Energy Efficiency in Electric Devices, Machines and Drives

Energy Efficiency in Electric Devices, Machines and Drives

Special Issue Editors

Gorazd Štumberger
Boštjan Polajžer

MDPI • Basel • Beijing • Wuhan • Barcelona • Belgrade • Manchester • Tokyo • Cluj • Tianjin

Special Issue Editors

Gorazd Štumberger
University of Maribor
Slovenia

Boštjan Polajžer
University of Maribor
Slovenia

Editorial Office
MDPI
St. Alban-Anlage 66
4052 Basel, Switzerland

This is a reprint of articles from the Special Issue published online in the open access journal *Energies* (ISSN 1996-1073) (available at: https://www.mdpi.com/journal/energies/special_issues/energy_efficiency_electric).

For citation purposes, cite each article independently as indicated on the article page online and as indicated below:

LastName, A.A.; LastName, B.B.; LastName, C.C. Article Title. *Journal Name* **Year**, *Article Number*, Page Range.

ISBN 978-3-03936-356-8 (Pbk)
ISBN 978-3-03936-357-5 (PDF)

Contents

About the Special Issue Editors

Gorazd Štumberger received B.Sc., M.Sc., and Ph.D. degrees from the University of Maribor, Maribor, Slovenia, in 1989, 1992 and 1996, respectively, all in electrical engineering. Since 1989, he has been with the Faculty of Electrical Engineering and Computer Science, University of Maribor, where he is currently a Professor of electrical engineering. In 1997 and 2001, he was a Visiting Researcher with the University of Wisconsin, Madison, and with the Catholic University Leuven, Leuven, Belgium, in 1998 and 1999. His current research interests include energy management, design, modeling, analysis, optimization and control of electrical machines and drives, and power system elements. Since 2019, he has been Dean of the Faculty of Electrical Engineering and Computer Science at the University of Maribor.

Boštjan Polajžer received B.Sc. and Ph.D. degrees in electrical engineering from the Faculty of Electrical Engineering and Computer Science, University of Maribor, Maribor, Slovenia, in 1997 and 2002, respectively. Since 1998, he has been with the Faculty of Electrical Engineering and Computer Science, University of Maribor, where he has been an Associate Professor since 2010. In 2000, he was a Visiting Scholar with the Catholic University Leuven, Leuven, Belgium, and the Graz University of Technology, Graz, Austria, in 2019. His research interests include electrical machines and devices, power quality, and power-system protection and control. Boštjan Polajžer is the head of the Power Engineering Institute at the Faculty of Electrical Engineering and Computer Science, University of Maribor.

Preface to "Energy Efficiency in Electric Devices, Machines and Drives"

Energy efficiency is one of the issues that cannot be avoided when dealing with electrical devices, machines, drives, and systems. Products that do not achieve the minimum efficiency levels specified in standards cannot be sold and every increase in energy efficiency reduces energy consumption and thus the cost of energy supply. Improvements in energy efficiency on a global scale reduce the energy demand and increase the energy supply, which indirectly reduces greenhouse gas emissions.

The key element in improving the energy efficiency of electrical devices, machines, and drives is the reduction of losses, which can be achieved using various methods. The development of new materials seems to be the most challenging. New solutions in design, design improvements, and optimization can contribute significantly to improving energy efficiency. When an electrical device or machine is manufactured, further improvements in energy efficiency can be achieved by an appropriate selection and matching of power electronic components with the inherited characteristics of machines or devices. Proper control of systems consisting of electrical devices or machines, and power electronics components based on advanced system models can further improve the overall energy efficiency of systems. When electric devices, machines, and drives operate within a system, a properly implemented energy management system, based on in-depth knowledge of the technological process, can improve energy efficiency at the system level.

Some cases of the aforementioned approaches to improving energy efficiency are discussed in this Special Issue.

Gorazd Štumberger, Boštjan Polajžer
Special Issue Editors

Article

Impact of the Winding Arrangement on Efficiency of the Resistance Spot Welding Transformer

Gašper Habjan * and Martin Petrun 

Faculty of Electrical Engineering and Computer Science, University of Maribor, 2000 Maribor, Slovenia; martin.petrun@um.si
* Correspondence: gasper.habjan@um.si

Received: 31 July 2019; Accepted: 23 September 2019; Published: 30 September 2019

Abstract: In this paper, the impact of the winding arrangement on the efficiency of the resistance spot welding (RSW) transformer is presented. First, the design and operation of the transformer inside a high power RSW system are analyzed. Based on the presented analysis, the generation of imbalanced excitation of the magnetic core is presented, which leads to unfavorable leakage magnetic fluxes inside the transformer. Such fluxes are linked to the dynamic power loss components that significantly decrease the efficiency of the transformer. Based on the presented analysis, design guidelines to reduce the unwanted leakage fluxes are pointed out. The presented theoretical analysis is confirmed by measurements using a laboratory experimental system. The presented experimental results confirm that the proposed improved winding arrangement increased the efficiency of the transformer in average for 6.27%.

Keywords: DC-DC converter; resistance spot welding; transformer; efficiency; dynamic power loss; design

---

1. Introduction

Resistance spot welding (RSW) systems have a very important role in modern industry. Considering the automotive industry alone, their importance is striking, as nowadays three new cars are produced every second worldwide [1,2]. An interesting fact is that three to five thousand welding spots are required to produce a contemporary personal car. Such a high amount of welding spots requires, on the one hand, the use of fully automated welding systems that are based on robot arms and on the other hand consumes large quantities of energy. Therefore welding systems should be designed as high power density devices where the efficiency of the whole system is crucial.

RSW systems can be generally divided in two groups—systems that produce AC and systems that produce DC welding currents. Nowadays the DC RSW systems are generally replacing the older AC RSW systems due to several advantages [3,4]. A typical contemporary high power RSW system consists of a AC-DC converter that operates at a frequency around 1 kHz, which is generally classified as a medium frequency RSW system [4–7]. A medium frequency RSW system with a DC welding current is shown in Figure 1.

The first part of the discussed system is the AC-AC converter, which consists of a passive input rectifier, a DC-link and H-bridge inverter. The full-wave input rectifier produces a DC voltage U_{dc} from phase voltages u_a, u_b and u_c, whereas the H-bridge inverter generates a pulse-width modulated (PWM) voltage with modulation frequency of 1 kHz. The presented AC-AC converter supplies the primary side of a welding transformer and is in high power RSW system usually not mounted on a robot arm.

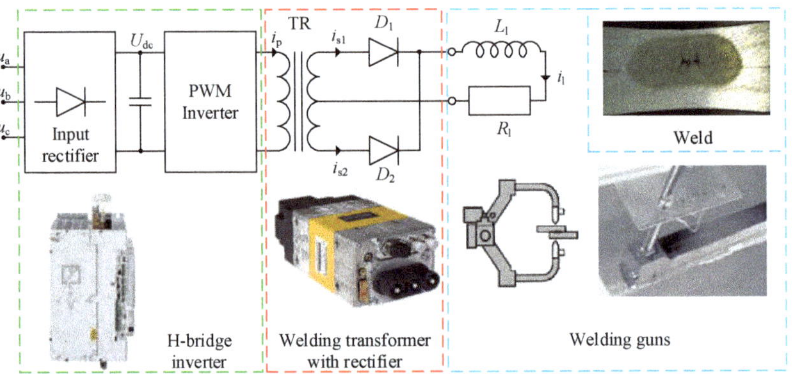

Figure 1. Schematic presentation of a typical resistance spot welding system [5].

The second part is a welding transformer, which consists of a three winding transformer with one primary and two secondary windings and an integrated center-tapped full-wave output rectifier. The main task of this transformer is to adequately increase the level of the primary current i_p, consequently a turn ration of 55:1 is used. The high alternating currents i_{s1} and i_{s2} in both secondary windings are rectified to a DC welding current i_l using a center-tapped output rectifier where two special high current diodes are used.

The third part of the RSW system is a welding gun that is connected to the output rectifier. The main task of a welding gun is to produce weld of prescribed quality, for what adequate electrical, mechanical and thermal conditions have to be fulfilled [2,4]. Due to very high welding currents in such systems, the welding transformer and the welding gun have to be placed together as close as possible to reduce high ohmic power losses [3]. Consequently both discussed parts are mounted on a moving robot arm, whereas the high power density of the welding transformer is one of the main goals in its design process.

In the RSW systems, the majority of the power loss occurs in the second and third part of the system due to very high currents. Therefore these parts are generally cooled using a cooling liquid [3,6]. By reducing the generated power loss in these parts consequently less cooling is required and the power density of the devices can be increased. In this paper the main focus was to analyze the impact of the winding arrangement on the power loss of the welding transformer. For this purpose, a laboratory experimental system was assembled that enabled comparisons of different winding arrangements. The obtained results have shown that an adequate winding layout reduced the power loss in the welding transformer significantly. The main contribution of presented research work is that presented experimental results are supported by a sound theoretical background, where also guidelines for the designing of high power transformers with center-tapped output rectifiers are discussed. The presented theoretical background and design guidelines are indispensable in the design process of the device, as they give an adequate starting point and can be applied to any contemporary design method (e.g., applying adequate winding layout in an finite element model of the device, where the dimensions of the winding can be furthermore optimized).

The paper is organized as follows. In the first section an introduction for the discussed problem is given. In the second section operation and the power loss inside a RSW system are discussed. In the third section the laboratory experimental system is presented, whereas in the fourth section the obtained results are presented and discussed. The fifth section gives a conclusion, where also the outline for future work is presented.

2. Design, Operation and Power Loss Inside a RSW Transformer

Design, operation and power loss inside a RSW transformer are inseparably interconnected. Power loss inside the transformer depends both on the operation as well as the design of the transformer, where the main focus of this paper is reduction of power loss by improving the design of the transformer.

2.1. Design of the Welding Transformer

A typical welding transformer is assembled from an iron core on which three windings are placed as shown in Figure 2.

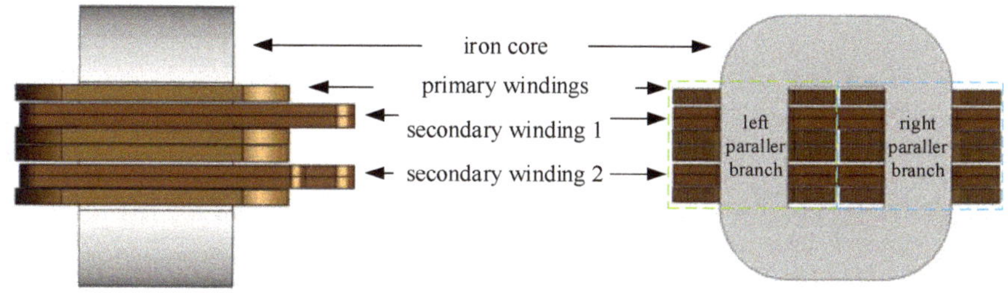

Figure 2. A typical design of a welding transformer [1].

The iron core consists of a wound sheet tape of a soft magnetic material that is wound on a model to obtain adequate geometry. The iron core is usually cut in two C-like segments to simplify the assembling process of the transformer [3,6,7]. The primary winding is generally obtained from a rectangular wire that is wound to adequate coils that fit on the iron core segments [3,7]. The primary winding is assembled using 4 such coils that are connected in series and have combined 55 winding turns. Due to high currents i_{s1} and i_{s2} the secondary windings consist of massive copper conductors that represent 1 winding turn and can be also equipped with adequate cooling channels [3,6]. In the presented case the massive secondary turns were furthermore divided into two halves of individual cross sections 17 mm × 2 mm, where the total cross section of a secondary winding was 17 mm × 4 mm. In the presented case both the primary as well as the secondary windings consist of two parallel branches that are placed on the opposite parts of the core. The final assembly is furthermore mechanically reinforced as the transformer is exposed to high mechanical stress due to high currents in the windings [1]. The presented transformer is designed for operation at the frequency of 1 kHz.

2.2. Operation of the Welding Transformer inside the RSW System

The presented transformer is only a part of the RSW system. The transformer is on the primary side connected to the H-bridge and on the secondary side to the output rectifier, as shown in Figure 1. Both connected parts determine the operation of the transformer, that is, how the electromagnetic variables inside the discussed transformer change with time. These variables generate power loss inside the transformer, therefore the analysis of operation is crucial. The operation of the RSW system can be broken down into 4 characteristic states that depend on the state of the H-bridge converter. These 4 states are shown in Figure 3.

In the presented case, T_p represents the time of one modulation period, T_{ON} is the time period in which relevant transistors are set in such a way that the transformer is supplied from the DC-link and T_{OFF} is the time period when all the transistors are not conducting and thus separate the transformer from the DC-link voltage. The typical 4 operation states hence correspond to conduction states of switches S_1, S_2, S_3 and S_4 of the H-bridge converter, which supplies the primary winding of the transformer either with voltage $u_p = U_{dc}$, $u_p = 0$ or $u_p = -U_{dc}$. The primary current i_p corresponds to the waveform of the generated voltage u_p, where the current increases or decreases according to the individual switching state. Each conducting switching state (states 1 and 3 in Figure 3) is followed by a short additional state (states 2a and 4a), where absolute value of i_p continuously decreases to 0 despite all transistors are already turned off. In these states the current i_p flows through the free-wheeling diodes of the H-bridge. However, the impact on power loss of these states is in this paper neglected due to a very short duration of these two states.

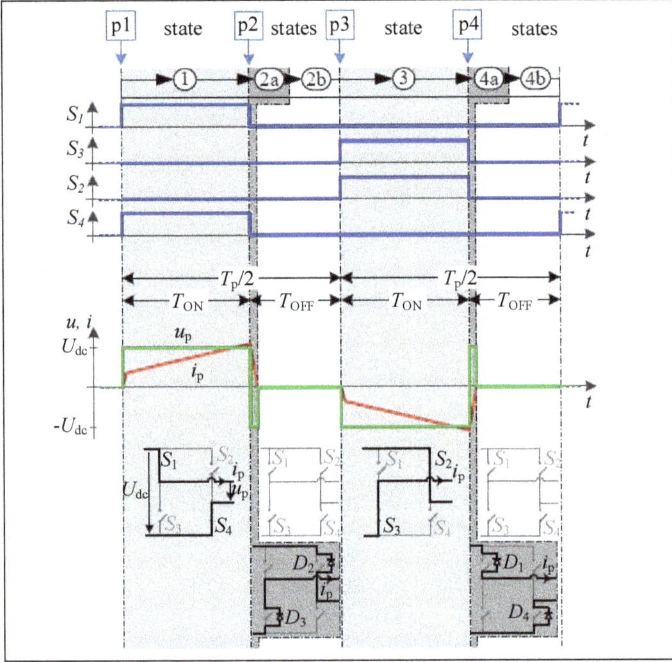

Figure 3. Pulse-width modulation with corresponding current and voltage on the primary side of the resistance spot welding (RSW) transformer [8].

The presented states in the primary winding generate also 4 main states in the secondary winding. These states depend on the output rectifier and are shown in Figure 4.

On the secondary side of the transformer the currents i_{s1} and i_{s2} flow only when individual rectifier diodes are polarized in a conducting way. Due to the inductive nature of the load, the presented converter operates in the continuous conducting mode on the secondary side of the transformer, that is, the current i_1 flows also in states 2 and 4, where it splits equally trough the two secondary windings [4,6]. Analogous to the primary side, there are also two brief switching states 2a and 4a that are in the presented analysis omitted. The generated continuous DC welding current i_1 is finally used to generate welds of adequate quality.

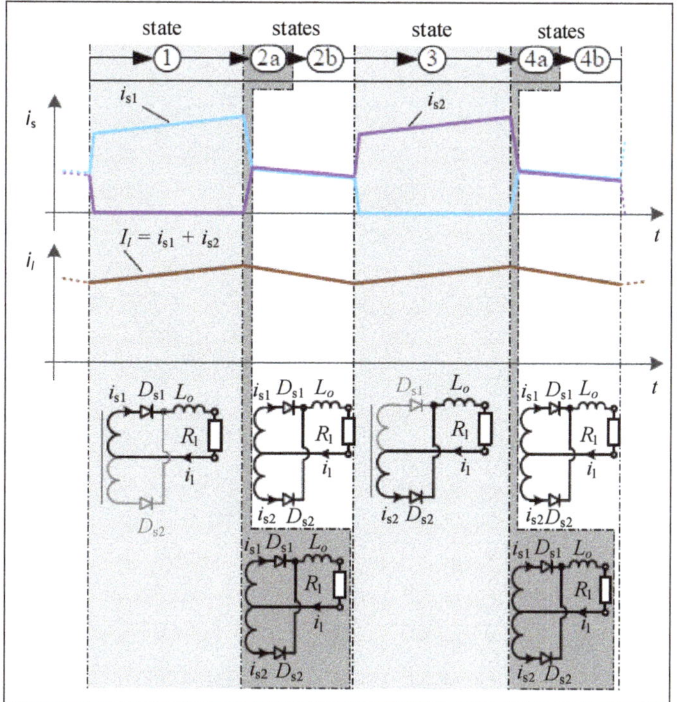

Figure 4. Typical states of the secondary side of the RSW transformer and output rectifier [8].

2.3. Power Loss Inside the Welding Transformer

Power loss inside the welding transformer can be in general separated in the iron core and winding power loss components. Both components are heavily dependent on used materials, design as well as the operation of the transformer [3,7].

The operation of the RSW system is for the needs of the presented analysis divided into 4 characteristic states as discussed in previous subsection. These 4 states can be furthermore combined into 3 characteristic conduction states of the transformer:

- State 1: The electrical current flows through the secondary winding 1 and the primary winding,
- State 3: The electrical current flows through the secondary winding 2 and the primary winding,
- State 2b and 4b: The electrical current flows only through the secondary windings.

These three states are determined by the direction of the currents and the activity of the windings in the transformer as shown in Figure 5.

In each of the presented state the excitation of the magnetic subsystem is substantially different; the currents in individual windings generate magneto-motive force (mmf) Θ along the iron core, which furthermore generates the magnetic flux and linkage between primary and secondary windings. The generated magnetic flux is, however, flowing not only in the magnetic core. Due to imperfections of the used materials (permeability of the iron core is not infinite, permeabilities of copper and air are not zero), magnetic flux is generated also in the areas around the core. Such areas include all the areas where mmf is generated, that is, areas where windings are located.

Due to the specific design, mmf in the iron core window acts predominantly along the y-direction. According to the presented conduction states, the distribution of the mmf $\Theta_y(x)$ in the iron core window can be easily approximated by considering that the density of current in the windings is uniform. According to the Ampere's Law, Θ_y is therefore changing linearly in respect to the length of the core window (*x*-axis) as shown in Figure 6a).

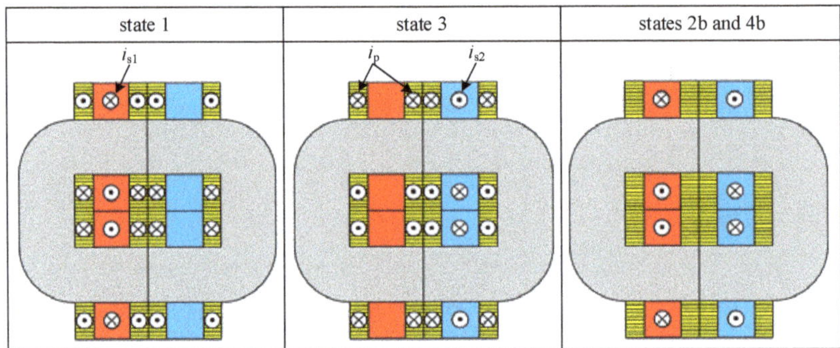

Figure 5. Current direction and winding activity for all four characteristic switching states of the RSW system [8].

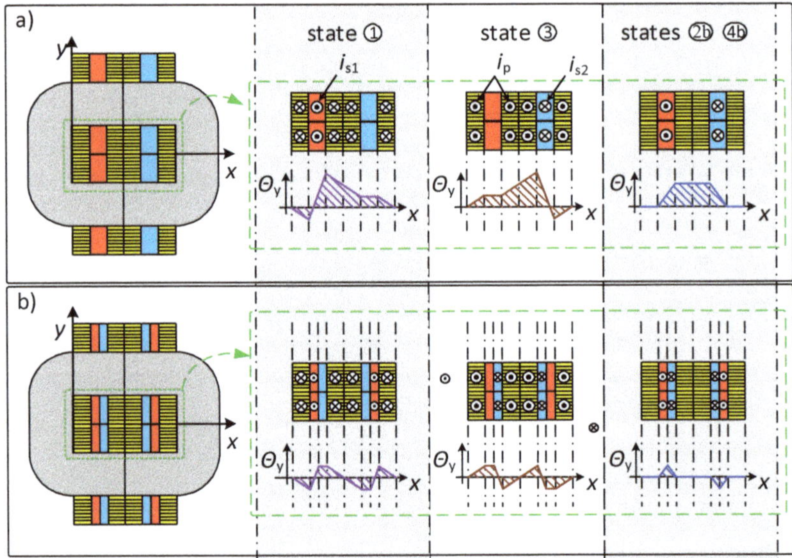

Figure 6. Two types of distribution of windings and a comparison of magneto-motive force for all states: (**a**) distribution of winding layout 1 and (**b**) distribution of winding layout 2 [8].

The obtained results have shown that the mmf distribution $\Theta_y(x)$ in different conducting states changes significantly. As the magnetic flux Φ_y is a direct consequence of $\Theta_y(x)$ and in the transformer window only linear magnetic materials (copper, air, insulation) are used, the distribution of the flux $\Phi_y(x)$ in this area corresponds to the $\Theta_y(x)$. The significant changes in $\Theta_y(x)$ consequently generate significant changes in $\Phi_y(x)$. As this leakage flux is changing mostly in electrically conducting materials (windings and iron core), the corresponding induced voltage generates unwanted eddy currents that increase the power loss in the device.

By analyzing the original winding layout shown in Figure 6a it was shown that the distribution of mmfs in individual conduction states is very imbalanced with relative high peak values. Moreover, the change of distribution $\Theta_y(x)$ between the states is significant. The generated mmf is acting in the y-direction, hence also the corresponding leakage flux will flow across the transformer window in this direction. This leakage flux changes significantly with time and induces voltages in the conducting materials of the transformer. The discussed induced voltage in the windings is causing the so called proximity effect, whereas the induced voltage in the iron core is causing additional eddy currents. Both effects are qualified as so called dynamic power loss, because they are generated due to changes of electromagnetic variables. These loss components increase the total loss significantly and should therefore be minimized. The imbalanced leakage magnetic field distribution $\Phi_y(x)$ furthermore leads to imbalanced flux in the iron core and can cause local iron core saturation in addition to increased power loss. This can again lead to increased power loss of the device, as the magnetizing current in the primary windings increases, which can be visible as characteristic current spikes in the primary current [6,9].

The key for discussed loss reduction was clear; the winding layout of the transformer should be designed in such a way that the distribution of the produced mmfs $\Theta_y(x)$ is in all conduction states balanced, whereas the peak values of Θ_y along the x axis should be held as low as possible. This can be achieved, for example, using the winding arrangement shown in Figure 6b. In comparison to original winding layout shown in Figure 6a, in this case the secondary windings are divided in two parallel branches that balance the mmf distribution in all states. Importantly also the peak values in $\Theta_y(x)$ are reduced for more than 50 %, therefore the changes between the distributions in individual conducting states are significantly smaller. In this way the dynamic loss components of the transformer were significantly reduced what was confirmed with the experimental analysis presented in following subsections.

3. Laboratory Experimental System

The presented simplified theoretical analysis was applied due to the very complex nature of the problem. In a real transformer, $\Theta_y(x)$ is due to possible skin and proximity effects not changing necessarily linearly in respect to x-axis. Consequently also $\Phi_y(x)$ in the core window can be distorted (however still imbalanced in a similar fashion as in the discussed linear analysis). Furthermore, the windings outside of the transformer core produce mmf that cannot be approximated using only the component in the y-axis. Lastly, the connections between the transformer and output rectifier should be considered as they impact the operation of the whole system significantly. Due to these facts simulation analysis would require very complex models (e.g., a 3D finite element model) where the theoretical background of the increased additional loss components could be overlooked. Such models furthermore cannot include all the effects, especially the increased eddy currents in the iron core that occur due to leakage magnetic fields that are perpendicular to core lamination. Consequently the analysis was carried out using a laboratory experimental systems that is presented below.

The laboratory experimental system was designed in such a way that changes of various winding arrangements could be performed. For this purpose a typical industrial RSW transformer was adjusted with additional bus system on the secondary side of the transformer, as shown in Figure 7. This system consisted of three buses; upper and lower buses (marked by 1 and 3 in Figure 7) were connected to both diodes of output rectifier, whereas the middle bus (marked by 2 in Figure 7) represents the center tap of the transformer and therefore the negative potential of the full wave rectifier.

The presented buses were designed in such a way that changing of the position of secondary and primary windings was possible. In order to enable the discussed analysis, also secondary windings of the transformer were adjusted. In a typical industrial RSW transformer, both secondary windings consist of single massive copper conductors that are equipped with cooling channels. In contrast to this, the secondary windings of the laboratory transformer were split into two halves as shown in Figure 2, where the cooling channels were omitted. In this way, comparable results for the presented analysis were obtained.

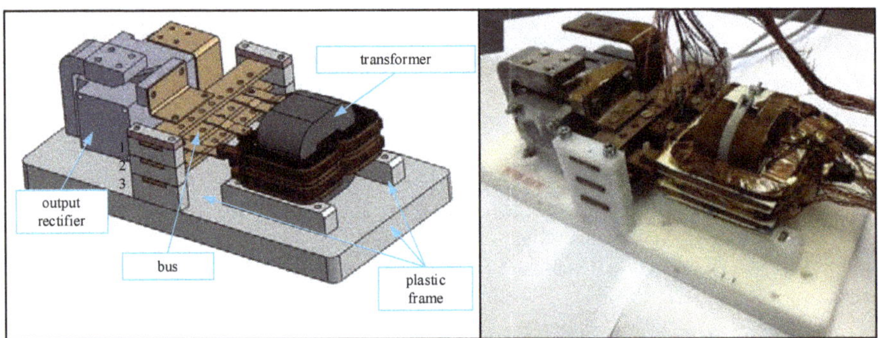

Figure 7. Design of the laboratory experimental system [1].

The drawback of the additional bus system was, however, that the total resistances on the secondary side of the transformer were increased in comparison to a compact industrial RSW transformer [5]. The measured total resistances for both winding arrangements are shown in Table 1. Measurements of resistances were performed using an adequate bridge based DC micro ohm meter which measured a DC resistance.

Table 1. Total resistance of individual windings for both winding arrangements.

	Primary Winding	**Secondary Winding 1**	**Secondary Winding 2**
distribution of winding layout 1	R_p = 23.7 mΩ	R_{s1} = 0.137 mΩ	R_{s2} = 0.184 mΩ
distribution of winding layout 2	R_p = 23.7 mΩ	R_{s1} = 0.106 mΩ	R_{s2} = 0.111 mΩ

From the obtained measurements it was shown that the total secondary resistance (windings combined with the bus system) was lower for proposed winding arrangement in comparison to the original arrangement. This was due to slightly shorter net path of the currents trough the winding combined with the connection between the transformer and the output rectifier. Furthermore, the total resistances of both secondary branches were better balanced compared to the original arrangement. An imbalance of secondary resistance can lead to drift of magnetic flux inside the core and consequently to its saturation [6,8,9]. Furthermore, the iron core of the transformer was equipped with several measuring coils as shown in Figure 8.

These measurement coils were used to calculate the density of magnetic flux inside the core, where in each measurement coil induced voltage $u_i(t)$ was measured. The densities of magnetic flux $B(t)$ were obtained by (1)

$$B(t) = \frac{1}{N_m S} \int u_i(t) dt, \tag{1}$$

where S represents the cross-section of the core and N_m represents the number of turns of the measurement winding. The laboratory experimental system was finally mechanically reinforced using an adequate plastic frame.

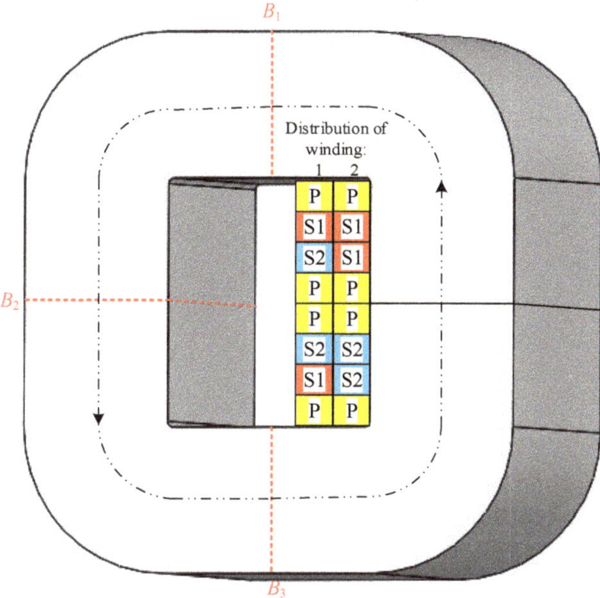

Figure 8. Placement of measuring coils on the core of laboratory RSW transformer [1].

4. Results

The discussed laboratory experimental system was tested for both winding arrangements using equal operation conditions. The tests were performed in such a way that the load resistance R_l and inductance L_l of the system were fixed, whereas the duty ratio of the H-bridge inverter was increased from 0.3 to 0.8 with a step of 0.1. The efficiency was determined by measuring the voltages and currents on the primary and secondary side of the transformer. Measurements were performed using the high performance measuring device DEWETRON DEWE-2600, where the voltages were measured directly and for the measurements of the currents special low frequency Rogowski's coils (CWT 3LFR and CWT 150LFR) were used.

Measured voltages for duty ratio of 0.6 are shown in Figure 9.

The obtained result have shown that the induced voltages in the secondary windings were slightly higher in the proposed winding arrangement in comparison to the original one. The RMS values of induced voltages in the secondary windings in the case of winding arrangement 1 were $U_{s1RMS} = 5.129$ V and $U_{s2RMS} = 5.225$ V. Compared to this, when winding arrangement 2 was applied, these values were $U_{s1RMS} = 5.249$ V and $U_{s2RMS} = 5.268$ V. This increase was attributed to lower leakage magnetic fields in the transformer window combined with lower total resistances on the secondary side of the transformer when improved winding arrangement was applied. Corresponding currents are shown in Figure 10.

The difference between all the currents in both analyzed winding arrangements was significant. The average steady state value of output welding current i_l was in the presented operation point for 1780 A higher, what accounts for 12.7 %. This increase was attributed to higher induced voltages in both secondary windings as well as lower total resistances on the secondary side of the RSW transformer.

Based on measured voltages $u(t)$ and currents $i(t)$ during the welding cycles, average input as well as output powers for different duty ratios were calculated. All the average power values P were determined using orthogonal decomposition technique in the time domain, where measured currents were decomposed into orthogonal and co-linear components in respect to adequate voltages [10,11]. In this way, from measured primary voltage u_p and current i_p the average input power of the transformer P_{in} was calculated and from measured secondary voltages $u_{\text{s}1}$ and $u_{\text{s}2}$ as well as currents $i_{\text{s}1}$ and $i_{\text{s}2}$ average output powers of the transformers $P_{\text{out}1}$ and $P_{\text{out}2}$ were calculated, respectively. The total output average power was determined by $P_{\text{out}} = P_{\text{out}1} + P_{\text{out}2}$. Based on the obtained P_{in} and P_{out}, total power loss P_{tr} and efficiency η_{tr} of the transformer were calculated by (2) and (3), respectively.

$$P_{\text{tr}} = P_{\text{in}} - P_{\text{out}} \tag{2}$$

$$\eta_{\text{tr}} = \frac{P_{\text{out}}}{P_{\text{in}}} \tag{3}$$

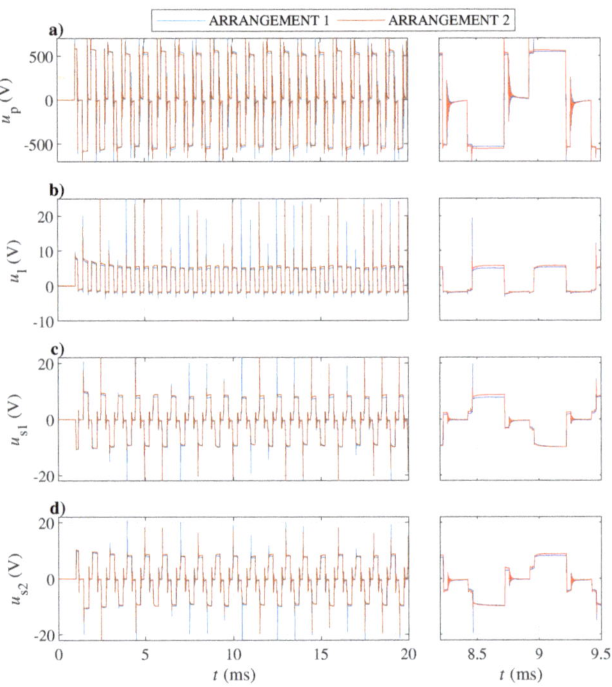

Figure 9. Comparison of measured primary and secondary voltages for both winding arrangements: (**a**) voltage in the primary winding u_{p}, (**b**) output voltage u_1, (**c**) voltage in the secondary winding 1 $u_{\text{s}1}$ and (**d**) voltage in the secondary winding 2 $u_{\text{s}2}$.

In addition to P_{tr}, also the total power loss of the output rectifier P_{rect} was determined. For this purpose furthermore the voltage u_1 and current i_1 on the output of the rectifier were measured. Based on these measurements, the average power supplied to the load P_{load} was determined. The power loss in the full-wave rectifier was obtained by (4)

$$P_{\text{rect}} = P_{\text{out}} - P_{\text{load}}, \tag{4}$$

whereas the efficiency of the output rectifier was calculated by (5) respectively.

$$\eta_{\text{rect}} = \frac{P_{\text{load}}}{P_{\text{out}}}. \tag{5}$$

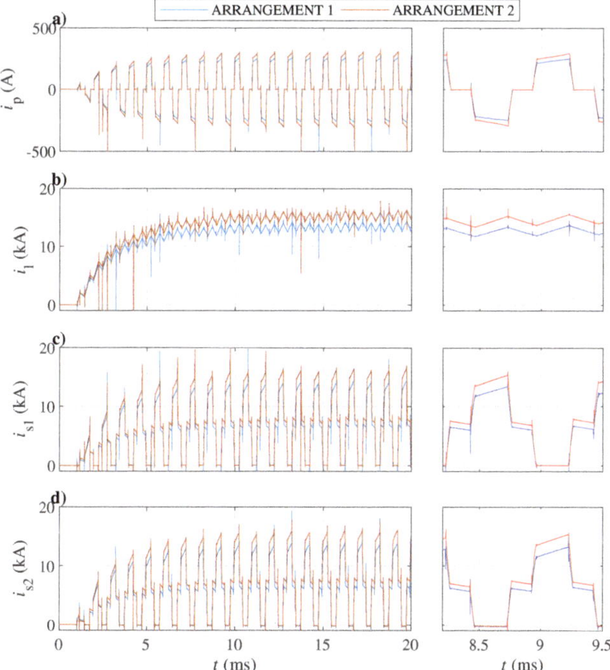

Figure 10. Comparison of measured primary and secondary currents for both winding arrangements: (**a**) current in the primary winding i_p, (**b**) welding current i_l, (**c**) current in the secondary winding 1 i_s1 and (**d**) current in the secondary winding 2 i_s2.

Due to the difference in the output currents for both winding arrangements, the obtained results were compared in respect to the output current instead of the duty ratio. The obtained results are presented in Figure 11.

The obtained results have shown that the efficiency of the transformer with improved winding arrangement was significantly higher; the efficiency improved for around 5% at low and around 6.7% at high welding currents, where the average efficiency difference between both distributions of windings was 6.27%. This increase corresponded to around 0.73 kW or 6 kW less total power loss in the transformer at low and high welding currents, respectively. In this way P_tr was decreased for 31.7–34.7%. This decrease of power loss was a direct consequence of the decreased leakage magnetic field inside the transformer window and generates the additional dynamic power loss component. Due to improved winding arrangement the proximity effect between all the winding coils was significantly reduced. Furthermore, as the leakage magnetic flux entered the iron core perpendicular to the lamination of iron core, the dynamic loss inside the iron core was reduced also. Changes of magnetic flux that enters the iron core perpendicular to lamination generate eddy currents that are due to unfavorable direction not limited by the lamination. This effect can increase the iron core power loss significantly [12]. The decrease of P_tr consequently enables to reduce the cooling of the transformer what furthermore increases the efficiency of the whole RSW system. In this way also weight of the transformer can be reduced, whereas the power density is improved.

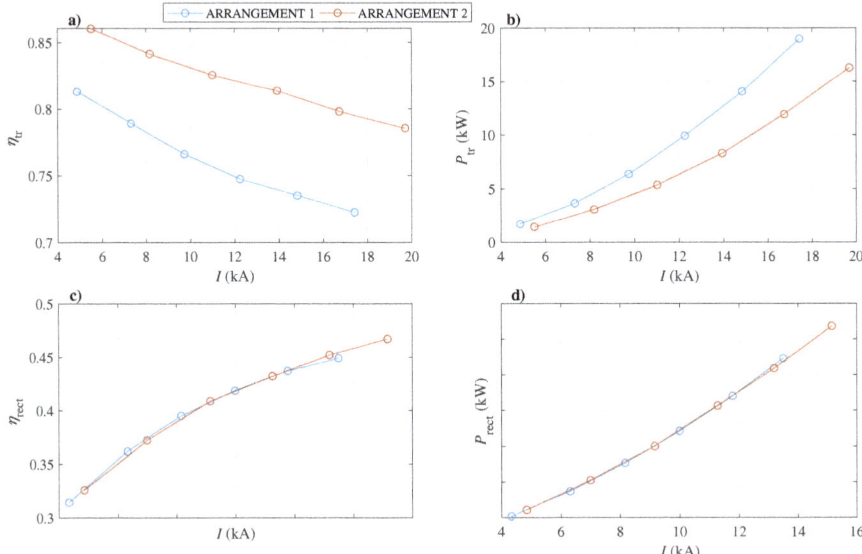

Figure 11. Comparison of determined power loss and efficiency of the transformer (**a**,**b**) and output rectifier (**c**,**d**) for both winding arrangements.

In contrast to the transformer, the efficiency and power loss of the rectifier depended only on the output current. Therefore the results were comparable for both winding arrangements. It is worthwhile to note that the rectifier power loss was even higher than power loss of the transformer.

Using the measurement coils on the core as presented in Figure 8 distribution of magnetic flux inside the iron core was determined. The obtained results are presented in Figure 12, where the imbalance of magnetic flux in the iron core was observed.

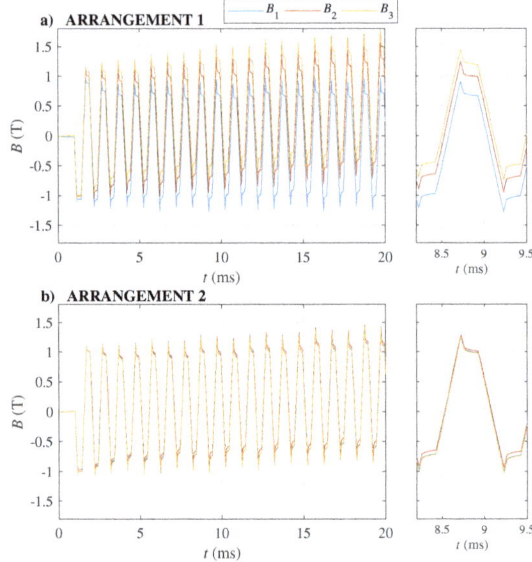

Figure 12. Results of the calculated densities of magnetic flux density B for: (**a**) arrangement 1 and (**b**) arrangement 2.

The calculated densities of magnetic flux were substantially different in all three core parts when original winding arrangement was analyzed (Figure 12a)). In contrast to this, all three densities of magnetic flux were balanced when the improved winding arrangement was analyzed. The observed non-uniform distribution of B was a direct consequence of imbalanced distribution $\Theta_y(x)$ trough all 4 characteristic states. As $\Theta_y(x)$ was balanced in all the states if improved winding arrangement was applied, consequently also a uniform distribution of B inside the core was achieved. A non-uniform distribution of B can lead to local iron core saturation and consequently to increased power loss and decreased operation reliability of the device.

5. Conclusions

In the paper, the impact of winding arrangements on dynamic power loss inside a RSW transformer was analyzed. The layout of the windings along the transformer's core has direct impact on amplitudes and distribution of leakage magnetic fields around these windings. First a sound theoretical background for improvement of winding arrangements is discussed, where the negative consequences of discussed phenomena were pointed out. The negative consequences include increased dynamic power loss inside the windings (proximity effect) and the iron core (increased eddy currents) as well as non-uniform distribution of magnetic flux density inside the core, which can lead to local saturation. The theoretical analysis was confirmed by experimental results. The obtained experimental results have shown that the losses inside the transformer were significantly decreased when the arrangement of the windings was improved with accordance with the presented theoretical background. The efficiency of the improved transformer was in average increased for 6.27 %. With the experimental analysis it was furthermore shown that the improved winding arrangement results in a better distribution of magnetic flux density inside the core. Consequently, local saturations of the core were avoided, whereas the magnetic material inside the device was better utilized. The presented analysis therefore points out basic guidelines when designing a high power DC-DC converter that utilizes a transformer with a center tapped rectifier. Future work will be focused on analysis of the non-uniform distribution of the magnetic field on the operation and losses of the whole system for RSW.

Author Contributions: Supervision, M.P.; Writing—original draft, G.H. and M.P.; Writing—review & editing, M.P.

Funding: This research received no external funding.

Conflicts of Interest: The authors declare no conflict of interest.

Abbreviations

The following abbreviations are used in this manuscript:

RSW	Resistance Spot Welding
DC	Direct Current
AC	Alternating Current
PWM	Pulse Width Modulation
mmf	magneto-motive force

References

1. Habjan, G. Nonhomogeneous Distribution of Magnetic Flux inside the core of a Resistance Spot Welding Transformer. Master's Thesis, Faculty of Electrical Engineering and Computer Science, University of Maribor, Maribor, Slovenia, 2018.
2. Kimchi, M.; Phillips, D.H. *Resistance Spot Welding: Fundamentals and Applications for the Automotive Industry*; Morgan & Claypool: San Rafael, USA, 2017; p. i-115
3. Popović Cukovic, J. Development of Winding design for Resistance Spot Welding Transformer. Ph.D. Thesis, Faculty of Electrical Engineering and Computer Science, University of Maribor, Maribor, Slovenia, 2013.
4. Zhou, K.; Yao, P. Review of Application of the Electrical Structure in Resistance Spot Welding. *IEEE Access* **2017**, *5*, 25741–25749. [CrossRef]

5. Brezovnik, R.; Černelič, J.; Petrun, M.; Dolinar, D.; Ritonja, J. Impact of the Switching Frequency on the Welding Current of a Spot-Welding System. *IEEE Trans. Ind. Electron.* **2017**, *64*, 9291–9301. [CrossRef]

6. Klopčič, B.; Dolinar, D.; Štumberger, G. Advanced Control of a Resistance Spot Welding System. *IEEE Trans. Power Electron.* **2008**, *23*, 144–152. [CrossRef]

7. Podlogar, V.; Klopcic, B.; Stumberger, G.; Dolinar, D. Magnetic Core Model of a Midfrequency Resistance Spot Welding Transformer. *IEEE Trans. Magn.* **2010**, *46*, 602–605. [CrossRef]

8. Petrun, M. Modeling and Analysis of Magnetic Components in High Power DC-DC Converters. Ph.D. Thesis, Faculty of Electrical Engineering and Computer Science, University of Maribor, Maribor, Slovenia, 2014.

9. Štumberger, G.; Klopčič, B.; Deželak, K.; Dolinar, D. Prevention of Iron Core Saturation in Multi-Winding Transformers for DC-DC Converters. *IEEE Trans. Magn.* **2010**, *46*, 582–585. [CrossRef]

10. Štumberger, G.; Polajžer, B.; Toman, M.; Dolinar, D. Orthogonal decomposition of currents, power definitions and energy transmission in three-phase systems treated in the time domain. In Proceedings of the International Conference on Renewable Energies and Power Quality (ICREPQ'06), Palma de Mallorca, Spain, 5–7 April 2006; Volume 1, pp. 263–267. [CrossRef]

11. Williems, J.L. A New Interpretation of the Akagi-Nabae Power Components for Nonsinusoidal Three-Phase Situations. *IEEE Trans. Instrum. Meas.* **1992**, *4*, 523–527. [CrossRef]

12. Cougo, B.; Tüysüz, A.; Mühlethaler, J.; Kolar, J.W. Increase of tape wound core losses due to interlamination short circuits and orthogonal flux components. In Proceedings of the IECON 2011—37th Annual Conference on IEEE Industrial Electronics Society, Melbourne, VIC, Australia, 7–10 November 2011; pp. 1372–1377.

Article

Modular Rotor Single Phase Field Excited Flux Switching Machine with Non-Overlapped Windings

Lutf Ur Rahman [1,*], Faisal Khan [1], Muhammad Afzal Khan [1], Naseer Ahmad [1], Hamid Ali Khan [1], Mohsin Shahzad [1], Siddique Ali [2] and Hazrat Ali [1]

[1] Department of Electrical and Computer Engineering, COMSATS University Islamabad, Abbottabad Campus, University Road Abbottabad, Abbottabad 22060, Pakistan; faisalkhan@cuiatd.edu.pk (F.K.); afzalk0370@gmail.com (M.A.K.); n.ahmadmwt@gmail.com (N.A.); hamidalims@gmail.com (H.A.K.); mohsinshahzad@cuiatd.edu.pk (M.S.); hazratali@cuiatd.edu.pk (H.A.)

[2] Department of Electrical Engineering, University of Engineering and Technology Peshawar, Peshawar 25120, Pakistan; siddique.ali@uetpeshawar.edu.pk

* Correspondence: lutfurrahman65@yahoo.com

Received: 15 December 2018; Accepted: 11 April 2019; Published: 25 April 2019

Abstract: This paper aims to propose and compare three new structures of single-phase field excited flux switching machine for pedestal fan application. Conventional six-slot/three-pole salient rotor design has better performance in terms of torque, whilst also having a higher back-EMF and unbalanced electromagnetic forces. Due to the alignment position of the rotor pole with stator teeth, the salient rotor design could not generate torque (called dead zone torque). A new structure having sub-part rotor design has the capability to eliminate dead zone torque. Both the conventional eight-slot/four-pole sub-part rotor design and six-slot/three-pole salient rotor design have an overlapped winding arrangement between armature coil and field excitation coil that depicts high copper losses as well as results in increased size of motor. Additionally, a field excited flux switching machine with a salient structure of the rotor has high flux strength in the stator-core that has considerable impact on high iron losses. Therefore, a novel topology in terms of modular rotor of single-phase field excited flux switching machine with eight-slot/six-pole configuration is proposed, which enable non-overlap arrangement between armature coil and FEC winding that facilitates reduction in the copper losses. The proposed modular rotor design acquires reduced iron losses as well as reduced active rotor mass comparatively to conventional rotor design. It is very persuasive to analyze the range of speed for these rotors to avoid cracks and deformation, the maximum tensile strength (can be measured with principal stress in research) of the rotor analysis is conducted using JMAG. A deterministic optimization technique is implemented to enhance the electromagnetic performance of eight-slot/six-pole modular rotor design. The electromagnetic performance of the conventional sub-part rotor design, doubly salient rotor design, and proposed novel-modular rotor design is analyzed by 3D-finite element analysis (3D-FEA), including flux linkage, flux distribution, flux strength, back-EMF, cogging torque, torque characteristics, iron losses, and efficiency.

Keywords: flux switching machine; modular rotor; non-overlap winding; magnetic flux analysis; iron losses; copper loss; stress analysis; finite element method

---

1. Introduction

In everyday applications, universal motors are mostly used in such devices as power tools, blenders, and fans. They are operated at high speed and deliver high starting torque as getting direct power from the ac-grid. At high speeds, universal motors cause noise due to their mechanical commutators, and they have a comparatively short maintenance period. To cope with these snags, research of a high-performance and low-cost brushless machine is greatly in demand [1,2]

Switched-flux brushless machines, a new class of electric machine were first presented in the 1950s [3]. Flux switching machines (FSMs), an unconventional machine, originated from the combination of principles among induction alternator and switched reluctance motor [4]. Distinct features of FSMs are their high torque density and robust rotor structure resulting from putting all excitation on the stator. In the past decade, various novel FSMs have been developed for several applications, confines from domestic appliances [5], automotive application [6,7], electric vehicles [8,9], wind power, and aerospace [10]. FSMs are categorized into permanent magnet flux switching machines (PMFSM), field excited flux switching machines (FEFSM), and hybrid excited flux switching machines (HEFSM). Permanent magnet FSMs and field excited FSMs have a permanent magnet (PM) and field excitation coil (FEC) for generation of flux source respectively, whilst both PM and FEC are generation sources of flux in HEFSM. The major advantage of FSMs is their simple/robust structure of rotor and easy management of temperature rise as all the excitation housed on stator. Recently, use of a permanent magnet as a primary source of excitation has dominated in flux switching research, due to their high torque/ high power density and optimum efficiency [11]. However, the maximum working temperature of PM is limited due to potential irreversible demagnetization. The use of PM in not always desirable due to high cost of rare earth material. For low cost applications, it is desirable to reduce the use of permanent magnets and hence they are replaced by DC-FEC. FEFSMs are capable of strengthening and weakening the generated flux as it is controlled by dc-current. FEFSMs have the disadvantage of less starting torque, fixed rotational direction, and high copper losses. The cumulative advantages of both FEC and PM are embedded in HEFSM having high torque capability/high torque density, HEFSMs also have high efficiency and flux weakening capability. However, the demerits of HEFSMs include a more complex structure, saturation of stator-core due to use of PM on stator, greater axial length, and high cost due to use of rare earth material. Therefore, FEFSMs could be considered a better alternative for requirements of low cost, wide speed controllability, high torque density, simple construction, less need of permanent magnet, and flux weakening operations as compared to other FSMs [12].

Numerous single-phase novel FS machines topologies has been developed for household appliances and different electric means. Single phase FSMs were first presented in [13] and further investigated in [14,15] by C. Pollock, they analyzed an 8S-4P doubly salient machine that offers high power density and low cost as shown in Figure 1. The FEC and armature has an overlapped winding arrangement resulting in longer end winding. To overcome the drawback of long end winding, the 12S-6P FSM has been developed that has same coil pitch as eight stator slots and four rotor poles but shorter end winding [16]. Figure 2 depicts how a 12-slot/6-pole machine has fully pitched winding arrangement as with C. Pollock's design. The end windings effect is even shorter by re-arranging the armature winding and FEC to different pitch of one and three slot pitches as shown in Figure 3. Both machines with F2-A2-six-pole and F1-A3-six-pole coil pitches have better copper consumption than a conventional machine (F2-A2-four-pole) for short axial length but has a disadvantage of higher iron loss due to more rotor poles [17]. The stator slots and rotor poles could be halved into F1-A3-3P machine as shown in Figure 4, that is more appropriate for high speeds due to a significant reduction in iron loss [16]. When the axial length is short, that is up to 25 mm, the average torque of both F2-A2/4P and six-pole machine is similar. However, F1-A3/3P exhibit higher average torque than F1-A3/6P machine at longer axial length of 60 mm. At the point when end winding is disregarded, a machine with fewer stator teeth and rotor poles has less average torque as compared to a machine with fewer rotor poles and stator slots for the same type of machine.

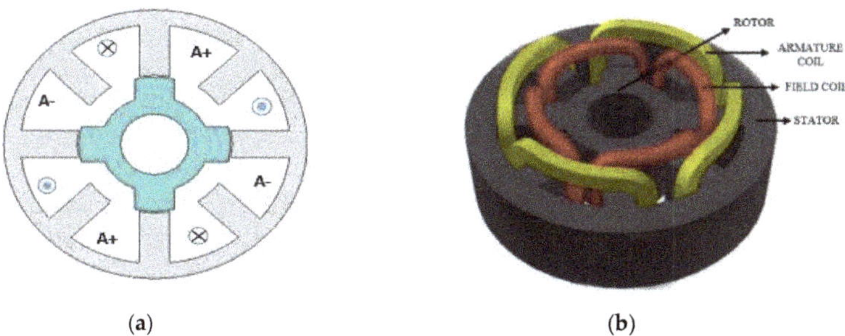

(**a**)   (**b**)

Figure 1. 8S/4P FEFS machine (F2-A2-4P). (**a**) Cross-sectional model; (**b**) 3D model [16].

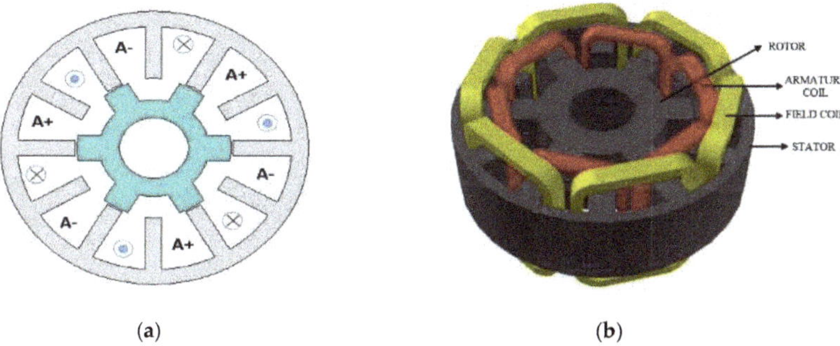

(**a**)   (**b**)

Figure 2. 12S/6P FEFS machine (F2-A2-6P). (**a**) Cross-sectional model; (**b**) 3D model [16].

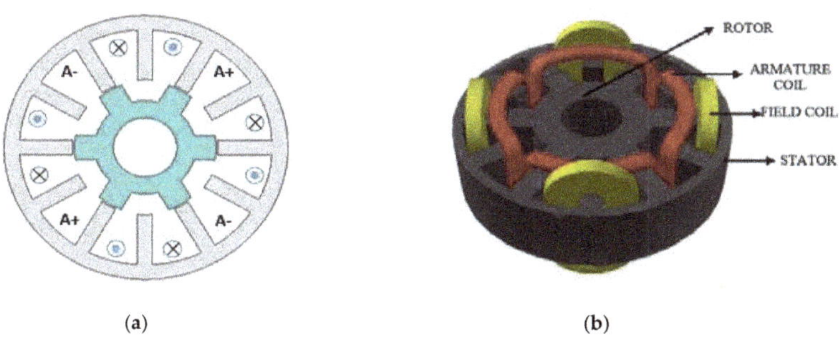

(**a**)   (**b**)

Figure 3. 12S/6P FEFS machine with rearranged winding (F1-A3-6P). (**a**) Cross-sectional model; (**b**) 3D model [16].

(**a**)   (**b**)

Figure 4. 6S/3P FEFS machine (F1-A3-3P). (**a**) Cross-sectional model; (**b**) 3D model [16].

In single phase FS machines, torque is generated with doubly salient structure due to the tendency of rotor to align itself into a minimum reluctance position as shown in Figure 5. When the stator slot and rotor pole are aligned at a minimum reluctance position, the motor cannot generate torque (called 'dead zone of torque') at aligned positions unless armature current direction is reversed. The dead zone of torque is eliminated in [18] with sub-part rotor structure having different pole arc lengths. The eight-slot/four-pole sub-part rotors are merged on the same face and pole axes are not parallel. However, sub-rotor poles cannot be allied with stator slots at a same-time, thus reluctance torque is generated at any rotor position. The single phase 8S/4P sub-part rotor FSM only applicable for a situation that requires a continuous unidirectional rotation. The conventional sub-part rotor design has demerit of overlapped winding arrangements between FEC and armature winding that results in higher copper consumption and higher iron losses due to salient rotor structure. A single phase sub-rotor FS machine minimizes the advantage of high speed, as it cannot operate at speed higher than normal level.

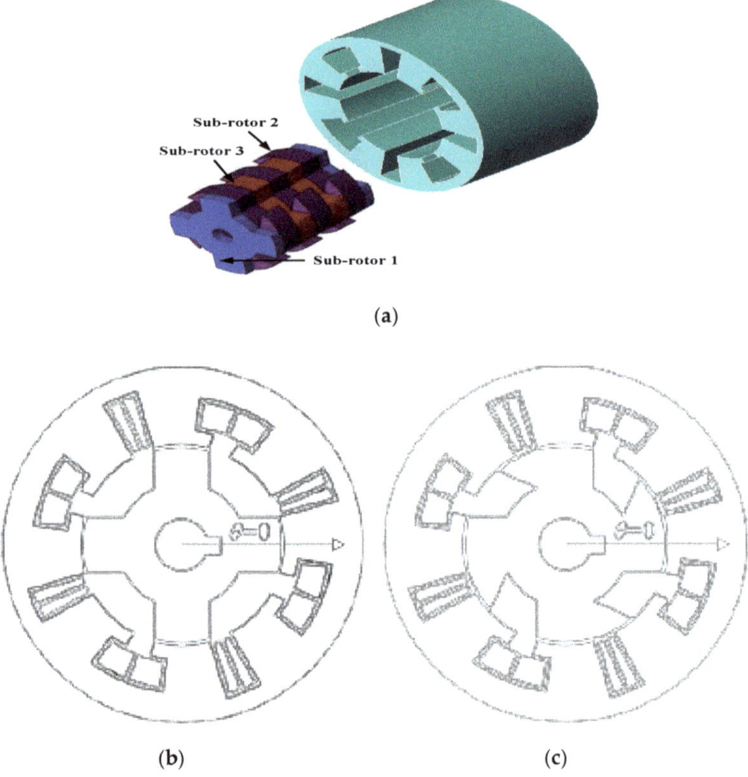

Figure 5. Sub-part rotor structure. (**a**) Manifestation of sub-part rotor; (**b**) pole arc of sub rotor-1; (**c**) pole arc of sub rotor-2.

This paper presents a novel-modular rotor structure for single phase FS machines as shown in Figure 6. The proposed design comprises of non-overlapped winding arrangements between armature winding and FEC, and modular rotor structure. The consumption of copper is much reduced due to non-overlapped winding arrangements. The modular rotor single phase FSM exhibit a significant reduction in iron losses, also reduces the rotor mass and lower the use of stator back-iron without diminishing output torque.

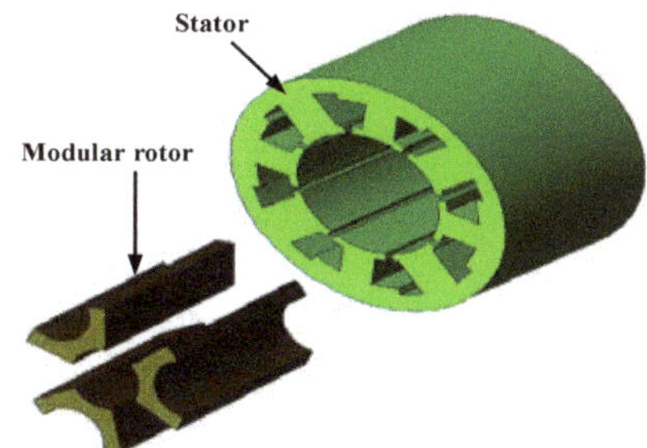

Figure 6. Modular rotor structure.

2. Design Methodology

The proposed novel modular rotor single phase eight-slot/six-pole FEFSM with non-overlapping winding arrangements is presented, as shown in Figure 7. To design the modular structure, JMAG designer ver.14.1 is used and the results obtained are validated by the 3D finite element analysis (3D-FEA). First, every section of motor such as stator, rotor, field excitation coil (FEC), and armature coil of modular design with eight stator slots and six rotor poles is designed in Geometry Editor. Then, the material, mesh properties, circuit, various properties, and conditions of the machine is selected and is simulated in the JMAG designer. The complete flow of the proposed design starts in Geometry Editor up to coil test analysis is shown in Figure 8. An electromagnetic steel sheet is used for the stator and rotor core. The design parameters and specifications of the modular design is illustrated in Table 1.

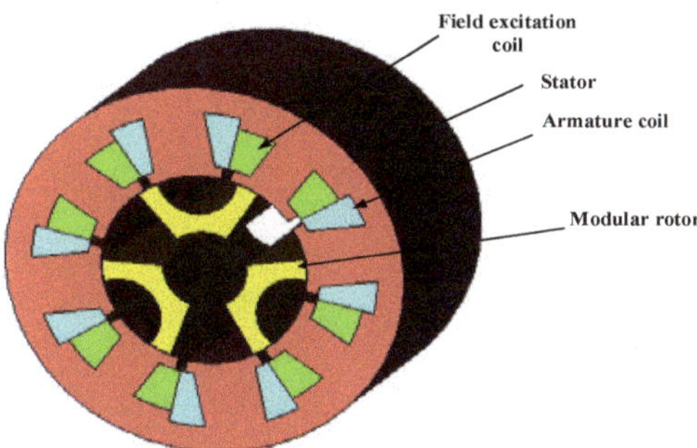

Figure 7. 8S-6P FEFSM with modular rotor.

Table 1. Design parameter of machines

Design Parameters	F1-A3-3P	Sub-Part Rotor Design	Modular Rotor Design
Number of phases	1	1	1
No. of slots	6	8	8
No. of pole	3	4	6
Stator outer diameter	96 mm	96 mm	96 mm
Rotor outer diameter	55.35 mm	53.55 mm	53.55 mm
Air-gap	0.45	0.45 mm	0.45 mm
Stator pole arc length	-	15.2 mm	5.5 mm
Teeth's arc of sub-rotor-1	-	15.2 mm	-
Teeth's arc of sub-rotor-2	-	26.9 mm	-
Rotor pole width	8 mm	-	5.6 mm
Stack length	60 mm	60 mm	60 mm
No. of turns per phase	120	30	30

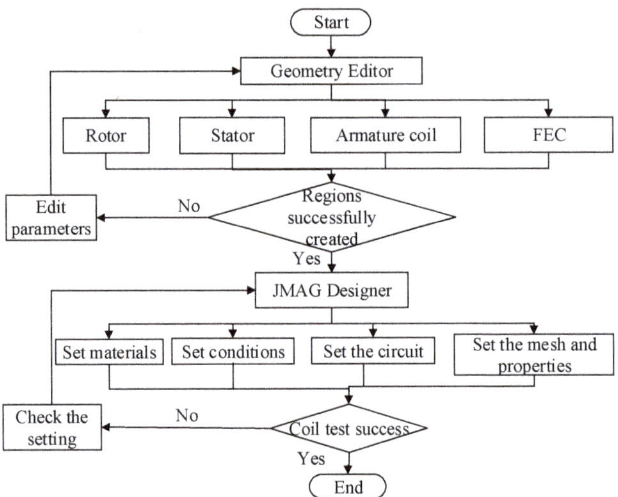

Figure 8. Design procedure.

3. Deterministic Optimization

The average torque analyses of eight-stator slots/six-rotor poles are examined. The maximum output torque obtained by the initial design is 0.88 Nm at speed of 400 rpm, which is much lower from the other designs. In order to improve the average torque characteristics, deterministic optimization is used. First optimization cycle consists of five steps, that is R_{IR}, θ, S_R, T_{WA}, T_{RA}, T_{WD}, and T_{RD}, as shown in Figure 9. Design free parameters R_{IR}, θ, S_R, T_{WA}, T_{RA}, T_{WD}, and T_{RD} are defined in rotor and stator part, as depicts in Figure 10 are optimized, while the outer radius of stator, air gap, and shaft of the motor are kept constant.

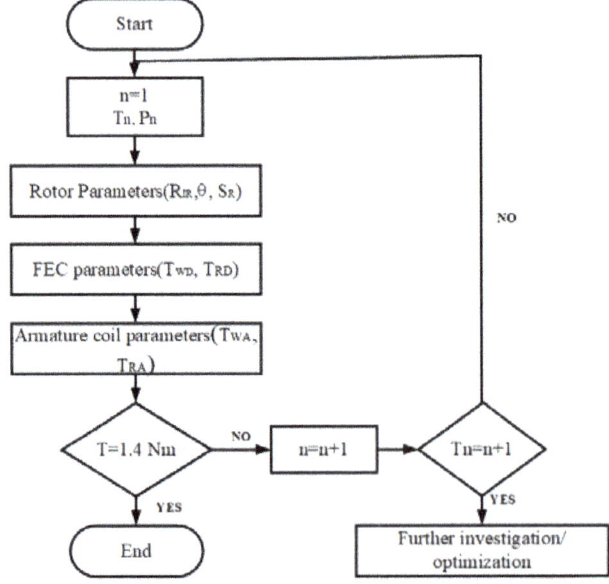

Figure 9. Optimization procedure.

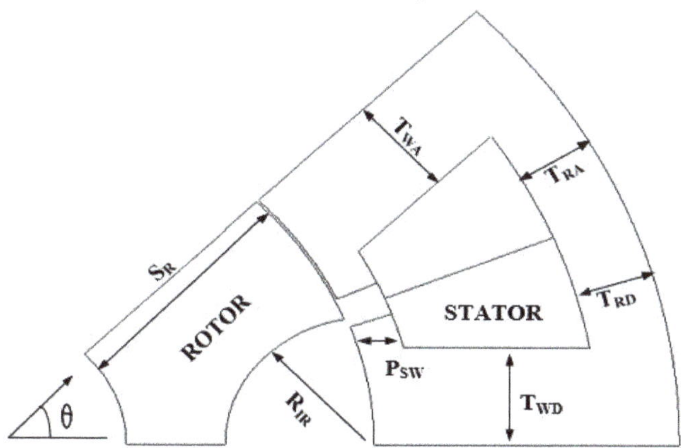

Figure 10. Design parameters of modular rotor design.

Initially, the design free parameters of rotor are updated, first of all, the inner rotor radius, R_{IR}, changes while the other parameters of stator and rotor remain constant. Then, rotor pole angle, θ, and split ratio, S_R, are varied and adjusted. The rotor pole angle, θ, is a dominant parameter in modular design to increase torque characteristics. Once the combination of promising values of rotor part for highest average output torque is determined, the next step is to refine the T_{WD} and T_{RD} of FEC, while rotor and armature slot parameter are kept constant. Finally, the essential armature slot is optimized by changing T_{WA} and T_{RA} while all other design parameters are preserved. To attain the highest average output torque, the above design optimization process is repeated. Figure 11 illustrates the highest average torque result after two cycles of optimization by updating several parameters that are already mentioned above. From Figure 11, it is also clear that during the first cycle the torque increases to a certain level by varying the above parameters of the machine and it becomes constant.

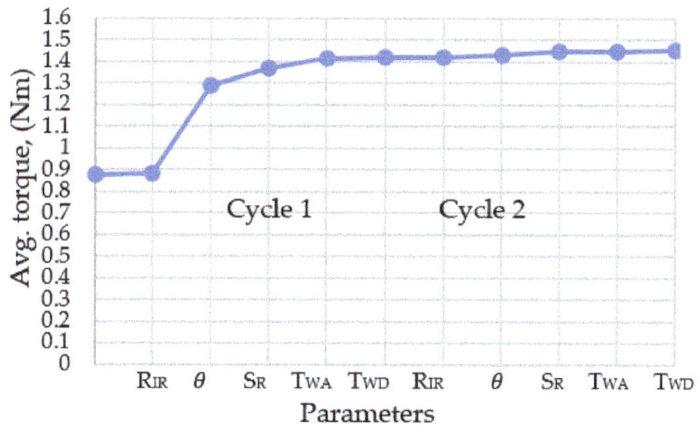

Figure 11. Effect of design parameters on average torque

During the first cycle, 32 percent of increase in the average output torque is achieved by refining the dominant parameter of rotor pole angle, θ, whilst other free design parameter adjustment shows less improvement in torque. In comparison with the initial design, the average output torque is improved by 40 percent after completion of second optimization cycle. The initial and optimized structure of

8S/6P modular design is illustrated in Figure 12. Additionally, the comparison of parameters of initial and final design is presented in Table 2.

Table 3 depicts comparison of cogging torque, flux linkage, back-EMF, average torque, and power of 3D-modular un-optimized and optimized design. The cogging torque and flux linkage of optimized designs is 0.3374 Nm and 0.2114 Wb respectively, which is 50% lower than the un-optimized cogging torque and flux linkage. Whilst, back-EMF of optimized modular 8S/4P is improved by 15%, that is still much lower than the applied input voltage of 220V. Furthermore, before optimization of modular design the maximum average output torque and power obtained is 9.77 Nm and 162.9 Watts respectively, at maximum FEC current density, J_e, is set to 10 A/mm^2 and 25 A/mm^2 is assigned to the armature coil, which is improved to 1.66 Nm and 288 Watts, respectively. Comparatively, average output torque and power is improved by 58.85% and 56.40%, respectively.

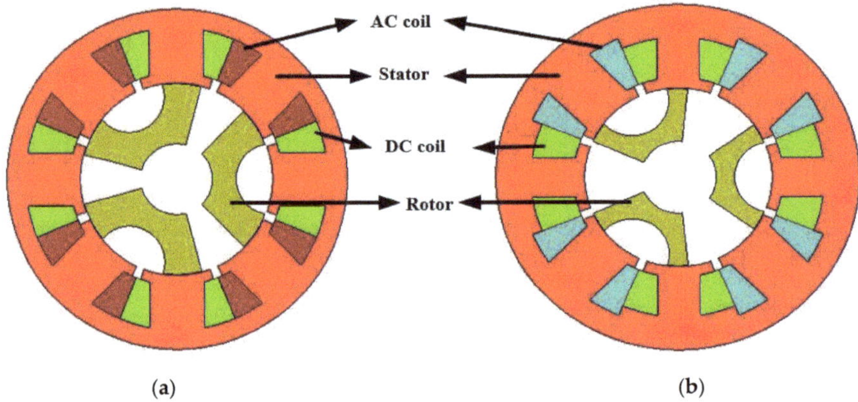

Figure 12. Structure of eight-slot/six-pole modular design.

Table 2. Initial and refined design parameters of Novel-modular rotor design

Parameters	Units	Initial Values	Optimized Values
Outer stator (OS)	mm	48	48
Inner stator (IS)	mm	27	24.7
Back-iron width of AC (T_{RA})	mm	42.5	41.4
Tooth width of AC (T_{WA})	mm	7.6	7.5
Back-iron width of DC (T_{RA})	mm	42.5	38.66
Tooth width of DC (T_{WD})	mm	7.6	5.5
Rotor inner circle radius (R_{IR})	mm	10	9.7
Pole shoe width (P_{SW})	mm	3	3
Rotor pole angle (θ)	deg	45	36
Split ratio (S_R)	-	0.55	0.5
Air gap	mm	0.45	0.45
Shaft radius	mm	10	10
Avg. torque	Nm	0.880	1.454

Table 3. Results comparison of optimized and un-optimized design

—	Cogging Torque (Nm)	Flux Linkage (Wb)	Back-EMF (Volt)	Avg. Torque (Nm)	Power (Watts)	Motor Mass (Kg)
Un-optimized design	0.67	0.01060	3.9	0.97775	162.986	4.02345
Optimized design	0.3374	0.02114	4.6	1.66148	288.967	3.84697

4. Result and Performance Based on 3D-FEA Finite Element Analysis (3D-FEA)

4.1. Flux Linkage

Comparison of flux linkages of three field excited FSM at no-load is examined by 3D-FEA [16,18]. To analyze the sinusoidal behavior of flux, the input current density of FEC, and armature coil is fixed to 10 A/mm^2 and 0 A/mm^2 respectively. Figure 13 shows that proposed modular design has peak flux of 0.021 Wb which is approximately equal to the peak flux of 15% of F1-A3-3P design. Similarly, sub-part rotor design has a 66% higher peak flux linkage than 8S/6P modular structure due to different pole arc length. The conventional F1-A3-3P design has the highest peak flux as compared to modular design, as well as sub-part rotor design due to the doubly salient structure.

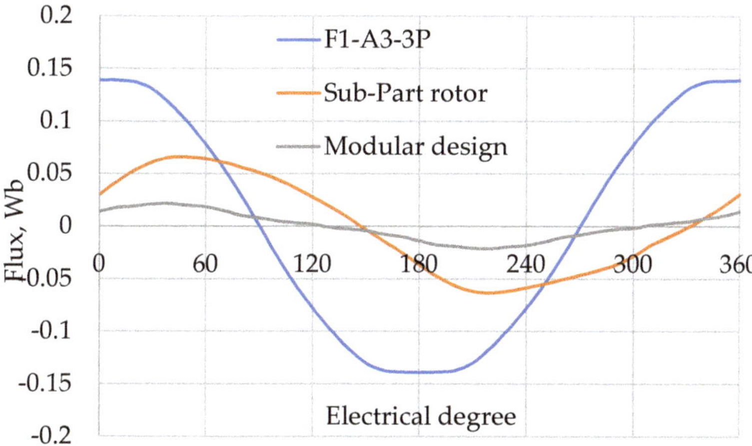

Figure 13. Comparison of U-flux linkages.

4.2. Flux Distribution

Flux density distribution generated by the DC coil in three FEFSM is shown in Figure 14. The red spot mention in Figure 14a–d show saturation of stator teeth and back-iron respectively of both conventional designs. F1-A3-3P design and sub-part rotor design has vector plot value of magnetic flux density distribution of 1.9953 and 1.9760 maximum, respectively. Whilst, the flux density distribution of modular design from the vector plot is 2.2528 maximum at 0° rotor position. Additionally, in comparison with 8S/4P sub-part rotor design and 6S/3P design, the proposed 8S/6P modular rotor design exhibits higher flux distribution. For completely utilizing flux in the proposed design, various parameters of the machine are optimized to enhance the flux distribution from the stator to the rotor and vice versa. The peak flux in the modular rotor pole at its lowest magnetic loading is 0.0453 Wb, which increases with increasing magnetic loading. Figure 15 illustrates flux density distribution at maximum armature current density of 25 A/mm^2.

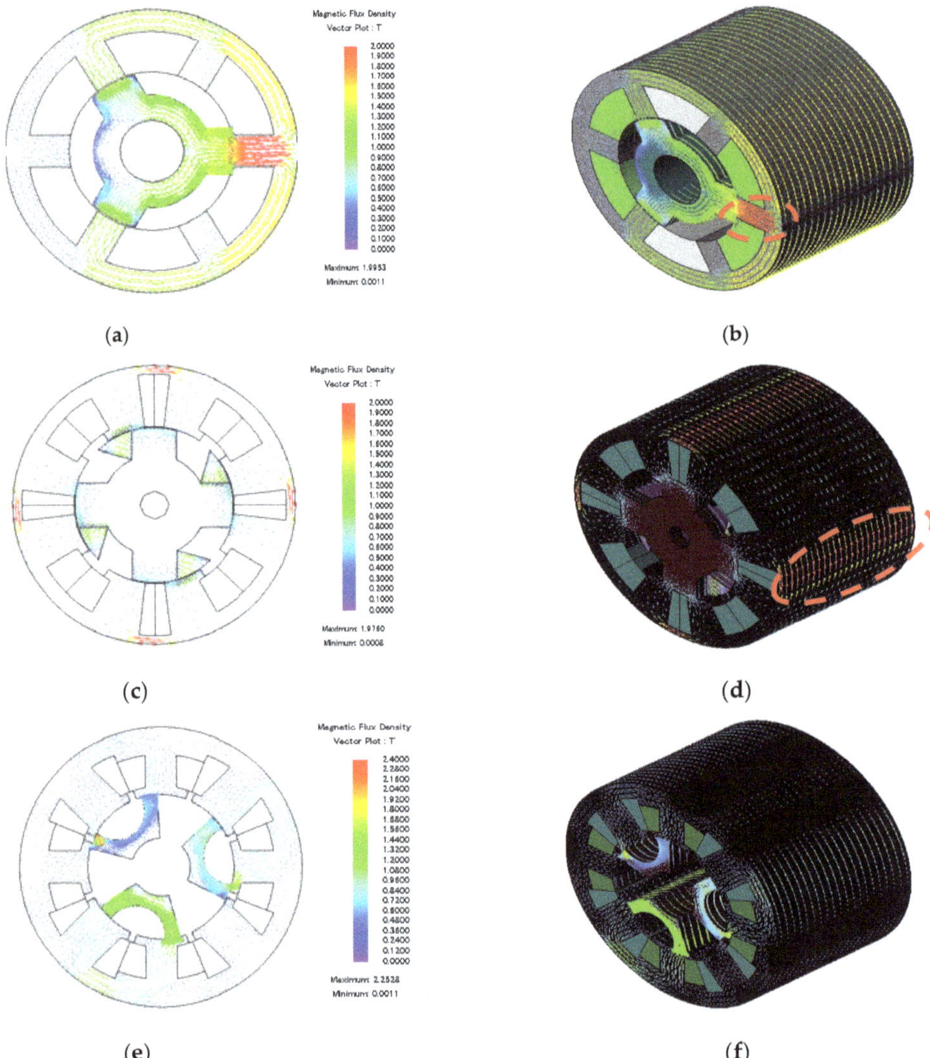

Figure 14. Flux distribution at no-load. (**a,b**) Flux distribution in conventional F1-A3-3P design; (**c,d**) Flux distribution in sub-part rotor design; (**e,f**) Flux distribution in proposed modular rotor design.

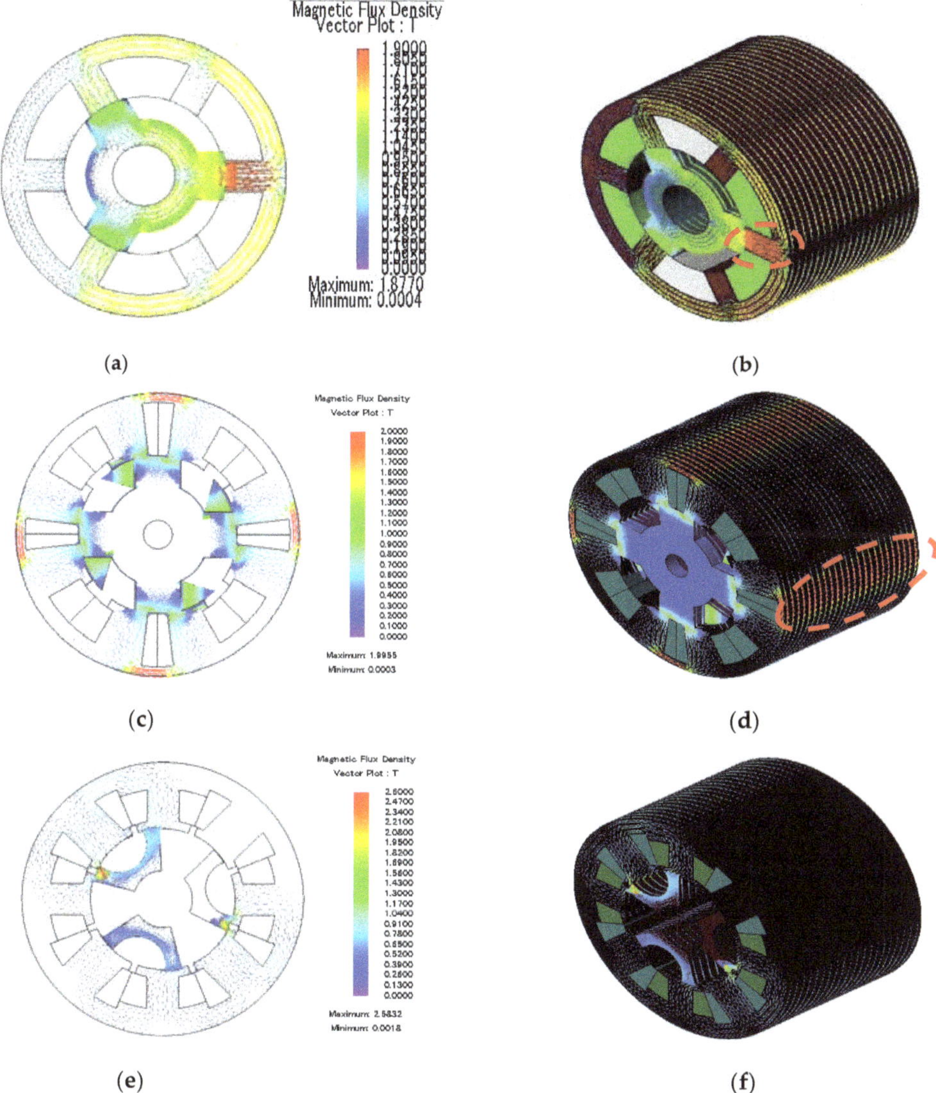

Figure 15. Flux distribution at maximum load. (**a**,**b**) Flux distribution in conventional F1-A3-3P design; (**c**,**d**) Flux distribution in sub-part rotor design; (**e**,**f**) Flux distribution in proposed modular rotor design.

4.3. Flux Strengthening

The effect of flux strength is analyzed by increasing current density; J_e of field excitation coil (FEC) is varied from 0 A/mm^2 to 20 A/mm^2, whilst armature current density; J_a is set 0 A/mm^2. The FEC input current is calculated from (1)

$$I_e = \frac{J_e \alpha S_e}{N_e} \tag{1}$$

where, I_e, J_e, α, S_e, and N_e are the input current of FEC, field current density, filling factor, slot area of FEC, and number of turns of field coil respectively. The analysis of coil test can be verified from the flux strengthening. Increasing the current densities of FEC, the pattern plot clearly shows a linear increase in flux until 0.027 Wb at J_e of 20 A/mm^2 as shown in Figure 16.

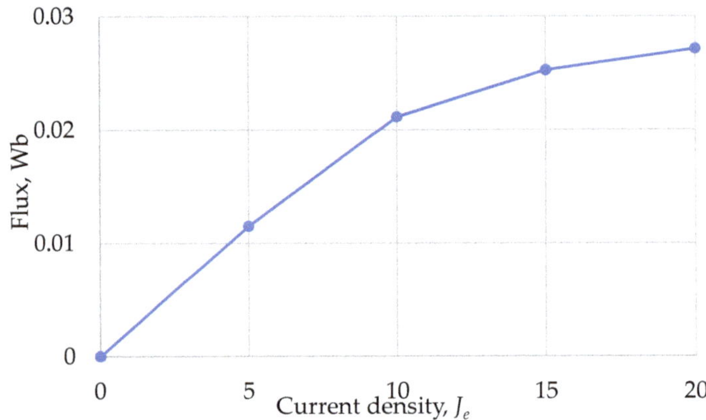

Figure 16. Peak flux strengthening with Modular rotor at various J_e.

4.4. Back-EMF Versus Speed

"Back-EMF is the induce voltages in the armature winding which opposes the change in current through which it is induced". The back-EMF (e_a) in the armature can be determined from the rate of change in armature flux or applying the co-energy concept [19]. For motor with N_r rotor poles

$$e_a = \omega\phi_f\frac{2KN_aN_r}{\pi} \tag{2}$$

where N_a, N_r, ω, ϕ_f and K is number of armature turns, number of rotor poles, rotational speed, field flux, and constant of field flux that are linked with the armature winding respectively. Substituting the ϕ_f (field flux) with

$$\phi_f = \frac{N_fI_f}{\mathcal{R}} \tag{3}$$

$$e_a = \frac{2KN_r}{\pi\mathcal{R}}N_aN_fI_f\omega \tag{4}$$

where N_f, I_f, and $\mathcal{R}$ is the number of field turns, the field current and reluctance of magnetic circuit. For the maximum conversion of electro-mechanical energy, armature current must flow in the opposite direction of the induced-EMF in the armature.

Figure 17 shows the 3D-FEA predicted induced-EMF of eight-slot/six-pole modular rotor structure at a fixed field current density (J_e; 10 A/mm^2) and various speeds. The induced-EMF increases linearly with increasing speed. The maximum induced voltage is 22 V at a maximum speed of 1600 rpm which is quite lower than the applied input voltage (220 V) which confirms the motor actioning of the machine.

4.5. Intantaneous Torque and Torque Ripple Calculation

Figure 18, investigates the instantaneous torque versus rotor mechanical revolution in electrical degrees of eight-slot/four-pole, six-slot/three-pole, and eight-slot/six-pole FESF machines. Six-slot/three-pole (F1-A3-3P) rotor design has high peak to peak torque as compared to eight-slot/six-pole (modular design) and eight-slot/four-pole (sub-part rotor) FESF machines. Figure 18 illustrates that characteristics of instantaneous torque at 10 A/mm^2 of six-slot/three-pole are better as compared to eight-slot/six-pole and eight-slot/four-pole FESF machines.

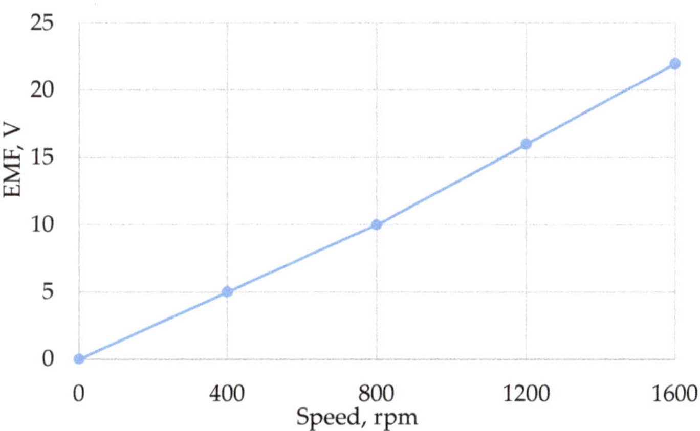

Figure 17. Maximum back-EMF at various speed.

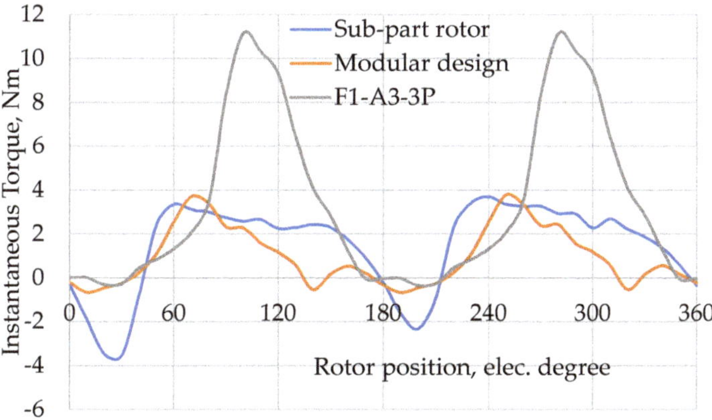

Figure 18. Comparison of instantaneous torque.

Whilst F1-A3-3P design exhibits the highest torque ripples comparatively to sub-part rotor design and modular rotor design. The proposed modular design has lower torque ripples than both the conventional designs, that is 29% and 60% lower than sub-part rotor design and F1-A3-3P design, respectively. Torque ripples are calculated from expression (5)

$$\left(\frac{\tau_{max} - \tau_{min}}{\tau_{avg}}\right) \times 100 \tag{5}$$

4.6. Total Harmonics Distortion (THD)

Total harmonics distortion is the ratio of the summation of all harmonic components to the fundamental frequency harmonics of the power or harmonics distortion that exists in flux. In electric machines, THD occurs due to harmonics present in flux. THD determines the electromagnetic performance of the machine as it is the representation of the harmonics in the machine. Mathematically, the THD of an electric machine can derived from equation (6)

$$THD = \frac{\sqrt{\sum_{n=1,2...}^{k=2n+1} \Phi_k^2}}{\Phi_1} \tag{6}$$

where k is odd number and Φ_k is the odd harmonics of flux. THD of proposed design is higher as compared to conventional design due to the modular structure of rotor. Figure 19 shows the THD of three FEFS machines. The graph shows that the THD of sub-part rotor design and F1-A3-3P design is 7% and 4% respectively, while THD of the proposed modular rotor design is 16.4%.

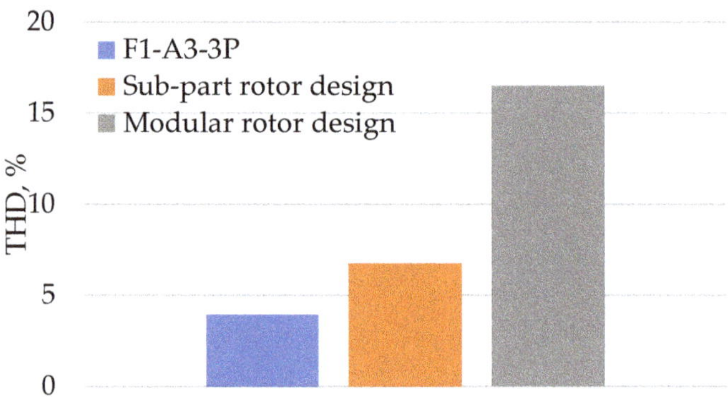

Figure 19. THD values of the conventional and proposed designs.

4.7. Cogging Torque

The interaction between the Stator excitation source (PM, excitation coil) and rotor pole of machine at no-load is called cogging torque. The magnetic circuit consists of an existing PM and coil having co-energy, the total co-energy is formulated as [20,21].

$$W_c = Ni\varphi_m + \frac{1}{2}\left(Li^2 + (\mathcal{R} + \mathcal{R}_m)\varphi_m{}^2\right) \tag{7}$$

where, N, i, $\mathcal{R}_m$, L, $\mathcal{R}$ and φ_m are the number of turns, current, magnetic flux, inductance of coil, magneto-motive force, and magnetic flux linkage respectively. The change in total co-energy with respect to the mechanical angle of the rotor determines the average torque of the machine.

$$T_e = \frac{\partial W_c}{\partial \theta} \; with \; i = constant \tag{8}$$

where, W_c and θ are total co-energy and mechanical rotor angle, respectively.

$$T_e = \frac{\partial\left(Ni\varphi_m + \frac{1}{2}\left(Li^2 + (\mathcal{R} + \mathcal{R}_m)\varphi_m{}^2\right)\right)}{\partial \theta} \quad T_e = Ni\frac{d\varphi_m}{d\theta} + \frac{1}{2}i^2\frac{dL}{d\theta} - \frac{1}{2}\varphi_m{}^2\frac{d\mathcal{R}}{d\theta} \tag{9}$$

The third term in Equation (9) changes in mmf with respect to the mechanical position of rotor causes cogging torque. The cogging torque produces unwanted noise and vibration. As Equation (9) shows that the cogging torque lead to a significant reduction in the average torque.

The cogging torque of F1-A3-3P, sub-part rotor and modular designs is comparing in Figure 16. The cogging of modular design is less than F1-A3-3P and sub-part rotor designs as depicts in Figure 16. Figure 20 illustrates that the cogging torque of modular design is 12% of F1-A3-3P and 53% of sub-part rotor. As a result, the modular design has less vibration and more average torque as to compare to F1-A3-3P and sub-part rotor design.

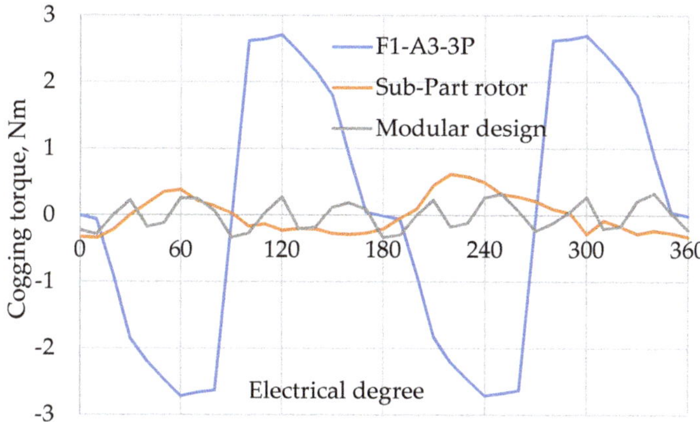

Figure 20. Comparison of cogging torque.

4.8. Copper Loss versus Torque

In field excited FSM, the copper consumption is the main constituent affecting the overall cost of the machine. As compared to other materials for FEFSM, copper is more expensive. The copper-loss of single phase FEFSM can be calculated from the formula as

$$P_{Cu} = I_a^2 R_a + I_f^2 R_f$$

where I_a, R_a, I_f, and R_f is the armature current, armature winding resistance, field current, field winding resistance, respectively. The comparison of copper loss-torque curve of three field excited FSM is shown in Figure 21. The average output torque modular design is almost similar to the sub-part rotor design but is much higher than the F1-A3-3P design. At fixed copper loss of 60 watts, the average torque of conventional sub-part rotor design, F1-A3-3P design, and proposed modular design is 1.6 Nm, 0.98 Nm, and 1.58 Nm, respectively. However, the plot clearly shows that modular design achieves a higher average torque under the constraint of maximum copper loss of 120 watts due to the short pitch coils.

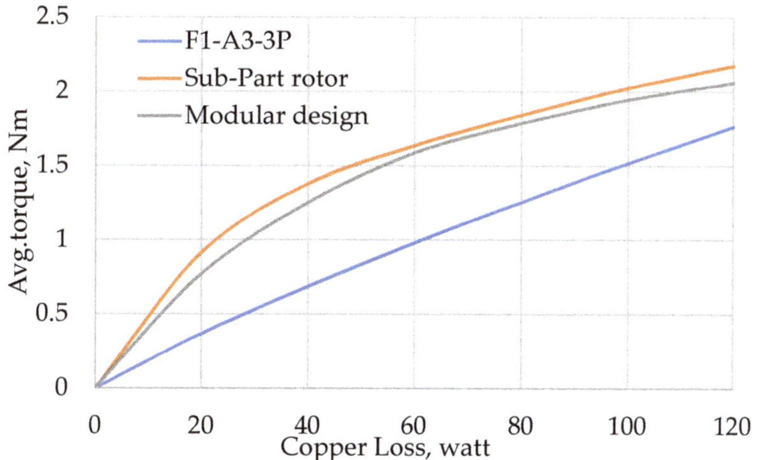

Figure 21. Comparison of average torque at fixed copper losses.

4.9. Torque versus Current Density

Torque versus current density of three FEFS machines is calculated at various current density and maximum current set to 25 A/mm^2. Figure 22 illustrates torque versus the current density of sub-part rotor design, F1-A3-3P design, and modular rotor design. Both conventional machines are saturated beyond 10 A/mm^2, while proposed machine has increasing torque profile, by increasing current density. At maximum 25 A/mm^2 current density F1-A3-3P design has higher average torque than modular rotor and sub-part rotor 65.6% and 63.05%, respectively. F1-A3-3P demonstrates high torque due to high flux linkage.

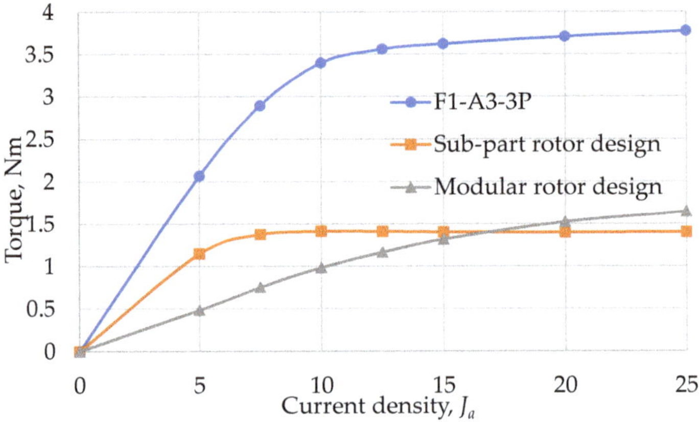

Figure 22. Comparison of average torque versus current densities.

4.10. Torque Density and Power Density

Torque density and power density of three FEFS machines is calculated at a fixed current density of 10 A/mm^2. Figure 23 illustrates torque densities of sub-part rotor design, F1-A3-3P design, and modular rotor design. Comparatively, the torque density of the F1-A3-3P design is 1.89 times higher than modular rotor design and 1.71 times higher than sub-part rotor design as shown in Figure 23. The proposed modular design has a reduced total mass of 23% and 44.8% as compared to sub-part rotor design and F1-A3-3P design, respectively, as shown in Table 4.

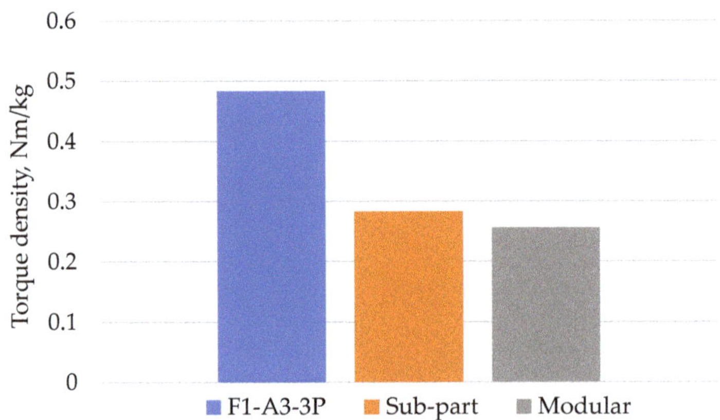

Figure 23. Torque density of three FEFS machines.

Table 4. Comparison of active mass

Design	Stator Mass (Kg)	Rotor Mass (Kg)	Copper Mass (Kg)	Total Mass (Kg)
F1-A3-3P	1.366	0.536	5.137	7.04
Sub-part rotor design	1.499	0.732	2.758	4.99
Modular rotor design	1.825	0.247	1.774	3.84

The power density of conventional and proposed design is expressed in Figure 24. Power density attains by modular rotor design is 0.0783 Watt/kg at current density of FEC, J_e, and armature current density, J_a, of 10 A/mm^2 as shown in Figure 24. High power density exhibits high efficiency and better electromagnetic performance. The proposed 8S/4P modular rotor design achieves 1.3 times and 1.9 times higher power densities as compared to F1-A3-3P design and sub-part rotor design respectively.

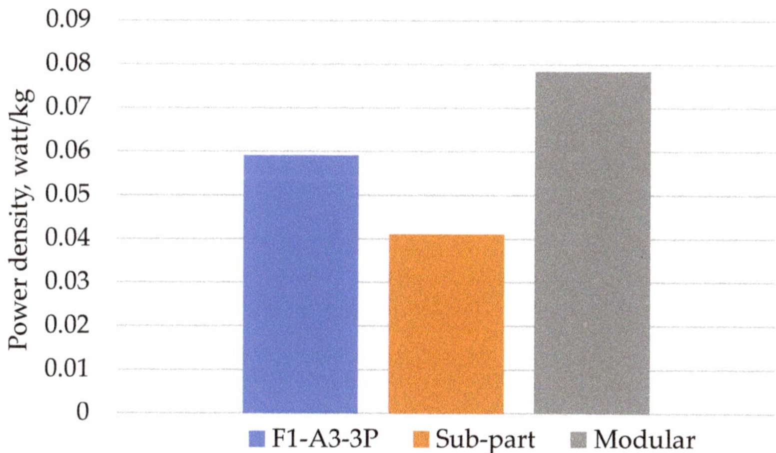

Figure 24. Power density of three FEFS machines.

4.11. Torque and Power Versus Speed Characteristics

The comparison of torque and power versus speed curve of three single phase FEFSM are illustrated in Figure 25. At a rated speed of 1664 rpm, the maximum average torque of the modular rotor design is 1.64 Nm which corresponds to the power generated by the proposed design at 286 W. Additionally, the average torque obtained by conventional 8S-4P sub part rotor design and 6S-3P salient rotor is 1.4 Nm and 3.77 Nm, at a base speed of 1389 rpm and 1053 rpm, respectively. The average torque of proposed design is higher as compared to the sub part rotor design. At a speed of 1600 rpm, the average torque of the proposed design is similar to 6S-3P design while being 19 percent higher than the 8S-4P design. Although the generated power of 8S-6P modular design is 28.4 percent higher than 8S-4P design, it is 31 percent lower than F1-A3-3P design. The pattern plot shows that, beyond rated speed, the average torque of the machine starts to decrease and power is decreased as well. The power of 6S-3P FEFSM decreases more rapidly due to an increase in iron loss above the rated speed.

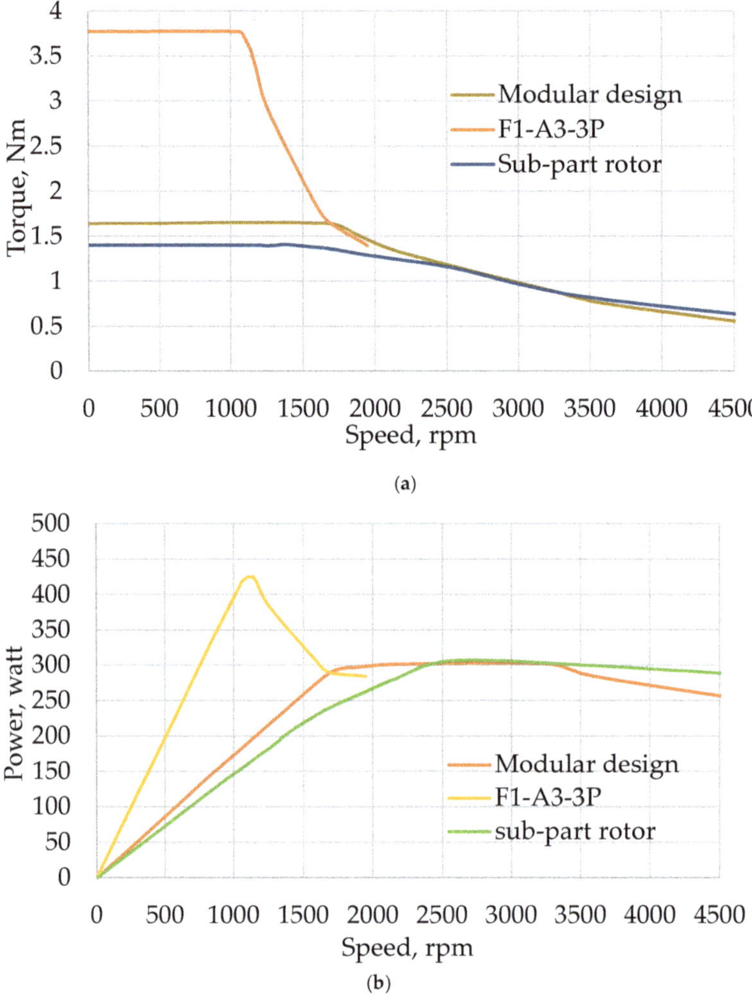

Figure 25. (**a**) Comparison of torque versus speed; (**b**) Comparison of power versus speed of three FEFSM.

4.12. Rotor Stress versus Speed

Rotor stress analysis is a technique to identify the principal stress, nodal force, and displacement occurred in the rotor structure in an ideal state after load is applied. Generally, the condition for mechanical stress of the rotor structure is accomplished by centrifugal force due to the longitudinal rotation of the rotor. Additionally, centrifugal force of the rotor is greatly affected by the speed. The rotor could highly withstand stress, if the principal stress of the rotor is higher. Principal stress is a crucial result in the analysis of stress. By increasing the angular velocity of the rotor, principal stress is increased exponentially. Thus, the rotor principal stress versus the speed of the three-field excited flux switching machines (rotor structure) is analyzed using 3D-FEA. The angular velocity varies from 0 rpm to 20,000 rpm for conventional three-pole salient rotor structure, four-pole sub-part rotor structure, and the proposed six-pole modular structure to analyze the maximum capability of mechanical stress. The constraints that coincide with the force acting on the rotor is faces, edges, and vertices. The maximum principal stress on each rotor at various speed is shown in Table 5.

Table 5. Stress analysis at various speed

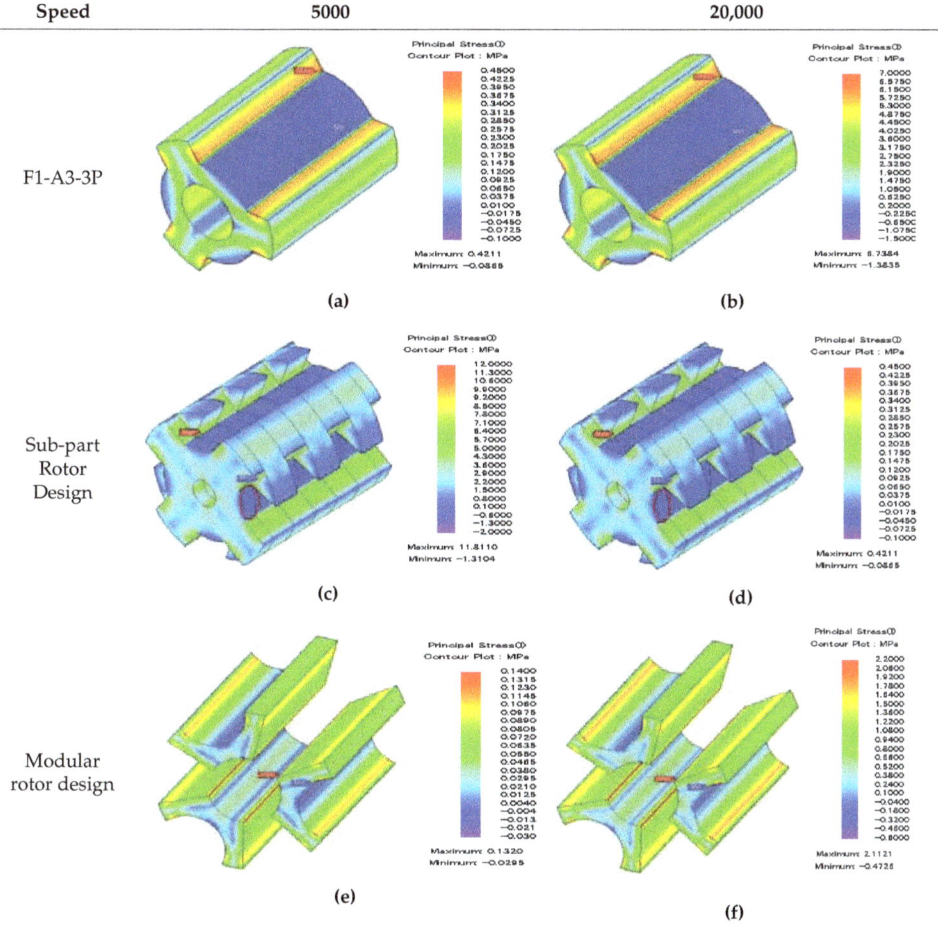

Speed	5000	20,000
F1-A3-3P	(a)	(b)
Sub-part Rotor Design	(c)	(d)
Modular rotor design	(e)	(f)

Figure 26 shows that comparison of principal stress of three different rotor structures versus speed. At a maximum speed of 20,000 rpm, the principal stress of salient rotor structure, sub-part rotor structure and modular rotor structure is 6.73 MPa, 11.61 MPa, and 2.11 MPa respectively. The pattern plot clearly shows that principal stress of proposed modular rotor structure is much lower as compared to the conventional rotor design. The maximum allowable principal stress of 35H210 electromagnetic steel is 300 MPa. All the three rotor structures are capable of high-speed applications, but the only salient rotor structure can be operated at high speeds due to the single piece rotor structure. Whilst the sub-part rotor and modular structure are only applicable for low-speed applications.

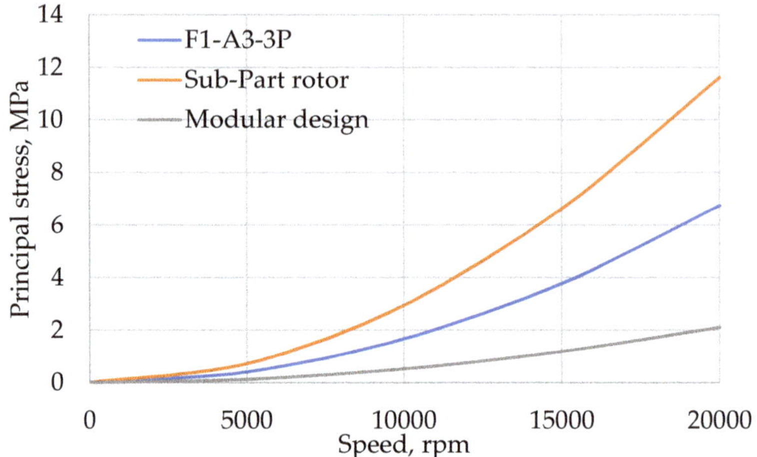

Figure 26. Principal stress against speed.

4.13. Copper Losses versus Je at Various Ja

Copper losses of three FEFSM at various armature current densities is shown in Figure 27. To analyze the total copper losses, FEC current density, J_e, is set to 10 A/mm^2, and armature current density is varied from 0 A/mm^2 to 25 A/mm^2. Figures 28–30, illustrate copper losses of both armature coil and FEC, in isolation at fixed J_e, whilst J_a changes to maximum. The pattern plot clearly showed that the copper losses are increased with increasing current densities. Comparatively, the proposed modular rotor design shows approximately 56% and 88% lower copper losses to sub-part rotor design and F1-A3-3P design respectively, at a maximum armature current density of 25 A/mm^2 as depicted in Figure 27. However, the proposed structure has reduced copper losses, indicating improved efficiency compared with the conventional designs.

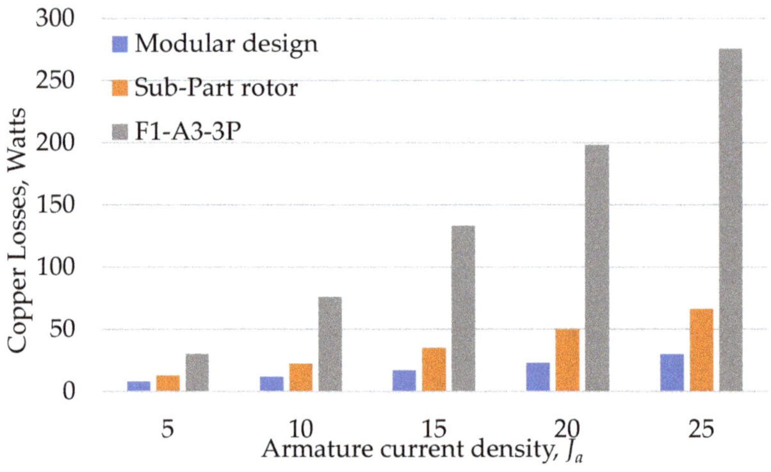

Figure 27. Copper losses versus various J_a.

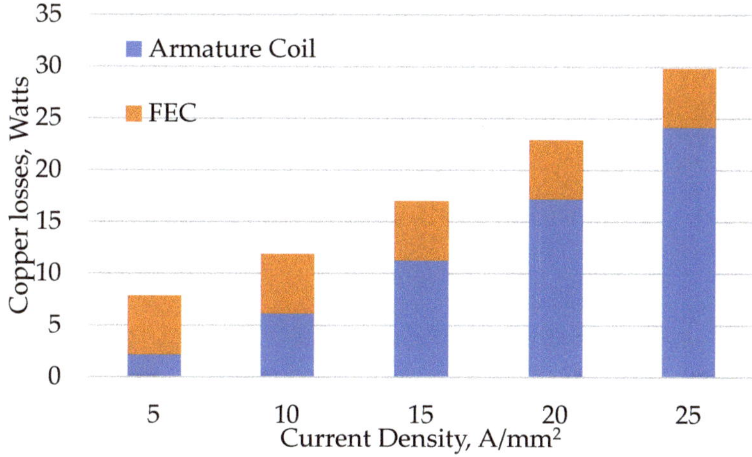

Figure 28. Copper losses of modular rotor design.

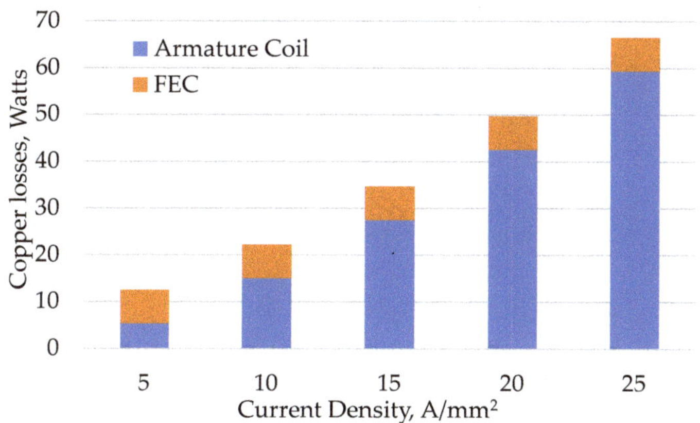

Figure 29. Copper losses of sub-part rotor design.

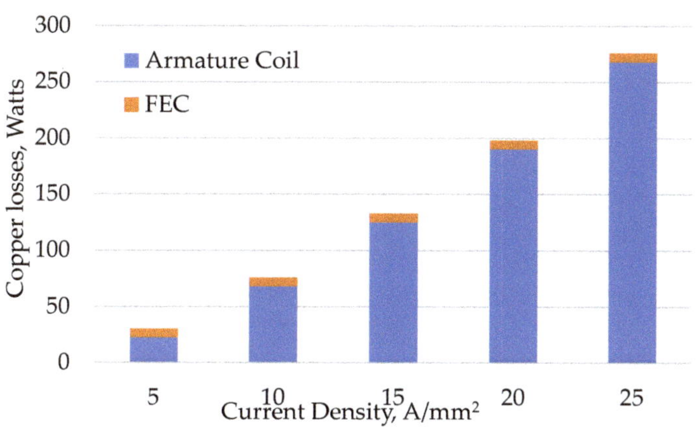

Figure 30. Copper losses of F1-A3-3P design.

4.14. Iron Loss versus Speed

Iron loss is a significant portion in the total losses of machine. Machine performance is greatly affected by iron losses due to flux emphasis of novel-modular topology in the stator, which generates a variation of flux densities in the rotor and stator core [17,22]. The flux density variation is expected to be reduced by implementing the novel-modular topology due to the reduction in utilization of the stator. Iron losses are increased with increasing electrical loading due to higher armature reaction [23]. The iron losses of the switched flux machine also vary greatly with speed at every part as shown in Figures 31–33. At low-speed, the machine dominates electromagnetic losses. The method of iron loss calculation can be found in [17,24].

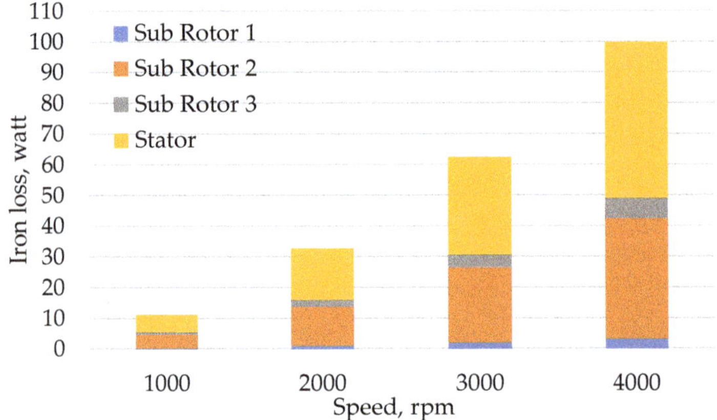

Figure 31. Iron loss at various parts of sub-part rotor design.

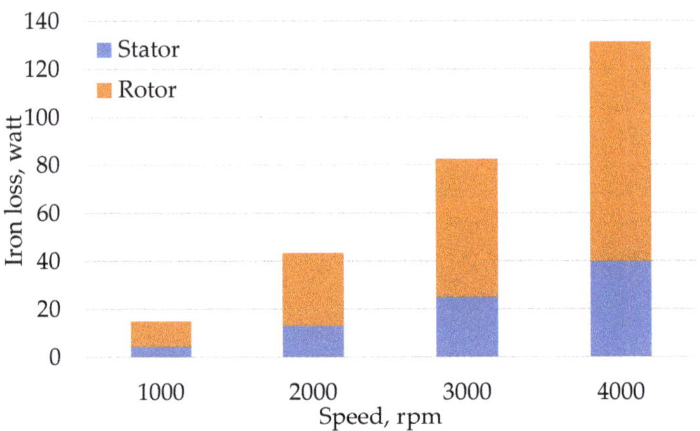

Figure 32. Iron loss at stator and rotor part of F1-A3-3P design.

The iron loss of each component of three-field excited FSM is calculated by 3D-FEA. In Figure 34, the plot clearly shows that the proposed modular rotor structure has lowest iron loss then the conventional sub-part rotor design and F1-A3-3P design. At a maximum speed of 4000 rpm, modular design reduces the iron losses of 29.44% and 7.22% compared with the conventional F1-A3-3P design and sub-part rotor design respectively. The reason for iron loss reduction in the stator is due to the modular rotor, variation of flux densities in the stator-core is investigated in [25].

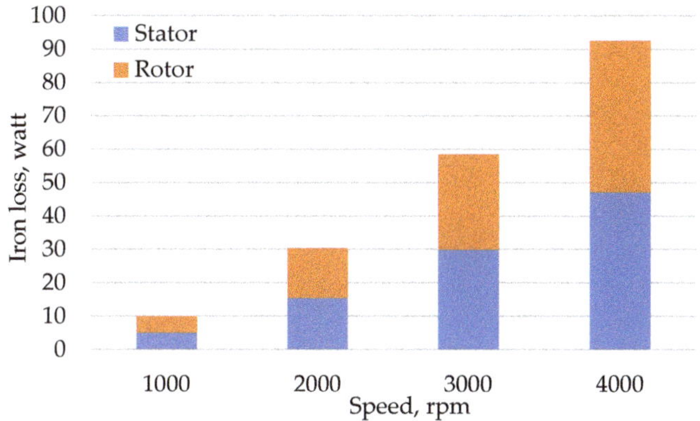

Figure 33. Iron loss at stator and rotor part of modular rotor design.

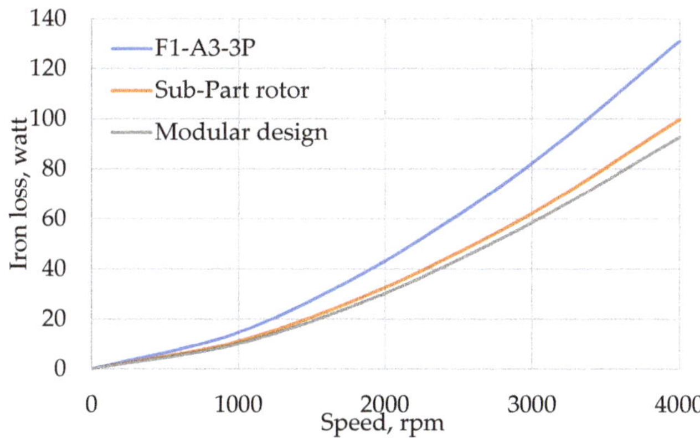

Figure 34. Comparison of iron losses at various speeds.

4.15. Motor Losses and Efficiency Analysis

The efficiencies of three FEFSMs are computed by 3D-FEA, considering all motor losses (iron losses in core laminations and copper losses in FEC and armature coil). Copper losses (P_{cu}) are calculated at fixed current densities of 10 A/mm^2, for both *FEC*, J_e, and armature coil, J_a, for all designs. Whilst, the iron losses are calculated at varying speed of 1000–4000 rpm. In single phase FEFS machines, copper losses can be illustrated as

$$P_{Cu} = I_a^2 R_a + I_f^2 R_f \tag{10}$$

where P_{cu}, I_f, R_f, I_a, and R_a are copper losses, field current, total field coil resistance, armature current, and total armature coil resistance, respectively. Figure 35a–c shows iron losses (P_i), copper losses (P_{cu}), output power (P_o), and efficiency at different speeds (range: 1000–4000 rpm) of sub-part rotor design, F1-A3-3P design, and modular rotor design, respectively. However, with increasing speed the iron losses increase in addition to further degrading efficiency. Furthermore, at every operating speed from 1000 rpm to 4000 rpm, the proposed design achieves comparatively higher efficiencies. At a max speed of 4000 rpm, the iron losses of the proposed modular rotor design are 9% and 30% lower than the conventional sub-part rotor design and F1-A3-3P design, respectively. However, reduction in iron losses shows a significant reduction in total machine losses, approximately 49% of F1-A3-3P design and 15% of sub-part rotor design. Furthermore, by adopting the modular structure, the proposed 8S/6P design achieves a higher average efficiency of approximately 12.8% and 11.4% higher than the conventional F1-A3-3P and sub-part rotor designs, respectively. Finally, it can be seen from Figure 36 that the efficiency of a single phase modular 8S/4P FEFS machine exhibit higher efficiency than other conventional FEFS machines.

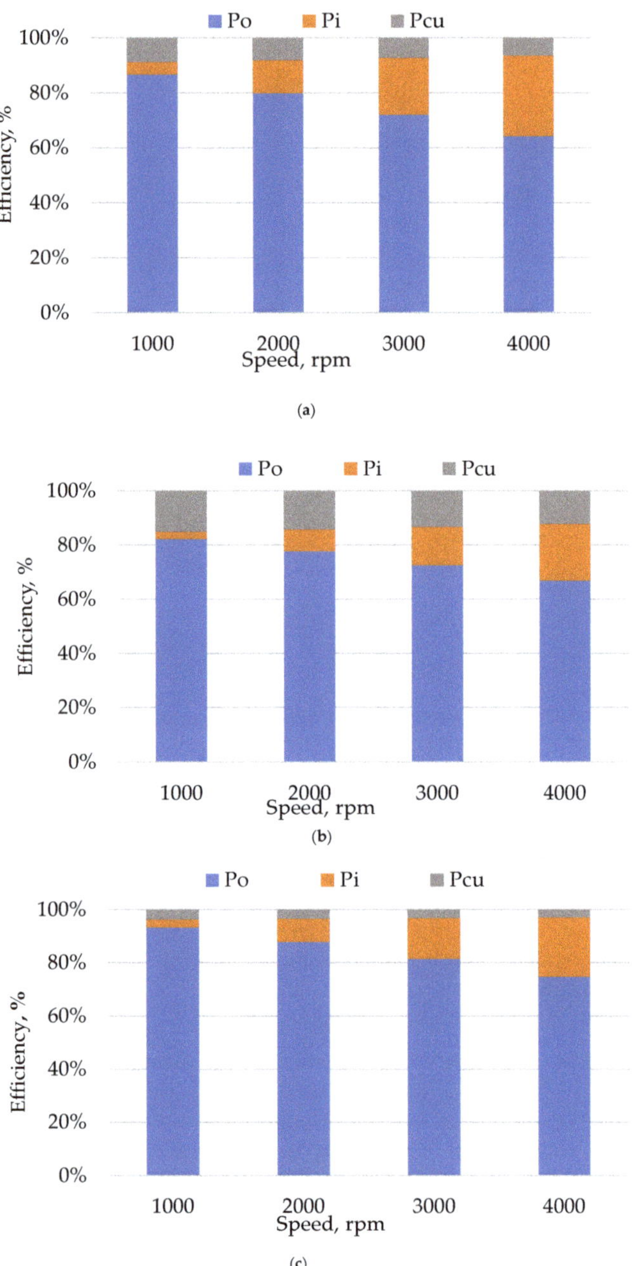

Figure 35. (**a**) Losses and efficiency of sub-part rotor design at various speeds. (**b**) Losses and efficiency of F1-A3-3P design at various speeds. (**c**) Losses and efficiency of modular rotor design at various speeds.

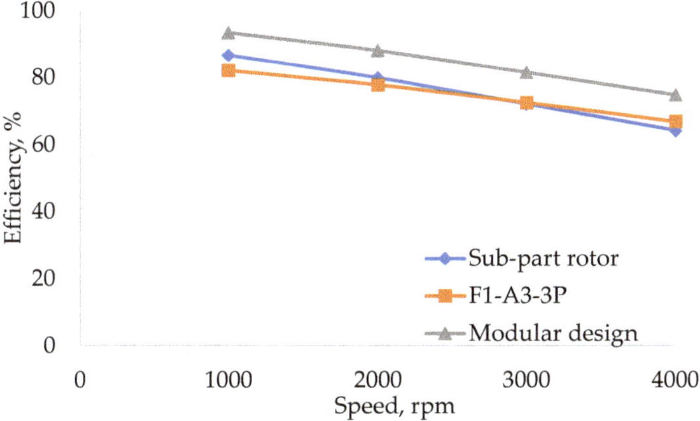

Figure 36. Comparison of efficiency at various rotor speeds.

5. Conclusions

A novel single-phase field excited topology of modular rotor flux switching machine is presented and the result is investigated by 3D-FEA.

In this paper, a comparison of single-phase eight-slot/four-pole sub rotor design and six-slot/three-pole salient rotor design with a novel modular rotor eight-slot/six-pole FSM is demonstrated. For comparison of flux linkage, cogging torque, average torque, and other different analyses of proposed FEFSMs, an optimal split ratio is set identical to the conventional designs.

The performance comparison of the three different types of single-phase eight-slot/four-pole sub rotor design and six-slot/three-pole salient rotor designs with a novel modular rotor eight-slot/six-pole FSM is demonstrated. he optimal split ratio is kept the same as the conventional designs for comparison of flux linkage, cogging torque, average torque, and other analyses of the proposed FSM. The initial design achieved inadequate power and torque production. Therefore, a deterministic optimization technique was adopted to improve the characteristics. The optimized design enhanced power, torque, and efficiency compared to existing eight-slot/four-pole and six-slot/three-pole FESF machines.

Novel modular 8S/6P single phase FSM with non-overlapped winding arrangement is designed. Copper consumption of modular rotor design is much lower than conventional designs that is 90% lower than F1-A3-3P and 56% than sub-part rotor design, at $J_a = 25$ A/mm^2, due to non-overlap winding between FEC and armature coil. The proposed design shows a higher average output torque when compared under constraints of fixed copper losses. Modular rotor structure also exhibits a significant reduction in iron losses, 30% as compared to F1-A3-3P and 9% reduced when compared with sub-part rotor design. Due to the modular structure of the rotor, the active rotor mass of the proposed design is reduced and the use of stator back-iron is lowered without diminishing torque output. This research also examines the principal rotor stress of the conventional rotor designs (sub-part rotor design and three-pole salient rotor design) and proposed (modular) rotor design with a different direction of constraints. Additionally, average efficiency of proposed modular design is increased by 12.8% compared with F1-A3-3P design and 11.4% compared with sub-part rotor design. Hence, the proposed motor is suitable for pedestal fan application by replacing induction machine. The proposed design has not yet been comprehensively analyzed and will be examined in our future work.

Author Contributions: Conceptualization, L.U.R. and F.K.; methodology, M.S.; software, H.A.K.; validation, N.A., L.U.R.; investigation, L.U.R.; resources, F.K.; data curation, H.A.K.; writing—original draft preparation, L.U.R.; writing—review and editing, L.U.R.; visualization, S.A.; supervision, F.K.; project administration, M.A.K.; funding acquisition, L.U.R., F.K., and H.A.

Funding: There is no external funding.

Acknowledgments: This work was supported by COMSATS University Islamabad, Abbottabad Campus and Higher Education Commission of Pakistan (No 8114/KPK/NRPU/R&D/HEC/2017).

References

1. You, Z.C.; Yang, S.M.; Yu, C.W.; Lee, Y.H.; Yang, S.C. Design of a High Starting Torque Single-Phase DC-Excited Flux Switching Machine. *IEEE Trans. Ind. Electron.* **2017**, *64*, 9905–9913. [CrossRef]
2. Fernando, N.; Nutkani, I.U.; Saha, S.; Niakinezhad, M. Flux switching machines: A review on design and applications. In Proceedings of the 2017 20th International Conference on Electrical Machines and Systems (ICEMS), Sydney, Australia, 11–14 August 2017; pp. 1–6.
3. Khan, F.; Sulaiman, E.; Ahmad, M.Z. A novel wound field flux switching machine with salient pole rotor and nonoverlapping windings. *Turk. J. Electr. Eng. Comput. Sci.* **2017**, *25*, 950–964. [CrossRef]
4. Omar, M.F.; Sulaiman, E.; Jenal, M.; Kumar, R.; Firdaus, R.N. Magnetic Flux Analysis of a New Field-Excitation Flux Switching Motor Using Segmental Rotor. *IEEE Trans. Magn.* **2017**, *53*, 1–4. [CrossRef]
5. Zhu, Z.Q.; Chen, J.T. Advanced flux-switching permanent magnet brushless machines. *IEEE Trans. Magn.* **2010**, *46*, 1447–1453. [CrossRef]
6. Zhao, X.; Niu, S.; Fu, W. Design and comparison of electrically excited double rotor flux switching motor drive systems for automotive applications. *CES Trans. Electr. Mach. Syst.* **2018**, *2*, 191–199. [CrossRef]
7. Taha, M.; Wale, J.; Greenwood, D. Design of a high power-low voltage multiphase permanent magnet flux switching machine for automotive applications. In Proceedings of the 2017 IEEE International Electric Machines and Drives Conference (IEMDC), Miami, FL, USA, 21–24 May 2017; pp. 1–8.
8. Sangdehi, S.; Kazemi, M.; Abdollahi, S.E.; Gholamian, S.A. A segmented rotor hybrid excited flux switching machine for electric vehicle application. In Proceedings of the Power Electronics, Drive Systems & Technologies Conference PEDSTC, Mashhad, Iran, 14–16 February 2017; pp. 347–352.
9. Zhang, G.; Hua, W.; Cheng, M.; Liao, J. Design and comparison of two six-phase hybrid-excited flux-switching machines for EV/HEV applications. *IEEE Trans. Ind. Electron.* **2016**, *63*, 481–493. [CrossRef]
10. Sanabria-Walter, C.; Polinder, H.; Ferreira, J.A.; Jänker, P.; Hofmann, M. Torque enhanced flux-switching PM machine for aerospace applications. In Proceedings of the IEEE International Conference in Electrical Machines ICEM, Marseille, France, 2–5 September 2012; pp. 2585–2595.
11. Chen, Y.; Zhu, Z.Q.; Howe, D. Three-dimensional lumped-parameter magnetic circuit analysis of single-phase flux-switching permanent-magnet motor. *IEEE Trans. Ind. Appl.* **2008**, *44*, 1701–1710. [CrossRef]
12. Tang, Y.; Paulides, J.J.H.; Motoasca, T.E.; Lomonova, E.A. Flux-switching machine with DC excitation. *IEEE Trans. Magn.* **2012**, *48*, 3583–3586. [CrossRef]
13. Rauch, S.E.; Johnson, L.J. Design principles of flux-switch alternators. *Trans. Am. Inst. Electr. Eng. Power Apparatus Syst.* **1955**, *74*, 1261–1268.
14. Pollock, H.; Pollock, C.; Walter, R.T.; Gorti, B.V. Low cost, high power density, flux switching machines and drives for power tools. *Ind. Appl. Conf.* **2003**, *3*, 1451–1457.
15. Pollock, C.; Pollock, H.; Brackley, M. Electronically controlled flux switching motors: A comparison with an induction motor driving an axial fan. *Ind. Electron. Soc. IECON* **2003**, *3*, 2465–2470.
16. Zhou, Y.J.; Zhu, Z.Q. Comparison of low-cost single-phase wound-field switched-flux machines. *IEEE Trans. Ind. Appl.* **2014**, *50*, 3335–3345. [CrossRef]
17. Pang, Y.; Zhu, Z.Q.; Howe, D.; Iwasaki, S.; Deodhar, R.; Pride, A. Investigation of iron loss in flux-switching PM machines. In Proceedings of the 4th IET International Conference on Power Electronics, Machines and Drives (PEMD), York, UK, 2–4 April 2008; pp. 460–464.
18. Zhang, Z.; Tang, X.; Wang, D.; Yang, Y.; Wang, X. Novel Rotor Design for Single-Phase Flux Switching Motor. *IEEE Trans. Energy Convers.* **2018**, *33*, 354–361. [CrossRef]
19. Zulu, A.; Mecrow, B.C.; Armstrong, M. A wound-field three-phase flux-switching synchronous motor with all excitation sources on the stator. *IEEE Trans. Ind. Appl.* **2010**, *46*, 2363–2371. [CrossRef]
20. Zhu, L.; Jiang, S.Z.; Zhu, Z.Q.; Chan, C.C. Analytical methods for minimizing cogging torque in permanent-magnet machines. *IEEE Trans. Magn.* **2009**, *45*, 2023–2031. [CrossRef]
21. Abdollahi, S.E.; Vaez-Zadeh, S. Reducing cogging torque in flux switching motors with segmented rotor. *IEEE Trans. Magn.* **2013**, *49*, 5304–5309. [CrossRef]

22. Hoang, E.; Mohamed, G.; Michel, L.; Bernard, M. Influence of magnetic losses on maximum power limits of synchronous permanent magnet drives in flux-weakening mode. In Proceedings of the Industry Applications Conference, Rome, Italy, 8–12 October 2000; pp. 299–303.
23. Chen, J.T.; Zhu, Z.Q.; Iwasaki, S.; Deodhar, R. Comparison of losses and efficiency in alternate flux-switching permanent magnet machines. In Proceedings of the 2010 XIX International Conference on Electrical Machines (ICEM), Rome, Italy, 6–8 September 2010; pp. 1–6.
24. Calverley, S.D.; Jewell, G.W.; Saunders, R.J. *Saunders. Design of a High Speed Switched Reluctance Machine for Automotive Turbo-Generator Applications*; No. 1999-01-2933; SAE Technical Paper; SAE International: Warrendale, PA, USA, 1999.
25. Thomas, A.S.; Zhu, Z.Q.; Wu, L.J. Novel modular-rotor switched-flux permanent magnet machines. *IEEE Trans. Ind. Appl.* **2012**, *48*, 2249–2258. [CrossRef]

Article

Analysis of Torque Ripple and Cogging Torque Reduction in Electric Vehicle Traction Platform Applying Rotor Notched Design

Myeong-Hwan Hwang [1,2], Hae-Sol Lee [1,3] and Hyun-Rok Cha [1,*]

[1]   EV Components & Materials R&D Group, Korea Institute of Industrial Technology,
      6 Cheomdan-gwagiro 208 beon-gil, Buk-gu, Gwangju 61012, Korea; han9215@kitech.re.kr (M.-H.H.);
      eddylee12@kitech.re.kr (H.-S.L.)
[2]   Department of Electrical Engineering, Chonnam National University, 77 Youngbong-ro, Buk-gu,
      Gwangju 61186, Korea
[3]   Robotics and Virtual Engineering, Korea University of Science and Technology, Daejeon 34113, Korea
*   Correspondence: hrcha@kitech.re.kr; Tel.: +82-62-600-6212

Received: 9 October 2018; Accepted: 29 October 2018; Published: 6 November 2018

Abstract: Drive motors, which are used in the drive modules of electric cars, are interior permanent magnet motors. These motors tend to have high cogging torque and torque ripple, which leads to the generation of high vibration and noise. Several studies have attempted to determine methods of reducing the cogging torque and torque ripple in interior permanent magnet motors. The primary methods of reducing the cogging torque involve either electric control or mechanical means. Herein, the authors focused on a mechanical method to reduce the cogging torque and torque ripple. Although various methods of reducing vibration and noise mechanically exist, there is no widely-known comparative analyses on reducing the vibration and noise by designing a notched rotor shape. Therefore, this paper proposes a method of reducing vibration and noise mechanically by designing a notched rotor shape. In the comparative analysis performed herein, the motor stator and rotor were set to be the same size, and electromagnetic field analysis was performed to determine a notch shape that is suitable for the rotor and that generates reasonable vibration and noise.

Keywords: interior permanent magnet synchronous motor; torque ripple; cogging torque; electric vehicle; notch

---

1. Introduction

Globally, the market for eco-friendly vehicles is continuing to grow, and a variety of electric car models are being released in the automotive market. The reason for this increased prevalence of electric cars in the market is because the laws concerning the average amount of carbon dioxide emissions from internal combustion vehicles are becoming increasingly stringent, and there is concern regarding fine particles being emitted into the atmosphere; thus, eco-friendly cars are becoming increasingly prevalent globally [1].

The electric motors that are used to drive electric cars influence the cars' performance considerably. The type of drive motor used can determine the car's mileage, efficiency, torque, vibration, maximum speed, and acceleration. One such drive motor that is widely used nowadays is the interior permanent magnet synchronous motor (IPMSM). The IPMSM involves a structure with a permanent magnet embedded in the interior of the rotor. It has a torque component that is caused by the interior magnet (alignment torque) and a torque component that is caused by the difference in the d–q axis magnetic reluctance (reluctance torque); thus, it can provide a large power density. Because the embedded permanent magnet's magnetic properties are similar to that of an air gap, marked differences in the d–q axis inductance distribution occur in the rotor interior. The motor can have a wide range of variable speed driving properties owing to weak field control in proportion with the saliency ratio. Therefore, these motors continue to be used as the drive motors of electric vehicles because they enable high-power and high-speed operation by ensuring high power density, a wide range of speeds, and mechanical strength, all of which are characteristics of the interior permanent magnet motor [2].

In the interior motor's structure, a magnet is inserted in the rotor, and magnet barriers exist at both ends of the rotor. The electrical device's properties vary based on the path of the magnetic flux produced by the magnet. By applying this property in the desired direction, it is possible to improve upon the loss, power, efficiency, cogging torque, and torque ripple [3]. This has various advantages; however, it also has several disadvantages. The magnetic distribution on the surface of the rotor is not uniform, and more cogging torque and torque ripple occur in this type of motor than with other forms of motors having the same magnetic circuits [4].

If the torque ripple is high, vibration and noise occur in the motor, and this may lead to motor drive system malfunctions. Unlike other motors, an IPMSM uses a permanent magnet to create magnetic field, and cogging torque generally occurs in the motor owing to the difference between the magnetic reluctance at the permanent magnet in the rotor interior and at the stator's slot structure. Cogging torque significantly affects the occurrence of noise and vibration in the motor; hence, it is necessary to reduce it as much as possible at the design stage. Torque ripple, which is related to the back electromotive force harmonics, must also be reduced as much as possible because "it increases the noise and vibration in the motor [5].

Thus, many studies have been conducted to reduce the vibration and noise in interior permanent magnet motors to ensure improved performance and reliability in electric car drive motors. A variety of papers have introduced methods that reduce the cogging torque and torque ripple. These methods involve adjusting the motor's arrangement, adjusting the width of slots and teeth, using permanent magnet skew, creating auxiliary teeth, using slot-less armatures, and using notches [6,7]. In contrast, this study aims to reduce cogging torque and torque ripple by changing the shape and arrangement of the permanent magnet [8,9].

The methods for reducing cogging torque and torque ripple that were used in other studies were mostly electromagnetic methods. However, this paper aims to reduce the cogging torque and torque ripple using a mechanical method. Existing electromagnetic methods have the disadvantage of also reducing the active magnetic flux, which lowers power and efficiency. Mechanical methods, however, reduce active magnetic flux far less than electromagnetic methods and enable high-efficiency high-power motors to be built. As a typical example of this, the T Company has built relatively high-efficiency motors for electric cars using a mechanical method similar to that introduced in this paper. Herein, a 3D finite element method is used to reduce the time required for obtaining an optimal design, and a variety of notches are introduced on the rotor to reduce the cogging torque and torque ripple. A notch shape for optimal torque ripple and cogging torque reduction through detailed shape design is determined. The results of this study are believed to offer a rotor notch design shape that can reduce the vibration and noise in future electric drive cars, and they are believed to contribute toward the efficiency gains of designing drive motors with reduced torque ripple.

This paper first describes the properties of the cogging torque and torque ripple in terms of theoretical equations, and the analysis model and its specifications. Furthermore, details of the comparative analysis performed according to the position and shape of the notch are presented, and the results and conclusions derived from the analysis are discussed.

2. Materials and Methods

2.1. Relevant Equations for Cogging Torque and Torque Ripple

2.1.1. Cogging Torque Equation

Cogging torque is the non-uniform torque of the stator, and it occurs inevitably in a motor that uses a permanent magnet. It is a radius-directed torque that is directed towards the position with the minimum magnetic energy, i.e., in an equilibrium state, in the motor. As shown in Equation (1), the cogging torque can be determined by deriving the drive motor's internal energy by differentiating the magnetic energy with respect to the synchronous motor's rotor position angle.

$$T_{Cogging}(\alpha) = -\frac{\partial W(\alpha)}{\partial \alpha} \tag{1}$$

In Equation (1), α is the rotor position angle and W is the motor's magnetic energy.

$$W_\alpha = \frac{1}{2\mu} \int_V B^2 dV \tag{2}$$

Here, B is the magnetic flux density, and μ is the permeability.

$$B = G(\theta, z)B(\theta, \alpha) \tag{3}$$

In Equation (3), $G(\theta, z)$ is the gap permeance function and $B(\theta, \alpha)$ is the gap magnetic flux density. Furthermore, θ is the angle along the circumference, and α is the rotation angle.

If Equation (3) is substituted into Equation (2):

$$\begin{aligned} W(\alpha) &= \frac{1}{2\mu} \int [G(\theta, z)B(\theta, \alpha)]^2 dV \\ &= \frac{1}{2\mu_0} \int_0^{L_s} \int_{R1}^{R2} \int_0^{2\pi} G^2(\theta, z)B^2(\theta, \alpha)d\theta dr dz \\ &= \frac{1}{2\mu} L_s \frac{1}{2}\left(R_2^2 - R_1^2\right) \int_0^{L_{ef}} \int_0^{2\pi} G^2(\theta, z)B^2(\theta, \alpha)d\theta dz \end{aligned} \tag{4}$$

Here, μ_0 is the air permeability, and L_s is the lamination layer length. R_1 is the inner radius of rotor, and R_2 is the outer radius of rotor.

In Equation (4), if a Fourier series expansion is performed on $G^2(\theta, z)$ and $B^2(\theta, \alpha)$, and the trigonometry function's orthogonality are used to solve Equation (4):

$$\begin{aligned} W(\alpha) &= \frac{L_s}{4\mu_0}\left(R_2^2 - R_1^2\right)\left[\sum_{n=0}^{\infty} G_{nN_L} B_{nN_L} \int_0^{2\pi} \cos nN_L\theta \cos nN_L(\theta + \alpha)\right] \\ &= \frac{L_s}{4\mu_0}\left(R_2^2 - R_1^2\right) \cdot 2\pi \cdot \sum_{n=0}^{\infty} G_{nN_L} B n N_L \cos nN_L\alpha \end{aligned} \tag{5}$$

In conclusion, for the cogging torque, the gap energy is partially differentiated by thve rotor's rotation angle, as shown in Equation (5), and it can be expressed as in Equation (6) [10–13]. Here, N_L is the least common multiple (LCM) between the number of rotor poles and stator slots:

$$\begin{aligned} T_{Cogging}(\alpha) &= -\frac{\partial W(\alpha)}{\partial \alpha} \\ &= \frac{L_s \pi}{2\mu_0}\left(R_2^2 - R_1^2\right) \sum_{n=0}^{\infty} G_{nNL} B_{nN_L} nN_L \sin nN_L\alpha \end{aligned} \tag{6}$$

2.1.2. Torque Ripple Equation

There are several reasons why torque ripple occurs in a permanent magnet synchronous motor, including the cogging torque generated because of the mechanical structure, offset and scale errors in the current sensor in terms of electrical control, fluctuations in the direct current (DC) link voltage, phase current distortions owing to the inherent properties of power switching elements, and dead time [14–17], and distortions in the back electromotive force [18–22].

A synchronous motor's power is determined by the maximum DC voltage and maximum current supplied by the inverter. The maximum current is denoted by I_{max}, and it is determined in conditions that satisfy the thermal rating of the inverter. The equations for current and voltage are as follows:

$$V_{ds}^2 + V_{qs}^2 \leq V_{max}^2$$
$$I_{ds}^2 + I_{qs}^2 \leq I_{max}^2 \tag{7}$$

Here, I_{ds} and I_{qs} are the d- and q-axis currents, respectively. V_{ds} and V_{qs} denote the d and q-axis terminal voltages, respectively.

The following are the voltage equations for a synchronous reference frame that sets the rotor that is rotating at a synchronous speed as the standard coordinate system.

$$v_{ds}^e = r_s i_{ds}^{ie} + \frac{d\lambda_{ds}^\varepsilon}{dt} - \omega_r \lambda_{qs}^e$$
$$v_{qs}^e = r_s i_{qs}^{ie} + \frac{d\lambda_{qs}^\varepsilon}{dt} - \omega_r \lambda_{ds}^e \tag{8}$$

As seen in Equation (8), if it is assumed that the voltage drop due to the stator's phase resistance is not large, it can be said that the terminal voltage is proportional to the speed, ω_r. In addition, the d–q axis magnetic flux part present in Equation (8), which takes into account the harmonic components, can be expressed as shown below. Here, λ_{df_har} and λ_{qf_har} are the d–q axis interlinked magnetic flux harmonic components caused by the permanent magnet.

$$\lambda_{ds}^e = L_d i_{ds}^e + \psi_f + \lambda_{df_{har}}$$
$$\lambda_{qs}^e = L_d i_{qs}^e + \lambda_{qf\_har} \tag{9}$$

The torque equation for the permanent magnet synchronous motor is as follows:

$$T_e = \frac{3P}{4}\left(\lambda_{ds}^e i_q - \lambda_{qs}^e i_d\right) \tag{10}$$

Equation (9) is substituted into Equation (10) to obtain Equation (11). From this, the torque ripple component, which is the torque component that occurs because of the interlinked magnetic flux's harmonic component in the synchronous motor's torque equation, can be derived, as in Equation (12):

$$T_e = \frac{3P}{4}\left\{\psi_f i_q + (L_d - L_q)i_d i_q + \left(\lambda_{df_{har}} i_q - \lambda_{qf\_har} i_d\right)\right\} \tag{11}$$

$$T_e = \frac{3P}{4}\left(\lambda_{df_{har}} i_q - \lambda_{qf\_har} i_d\right) \tag{12}$$

2.2. Analysis Model and Specifications

Figure 1 shows the shapes of the rotor and stator of the basic model employed in this study. A V-shaped magnet configuration was selected from several rotor shapes owing to its excellent speed-versus-torque characteristics and high allowable radial direction force [23–25]. Table 1 shows basic specifications of the interior permanent magnet synchronous motor, used in this paper.

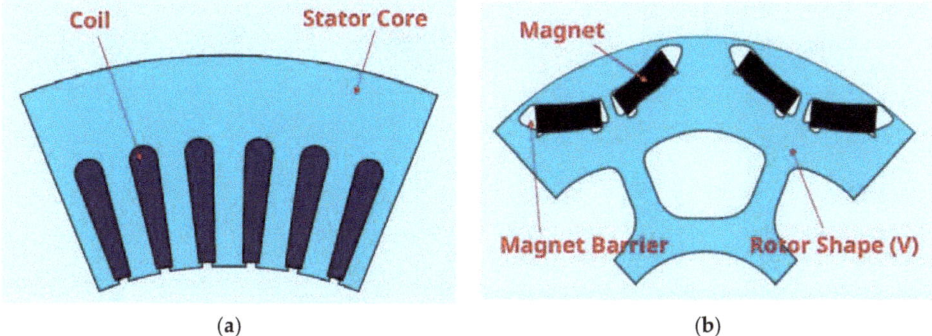

(**a**)

(**b**)

Figure 1. Slot stator and 8-pole rotor core: (**a**) stator core; (**b**) rotor core.

Table 1. Interior permanent magnet synchronous motor (IPMSM) basic specifications.

Parameters	Unit	Value
Number of Slots	EA	48
Number of poles	EA	8
Capacity	kW	35
Rated Speed	rpm	10,000–11,000
Outside Diameter	mm	200
Inside Diameter	mm	122
Stack Length	mm	50
Air Gap	mm	0.7

Thereafter, notches were added at various positions in accordance with the basic model specifications listed above, and a comparative analysis was performed. According to the energy method, the cogging torque is a change in the magnetostatic energy that occurs because of the rotation of the motor. Changes in magnetostatic energy mostly occur when a pole transition occurs through the gap between the stator and the rotor. Consequently, cogging torque occurs at this time. Therefore, when a notch is added to the rotor's surface, the notch acts in the same way as the air gap. As a result of acting as the air gap, it changes the distribution of the gap permeance function, $G(\theta, z)$, and as the number of active slots is changed, G_{nNL} is also changed, which results in reducing the cogging torque This has the effect of reducing the energy changes due to the rotation of the motor, thereby reducing the cogging torque and torque ripple [26,27].

2.3. Test Method

The aforementioned V-shape was selected as a typical shape for an interior permanent magnet. Notches were designed for the drive motor rotor magnet using a mechanical method rather than the existing electromagnetic method, and a comparative analysis was performed on the position with the smallest cogging torque and torque ripple. In each model, a notch was placed on the surface of the rotor to reduce the interior permanent magnet motor's cogging torque, and a comparative analysis was performed. The magnetic flux density distribution of the gap and the cogging torque properties change according to the position of the notch, and thus an optimal notch shape design is necessary. Figure 2 shows the positions of the notches in the analysis. In (a), (b), and (c), respectively, the notches are located on the inner, outer, and central parts of the rotor's surface. In (d), a central notch is added to the (a) notch shape. Finally, (e) shows the basic shape without any notches. A comparative analysis of the shapes according to the five notch positions was performed. After the optimal position was determined, a more detailed design was created accordingly.

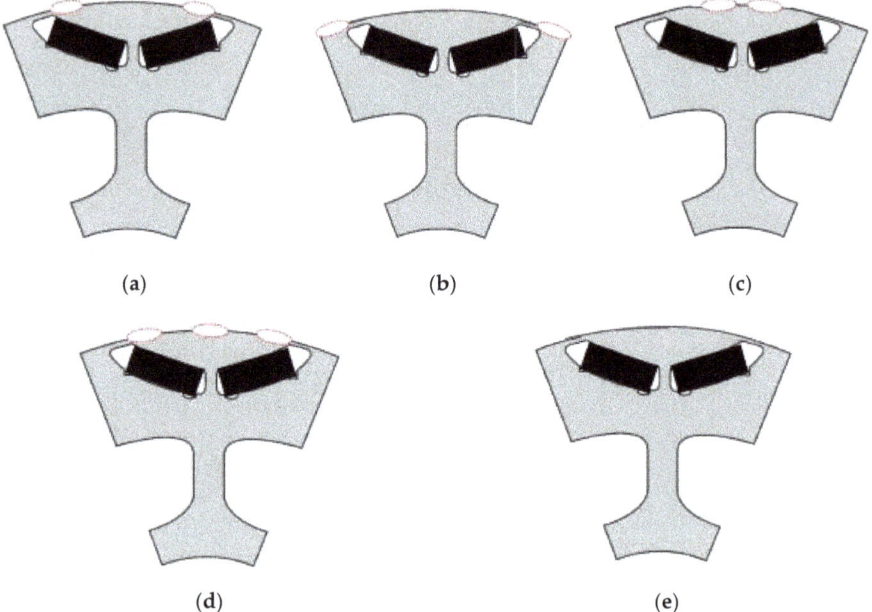

Figure 2. Motor rotor notch design positions: (**a**) magnet's inner notches; (**b**) magnet's outer notches; (**c**) magnet's central notches; (**d**) magnet's additional central notches; (**e**) no notches (basic form).

3. Comparative Analysis and Results

3.1. Comparative Analysis of Cogging Torque and Torque Ripple According to Notch Position

For analysis, we used the motor electromagnetic field analysis tool program, "JMAG" (ver. 14.1), made from Nagoya, Japan, with data produced through simulation analysis. The cogging torque is calculated by dividing the difference between the maximum and minimum value of torque by the mean value of torque and presenting the result in percentile values. The back electromotive force voltage and cogging torque analysis speed were analyzed at 1000 rpm. For the torque ripple, the analysis was performed at 4000 rpm, which is the rotational speed corresponding to the maximum power output. Figure 3 and Table 2 show the corresponding analysis results. The back electromotive force voltage was found to be low for the V-shape (a) (20.68 V) and V-shape (b) (20.87 V). These are thus considered shapes that can ensure a high rotational speed when the motor is under no load. The cogging torque was 1.7 N m for V-shape (a) and 1.73 N m for V-shape (b). When there are notches on the inner and outer parts of the magnet, a rotor with a low cogging torque value can be designed. The torque ripple of V-shape (b) was found to be 8.04%, which is 7.86% lower than that of V-shape (a). The cogging torque of V-shape (b) was 0.03 N m higher than that of V-shape (a); however, its cogging torque value was lower than V-shape (a) against its motor power and size. These results show that the torque ripple and cogging torque can be lowered simultaneously when notches are placed on the outer edge of the magnet, as in V-shape (b).

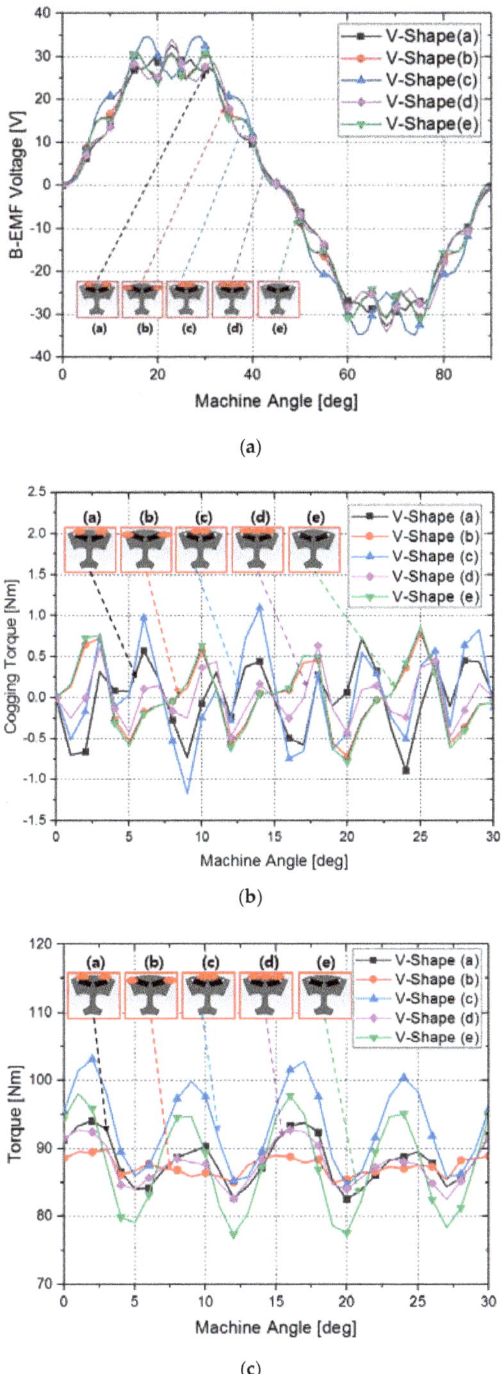

Figure 3. Back electromotive force voltage, cogging torque, and torque ripple according to notch position: (**a**) back electromotive force voltage; (**b**) cogging torque; (**c**) comparative analysis of torque ripple.

Table 2. Comparative analysis of cogging torque and torque ripple according to notch position.

Characteristic	V-Shape (a)	V-Shape (b)	V-Shape (c)	V-Shape (d)	V-Shape (e)
B-EMF Voltage [V_rms]	20.68	20.87	23.03	26.45	24.08
Cogging Torque [Nm]	1.7	1.73	2.6	1.21	1.882
Torque [Nm]	88	87	93	87.66	87.62
Torque Ripple [%]	15.9	8.04	22.58	12.9	24.41
Power [kW]	36.86	36.44	38.95	36.71	36.70

3.2. Comparative Analysis of Cogging Torque and Torque Ripple According to Notch Shape

The comparison results in Table 2 show that the shape with notches on the outer part of the magnet had low torque ripple and cogging torque. Therefore, a method is proposed that can optimize this shape to obtain the lowest torque ripple. Figure 4 shows diagrams of the notch optimization position and the detailed design. In Figure 4b, the radius refers to the length of the radius of a circle drawn from an arbitrary point on the stator surface. The shape and size of the notch is changed by changing the radius.

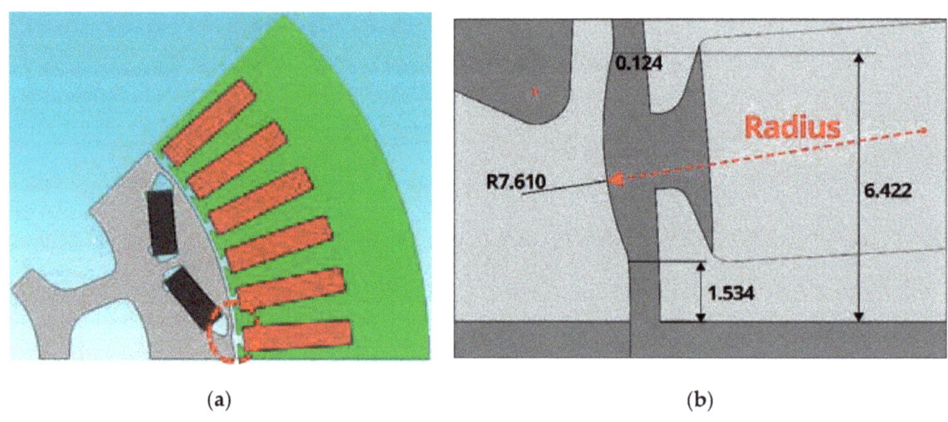

(a)   (b)

Figure 4. Notch shape optimization: (**a**) notch optimization position; (**b**) notch optimization diagram.

3.2.1. Comparative Analysis of Cogging Torque and Torque Ripple According to Changes in Notched Shape

In the first optimized shape, the standard point radius was 7.61 mm, and shape optimizations within a range of ±3 mm of the radius were analyzed. Table 3 lists the changes in notch width according to changes in the radius.

Table 3. Rotor notch radius parameter.

Analysis Case	Width [mm]	Radius(2) [mm]
1	0.75	5
2	0.82	6
3	0.9	7
4	0.93	7.61
5	0.94	8
6	0.96	9
7	0.98	10

The cogging torque analysis results shown in Figure 5 show that the cogging torque was 4.14 N m at the initial standard point radius of 7.61 mm, and it was 3.88 N m when the radius was 5 mm, which corresponds to the lowest cogging torque. As the radius increased, the cogging torque increased.

An interior permanent magnet (IPM) motor delays the current phase angle and uses the magnet torque and reluctance torque together; thus, the torque ripple and cogging torque according to the current phase angle were analyzed. Figures 6 and 7, respectively, show the torque ripple and torque, according to the radius, for various current phase angles. When the radius was 5 mm, the torque ripple was low, at 9.7%. As the radius increased, the power increased, and the torque ripple increased. When the radius was 10 mm, and the current source was $0°$, the torque ripple was 15%.

The comparative analysis results in Figure 7 shows that when the notch radius changes from 5 to 10 mm, the difference in the torque is less than 1 N m. Overall, it was determined that a smaller radius can lower the cogging torque and torque ripple considerably. A large notch shape does not reduce the cogging torque. Consequently, an additional analysis was performed to obtain a design to minimize the notch size and reduce the cogging torque and torque ripple.

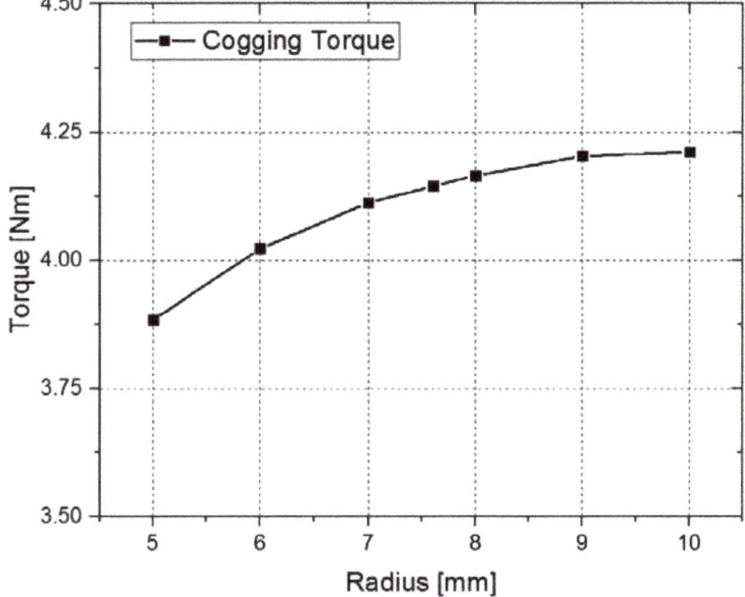

Figure 5. Comparison of cogging torque according to notch radius.

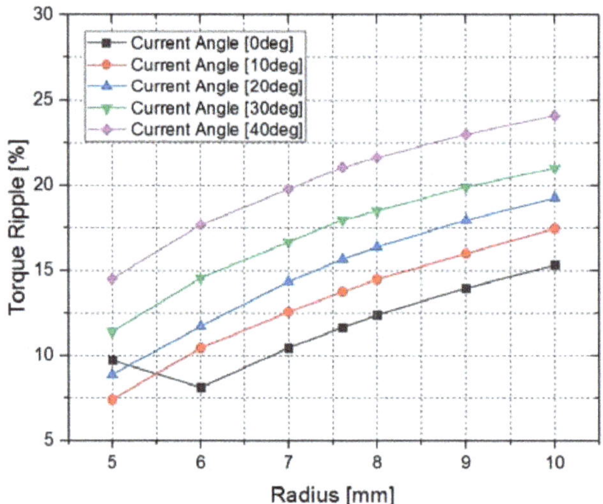

Figure 6. Comparative analysis of torque ripple according to notch radius.

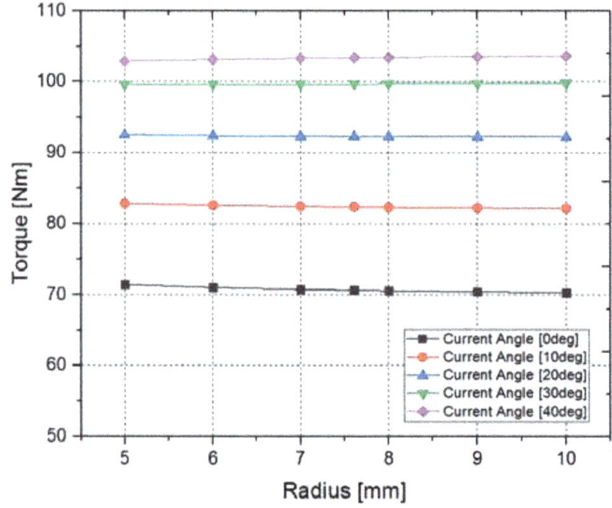

Figure 7. Comparative analysis of torque according to notch radius.

3.2.2. Comparative Analysis of Notch Dimensions for Optimal Design

According to the previous discussion, new parameters were set to obtain a notch design to reduce the cogging torque and torque ripple, as shown in Figure 8. Distance (1) was fixed at 0.5 mm, and it is the distance from an end of the notch's arc, when its radius (3) is 2 mm. Distance (1) was fixed to observe the analysis result of torque and torque ripple with the width difference. The changes in radius (3) were analyzed according to changes in notch depth, and not the width, and distance (2) indicates the distance from the magnet barrier to the notch, and it was measured because it is related to the flow of the magnetic flux.

Table 4 lists the values of distance (2) when distance (1) was fixed at 0.5 mm and radius (3) was changed.

To obtain the results shown in Figure 9, distance (1) was fixed at 0.5 mm, and a comparative analysis was performed on the cogging torque. The cogging torque was 3.7 N m at a distance (2) of 0.5 mm and a radius of 2.1 mm. At the deepest point, the cogging torque was 4.29 N m, which indicates a difference of 0.59 N m.

Figure 10 shows the results of analysis of the torque ripple according to radius (2) for various current phase angles. When the current phase angle was 0°, 10°, or 20°, the torque ripple was large if radius (2) was small. However, when the current phase angle was controlled to be 30° or 40°, the torque ripple tended to decrease as radius (2) decreased.

Table 4. Changes in distance (2) and radius (3) when distance (1) is fixed at 0.5 mm.

	Distance (2) [mm]	Radius (3) [mm]
1	0.5	2.1
2	0.78	2.5
3	0.93	3.1
4	1	3.5
5	1	4
6	1.05	4.451

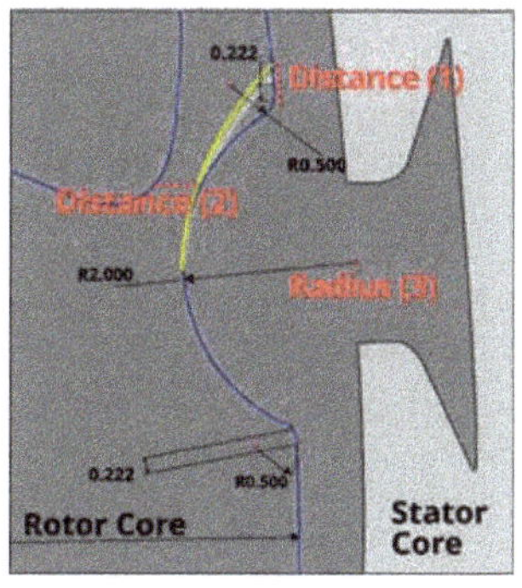

Figure 8. Notch optimization diagram with length parameter.

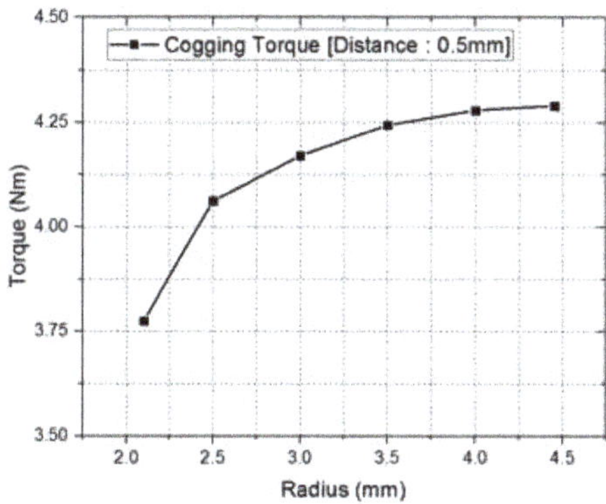

Figure 9. Comparative analysis of cogging torque when notch shape distance (1) is 0.5 mm.

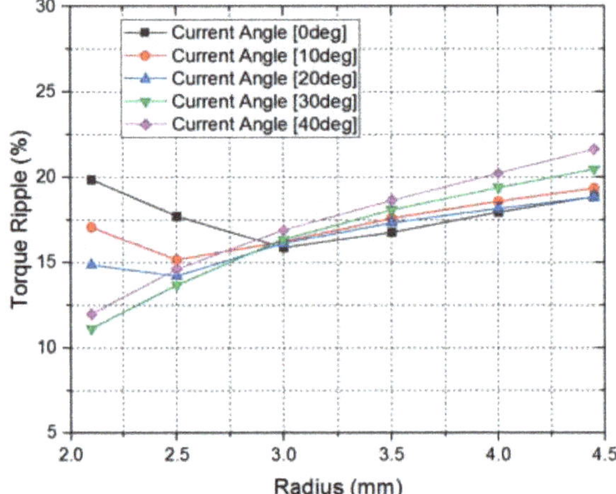

Figure 10. Comparative analysis of torque ripple when notch shape distance (1) is 0.5 mm.

4. Discussion

In this study, a finite element method was used to analyze the properties of interior permanent magnet synchronous motors that are used as drive motors in electric cars. The external diameters of the stator and the rotor were set to be the same in order to compare various parameters such as the cogging torque, torque ripple, and back electromotive force voltage according to the position of a notch on the rotor shape. In the electromagnetic field analysis, the input voltage and input current were set to be the same. The motor's performance characteristics were analyzed according to the notch position on the rotor shape.

The aim of this work was to determine an approach to obtain the design a rotor shape that can minimize the cogging torque and torque ripple so that a low-noise drive motor can be designed. The application target was an IPM motor; thus, current phase angle control was required to ensure that various characteristics of the motor, such as efficiency, power, and torque, do not degrade. Therefore, the point with the maximum torque was found via phase angle control, and the torque ripple, cogging torque, and maximum power were determined.

First, the notch position on the rotor shape was analyzed. Notches can be placed in a variety of positions, and thus, to determine the optimum position, notches were placed in five locations on the rotor and a comparative analysis of the resulting characteristics was performed. The results confirmed that placing the notches on the outer part of the magnet produced the best properties, and this design was then further optimized to obtain the best possible design.

Next, in the design phase, the distance between the rotor and the magnet barrier was fixed, and the original size of the notch was changed. When the overall analysis results were considered, V-shape (a) was found to be the most suitable notch shape in terms of reducing the noise and vibration, owing to the low values of the configuration's torque ripple and cogging torque in the analysis in which distance (1) was 1 mm.

This paper proposed a notch shape optimization method for reducing the vibration and noise in vehicles. Vibration and noise are considered to be particularly important factors in electric cars. Luxury cars experience many problems pertaining to vibration and noise, and these factors can be reduced by designing the drive module via the proposed method. The proposed method can be widely used in the design of electric car drive motors to ensure an appropriate noise level.

Author Contributions: Conceptualization, M.-H.H. and H.-S.L.; Data curation, M.-H.H.; Formal analysis, M.-H.H.; Methodology, H.-S.L.; Supervision, H.-R.C.; Validation, H.-S.L.; Visualization, M.-H.H.; Writing—original draft, M.-H.H. and H.-S.L.; Writing—review and editing, H.-R.C.

Funding: This research was funded by the support of the Korea Institute of Industrial Technology as "Variable Architecture Powertrain Platform and Self-driving Factor Technology Development for Industrial EV Self-driving Vehicle" KITECH EO-18-0020.

Conflicts of Interest: The authors declare no conflicts of interest.

References

1. Miller, T.J.E. *Design of Brushless Permanent Magnet Motor*; Clarendon Press: Oxford, UK, 1994.
2. Pyrhonen, O.; Niemela, M.; Pyrhonen, J.; Kaukonen, J. Excitation control of DTC controlled salient pole synchronous motor in field weakening range. In Proceedings of the 5th International Workshop Advanced Motion Control (AMC), Coimbra, Portugal, 29 June–1 July 1998; pp. 294–298.
3. Aihara, T.; Toba, A.; Yanase, T.; Mashimo, A.; Endo, K. Sensorless torque control of salient-pole synchronous motor at zero-speed operation. *IEEE Trans. Power Electron.* **1999**, *14*, 202–208. [CrossRef]
4. Ko, H.S.; Kim, K.J. Characterization of Noise and Vibration Sources in Interior Permanent-Magnet Brushless DC Motors. *IEEE Trans. Magn.* **2004**, *40*, 3482–3489. [CrossRef]
5. Hosinger, V.B. Performance of Polyphase Permanent Magnet AC Motor. *IEEE Trans. Power Appar. Syst.* **1980**, *99*, 1510–1518. [CrossRef]
6. Sitapati, K.; Krishnan, R. Performance Comparisons of Radial and Axial Field, Permanent-Magnet, Brushless Machines. *Proc. IEEE Ind. Appl.* **2001**, *37*, 1219–1226. [CrossRef]
7. Hwang, S.M.; Eom, J.B.; Jung, Y.H.; Lee, D.W.; Kang, B.S. Various design techniques to reduce cogging torque by controlling energy variation in permanent magnet motors. *IEEE Trans. Magn.* **2001**, *37*, 2806–2809. [CrossRef]
8. Mi, C.; Filippa, M.; Liu, W.; Ma, R. Analytical method for prediction the air-gap flux of interior-type permanent magnet machine. *IEEE Trans. Magn.* **2004**, *4*, 50–58. [CrossRef]
9. Im, C.Y.; Son, J.W.; Jung, S.Y. *Characteristic Analysis of Interior Permanent Magnet Synchronous Motor for Electric Vehicles Propulsion with High-Speed Operation*; The Korean Institute of Electrical Engineers: Seoul, Korea, 2011; pp. 97–99.
10. Jin, Y.S.; Cho, K.Y.; Kim, H.W.; Lim, B.K.; Han, B.M. Torque ripple reduction based on flux linkage harmonics observer for an interior PM synchronous motor including back EMF harmonics. *Trans. Korean Inst. Power Electron.* **2013**, *18*, 367–375. [CrossRef]
11. Lee, S.H.; Hong, I.P.; Park, S.J.; Kim, C.U. Torque ripple minimization for IPMSM with non-sinusoidal back-EMF. *Trans. Korean Inst. Power Electron.* **2002**, *7*, 91–100.
12. Lee, D.H.; Kim, C.H.; Kwon, Y.A. Reduction of torque ripple of PMSM using iterative flux estimation. *Trans. Korean Inst. Power Electron.* **2001**, *6*, 346–350.
13. Hwang, S.M.; Eom, J.B.; Hwang, G.B.; Jeong, W.B.; Jung, Y.H. Cogging torque and acoustic noise reduction in permanent magnet motors by teeth paring. *IEEE Trans. Magn.* **2000**, *36*, 3144–3146. [CrossRef]
14. De la Ree, J.; Boules, N. Torque production impermanent-magnet synchronous motors. *IEEE Trans. Ind. Appl.* **1989**, *25*, 107–112. [CrossRef]
15. Kwon, S.O.; Lee, J.J.; Lee, G.H.; Hong, J.P. Torque ripple reduction for permanent magnet synchronous motor using harmonic current injection. *Trans. Korean Inst. Power Electron.* **2009**, *58*, 1930–1935.
16. Chen, S.; Song, A.; Sekiguchi, T. High efficiency and low torque ripple control of permanent magnet synchronous motor based on the current tracking vector of electromotive force. *Conf. Rec. IEEE Ind. Appl. Soc.* **2000**, *3*, 1725–1729. [CrossRef]
17. Yang, Y.; Castano, S.; Yang, R.; Kasprzak, M.; Bilgin, B.; Sathyan, A.; Dadkhah, H.; Emadi, A. Design and Comparison of Interior Permanent Magnet Motor Topologies for Traction Application. *IEEE Trans. Transp. Electr.* **2017**, *3*, 4–9. [CrossRef]
18. Kwak, J.; Min, S.; Hong, J.P. Optimal Stator Design of Interior Permanent Magnet Motor to Reduce Torque Ripple Using Level Set Method. *IEEE Trans. Magn.* **2010**, *46*, 2108–2111. [CrossRef]
19. Lee, S.H.; Hong, J.P.; Hwang, S.M. Optimal Design for Noise Reduction in Interior Permanent-Magnet Motor. *IEEE Trans. Ind. Appl.* **2009**, *45*, 1945–1960.
20. Kioumarsi, A.; Moallem, M.; Fahimi, B. Mitigation of Torque Ripple in Interior Permanent Magnet Motors by Optimal Shape Design. *IEEE Trans. Magn.* **2006**, *42*, 3706–3711. [CrossRef]

21. Markovic, M.; Jufer, M.; Perriard, Y. Determination of tooth cogging for hard-disk Brushless DC motor. *IEEE Trans. Magn.* **2005**, *41*, 4421–4426. [CrossRef]

22. Lee, J.; Bae, J.; Oh, S.-Y.; Kim, S.-J. Optimal design of a Magnetization Fixture for Cogging Torque Reduction of ODD Spindle Motor. In Proceedings of the 31st International Telecommunications Energy Conference, Incheon, Korea, 18–22 October 2009; pp. 1–3.

23. Zhu, Z.Q.; Howe, D. Analytical prediction of the cogging torque in radial-field permanent magnet brushless motors. *IEEE Trans. Magn.* **1992**, *28*, 1371–1374. [CrossRef]

24. Kurnia, A.; de Larminat, R.; Desmond, P.; O'Gorman, T. A low torque ripple PMSM drive for EPS applications. In Proceedings of the Nineteenth Annual IEEE Applied Power Electronics Conference & Exposition (APEC), Anaheim, CA, USA, 22–26 February 2004; Volume 2, pp. 1130–1136.

25. Tamura, H.; Ajima, T.; Noto, Y. A torque ripple reduction method by current sensor offset error compensation. In Proceedings of the 15th European conference on Power Electronics and Application (EPE), Lille, France, 2–6 September 2013; pp. 1–10.

26. Cho, K.Y.; Lee, Y.K.; Mok, H.S.; Kim, H.W. Torque ripple reduction of a PM synchronous motor for electric power steering using a low resolution position sensor. *J. Power Electron.* **2010**, *10*, 709–716. [CrossRef]

27. Yoon, D.Y.; Hong, S.C. Reduction of torque ripple due to current-sensing errors in inverter-fed AC motor systems. *Trans. Korean Inst. Power Electron.* **1998**, *3*, 211–217.

Article

An Analytical Subdomain Model of Torque Dense Halbach Array Motors

Moadh Mallek [1],*, Yingjie Tang [1], Jaecheol Lee [1], Taoufik Wassar [1], Matthew A. Franchek [1] and Jay Pickett [2]

[1] Department of Mechanical Engineering, University of Houston, Houston, TX 77004, USA; ytang23@Central.UH.EDU (Y.T.); jlee84@uh.edu (J.L.); twassar@Central.UH.EDU (T.W.); MFranchek@Central.UH.EDU (M.A.F.)
[2] Technology Development, NOV Corporate Engineering, Houston, TX 77036, USA; Jay.pickett@nov.com
* Correspondence: mmallek@uh.edu; Tel.: +1-832-991-0491

Received: 30 October 2018; Accepted: 20 November 2018; Published: 22 November 2018

Abstract: A two-dimensional mathematical model estimating the torque of a Halbach Array surface permanent magnet (SPM) motor with a non-overlapping winding layout is developed. The magnetic field domain for the two-dimensional (2-D) motor model is divided into five regions: slots, slot openings, air gap, rotor magnets and rotor back iron. Applying the separation of variable method, an expression of magnetic vector potential distribution can be represented as Fourier series. By considering the interface and boundary conditions connecting the proposed regions, the Fourier series constants are determined. The proposed model offers a computationally efficient approach to analyze SPM motor designs including those having a Halbach Array. Since the tooth-tip and slots parameters are included in the model, the electromagnetic performance of an SPM motor, described using the cogging torque, back-EMF and electromagnetic torque, can be calculated as function of the slots and tooth-tips effects. The proposed analytical predictions are compared with results obtained from finite-element analysis. Finally, a performance comparison between a conventional and Halbach Array SPM motor is performed.

Keywords: mathematical model; Halbach Array; surface permanent magnet; magnetic vector potential; torque

1. Introduction

Permanent magnet (PM) motors have been deployed on machine electromechanical machinery for decades, owing to their reliable performance, electrical stability and durability [1]. These motors have been the industry gold standard, finding their way into both rotary and linear applications. PM machines are largely classified into three categories based on the placement of permanent magnets, namely the: (i) inset PM machine, (ii) internal PM machine and (iii) surface PM machine. To improve the electromagnetic torque density, investigations have focused on improving permanent magnet performance within the rotor. Recently, motor designs have moved from samaraium cobalt (SmCo) to neodymium iron boron (NdFeB). While NdFeB has higher remanence, achieving torque enhancements, such embodiments suffer from less coercivity and are prone to demagnetization. As a result, motor designs have created intricate ways to remove heat from this class of machines whereby NdFeB is applicable. Different laminate materials have also been investigated with the use of water-cooling and different magnet arrangements to improve performance. All these advancements are valuable and have improved the torque density of rotating machines. Presented in this manuscript is the application of these integrated principles to discover new magnet arrangements based on the Halbach Array.

The Halbach Array was originally discovered in the 1970s and replicated by Hans Halbach in the mid-1980s. This new magnetic array arrangement oriented the magnet poles such that the flux is magnified in a specific direction while limiting the flux in the opposite direction. Despite concentrating the magnetic flux is a desired direction; a Halbach Array may not necessarily improvement torque density. In fact, it is possible to create an adverse effect on torque density. For that reason and the associated manufacturing costs, Halbach Arrays have not found a niche in electric motor applications.

Halbach Array magnetization was proposed in many applications [2] and has become increasingly popular [1] with various topologies and applications: radial [3] and axial-field [4], slotted and slotless [5–8], tubular and planar for rotary [9,10] or linear [11] machines. Halbach Array configurations exhibit several attractive features including a sinusoidal field distribution in air gap that results in a minimal cogging torque and a more sinusoidal back EMF waveform [12]. Thus, using Halbach Array SPM eliminates the conventional design techniques such as skewing of the stator/rotor [13], optimization of the magnet pole-arc [14] and distributed stator windings [15]. The Halbach Array magnets configuration generates a steady magnetic field in the active region which yields many benefits including: (i) compact form with high torque density up to 30%, (ii) less weight due to less rotor back iron volume and (iii) lower rotor inertia [16]. Adopting a Halbach Array does not always assure torque improvement in motors. To improve torque performance, the magnetization pattern of the PMs in the array must be cautiously designed. Furthermore, the theoretical electro-magnetic modeling is very advantageous to improve the control applications [17,18]. Torque-dense motor design optimization requires a computationally efficient mathematical model parameterized by design parameters including geometric, electric and magnetic motor properties. Many studies have been dedicated to the analytical calculation of the electromagnetic devices and Halbach Array machines [19].

Liu et. al proposed a method to divide and establish nonlinear adaptive lumped equivalent magnetic circuit (LPMC) that included the slot effect [20]. Using Kirchhoff's Laws, the model predicts the electromagnetic performance through an iterative process. The accuracy of these predictions were found to be lower than that of the subdomain analytical solutions [21]. In addition, transfer relations, such as Melcher's method, analytically predict the electromagnetic characteristics of a tubular linear actuator with Halbach Array [22]. Their work derives the generalized vector potentials due to permanent magnets and single-phase winding current using transfer relations. However, the influences of stator slot and tooth tips effects were neglected. Shen et al. applied the subdomain method for Halbach Array slotless machine [3]. The motor performance is derived through a scalar potential calculation over three regions: rotor back iron, permanent magnets and air gap. The major limitation of this approach is that the model does not account for the slot and tooth-tips effects for different winding layouts.

A 2-D model for SPM machines was developed in for conventional permanent magnets magnetizations [23]. A subdomain model incorporating the influence of slot and tooth-tips was developed to predict the armature reaction field for conventional SPM [24]. This particular model is valid for Halbach Array magnetization. Broadening the class of Halbach Array magnetic parameterization, a slotless motor model was developed having different pole-arc [25,26]. This feature is employed in the developed model by introducing the so-called *magnet ratio* parameter that improved the electromagnetic performance over an equal pole-arc machine (conventional SPM).

The contribution of this paper is the development of a subdomain motor model for a slotted Halbach Array SPM motor incorporating both slot and tooth-tip effects. Torque maximization is accomplished by optimizing Halbach Array magnets configuration through the introduction of the magnet ratio. The objective is to predict the electromagnetic characteristics quantified by cogging torque, back-EMF and electromagnetic torque. The motor embodiment is divided into five physical regions; rotor back iron, magnets region, air-gap, tooth opening and slot region. Maxwell Equations are developed for each subdomain vector potential from which the flux density is derived. The governing equations for vector potential distribution are solved using the Fourier method. This solution provides the vector potential in each subdomain expressed as a Fourier series that is a function of unknown constants. The boundary conditions are emphasized at this stage, the vector potential and magnetic field should be continuous at the interface between two subdomains. A resultant linear system of analytical independent equations relating all the unknown constants is formulated. Then, the permanent magnet flux linkage is computed as an integral of the normal flux density along the stator bore. The coil is considered as either a punctual source of current [27,28] or a current sheet over the slot opening [29] where the slotting effect is approximated. The flux linkage is calculated by an integral of vector potential over the slot area. The flux passing through the conductors is accounted for using this method. Finally, having the flux linkage, the back-EMF and electromagnetic torque will be derived.

The remaining of the paper is organized as follows. In Section 2, the analytical field modeling and derivation with the subdomain method is developed. In Section 3, the electromagnetic performance, including back-EMF waveforms, electromagnetic torque and cogging torque are compared with the FE results, and the impact of magnet ratio on torque improvement is analyzed. Thermal analysis is carried out in Section 4 and conclusions are provided in Section 5.

2. Methods and Mathematical Modeling

For the purposes of model development, consider the schematic view of a 10-pole/12-slot PM machine Figure 1. The motor has a two-segmented Halbach Array with non-overlapping windings. The 2-D mathematical model developed in this section is based on the following assumptions: (1) infinite permeable iron materials, (2) negligible end effect, (3) linear demagnetization characteristic and full magnetization in the direction of magnetization, (4) non-conductive stator/rotor laminations, and (5) the gaps between magnets have the same constant relative permeability as magnets [23].

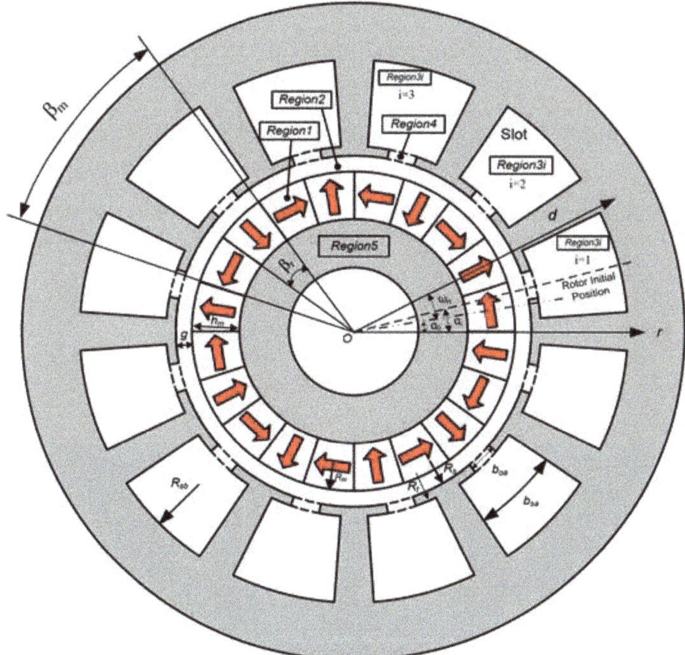

Figure 1. Symbols and Regions of Subdomain Model with Tooth-tips.

2.1. Vector Potential Distribution

The magnetic flux density can be expressed as [19]:

$$B = \mu_0(\mu_r H + M).\tag{1}$$

and μ_r (H/m) is the relative permeability, μ_0 (H/m) is the permeability of vacuum, M (A/m) is the magnetization, and H [A/m] is the magnetic field intensity. The vector potential can be expressed using the magnetic field as:

$$\nabla B = -\nabla^2 A. \tag{2}$$

Substituting B from (1) into (2) gives:

$$\nabla^2 A = -\mu_0 \mu_r \nabla \times H - \mu_0 \nabla \times M = -\mu_0 \mu_r J - \mu_0 \nabla \times M \tag{3}$$

where J (A/m^2) is the current density. Provided the eddy current does not influence the field distribution, the vector potential equation in the magnets becomes:

$$\nabla^2 A = -\mu_0 \nabla \times M. \tag{4}$$

The 2-D field vector potential has only one component along the z-axis that must satisfy the following equations based on (4):

1) Magnet Region:

$$\frac{\partial^2 A_{z1}}{\partial r^2} + \frac{1}{r}\frac{\partial A_{z1}}{\partial r} + \frac{1}{r^2}\frac{\partial^2 A_{z1}}{\partial^2 \alpha^2} = -\mu_0 \nabla \times M = -\frac{\mu_0}{r}\left(M_\alpha - \frac{\partial M_r}{\partial \alpha}\right) \tag{5}$$

2) Slot Region:

$$\frac{\partial^2 A_{z3}}{\partial r^2} + \frac{1}{r}\frac{\partial A_{z3}}{\partial r} + \frac{1}{r^2}\frac{\partial^2 A_{z3}}{\partial^2 \alpha^2} = -\mu_0 J \tag{6}$$

3) Air-Gap, Slot Opening and Rotor Back Iron Regions:

$$\frac{\partial^2 A_{z2,4,5}}{\partial r^2} + \frac{1}{r}\frac{\partial A_{z2,4,5}}{\partial r} + \frac{1}{r^2}\frac{\partial^2 A_{z2,4,5}}{\partial^2 \alpha^2} = 0 \tag{7}$$

where r is the radial position and α is the circumferential position where the d-axis shown in Figure 1 is the origin and anti-clockwise is the forward direction. The radial and circumferential components M_r Figure 2 and M_α Figure 3 are:

$$M_r = \sum_{k=1,3,5,\ldots} M_{rk} \cos(kp\alpha) \tag{8}$$

and:

$$M_\alpha = \sum_{k=1,3,5,\ldots} M_{\alpha k} \sin(kp\alpha). \tag{9}$$

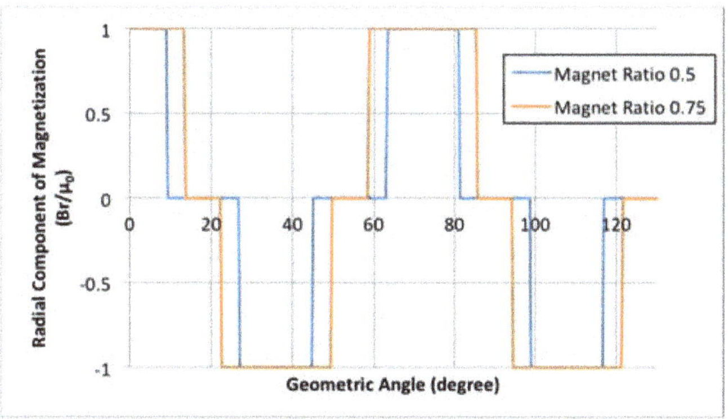

Figure 2. Radial Component of Magnetization for Two Magnet Ratios (0.5 and 0.75).

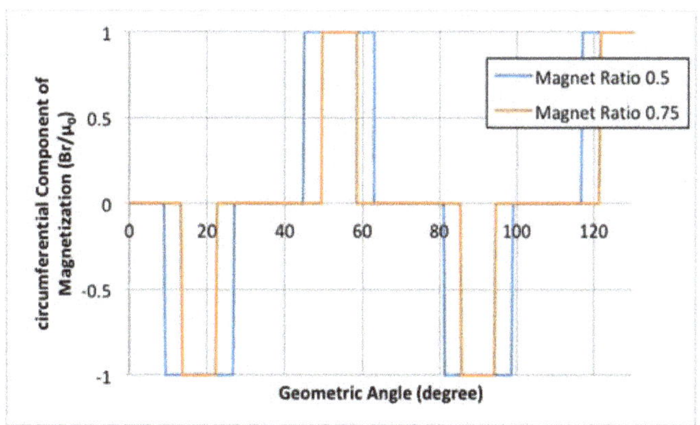

Figure 3. Circumferential Component of Magnetization for Two Magnet Ratios (0.5 and 0.75).

All the series expansions in this paper are infinite but there will be an introduction of finite harmonic orders in Section 3 for the sake of calculations and matrices inversions. Using the Halbach Array representation, the Fourier series coefficients are defined as:

$$M_{rk} = 2\frac{B_r}{\mu_0} R_{mp} \frac{\sin\left(R_{mp}\frac{k\pi}{2}\right)}{R_{mp}\frac{k\pi}{2}} \tag{10}$$

and:

$$M_{\alpha k} = -2\frac{B_r}{\mu_0} R_{mp} \frac{\cos\left(R_{mp}\frac{k\pi}{2}\right)}{R_{mp}\frac{k\pi}{2}} \tag{11}$$

where B_r is residual flux density of magnet and R_{mp} is the *magnet ratio* defined as the ratio of the pole arc β_r to pole pitch of a single pole β_m (Figure 4):

$$R_{mp} = \frac{\beta_r}{\beta_m} \tag{12}$$

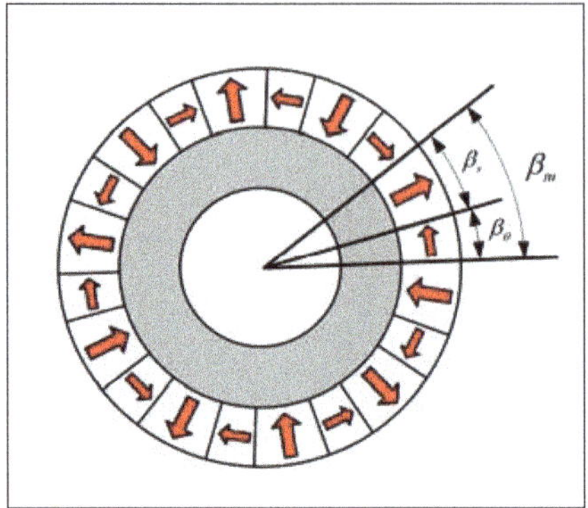

Figure 4. Halbach Array Magnet Ring.

The current density, J, can be expressed as:

$$J = J_{i0} + \sum_n J_{in} \cos\left[E_n\left(\alpha + \frac{b_{sa}}{2} - \alpha_i\right)\right], \tag{13}$$

$$J_{i0} = \frac{d(J_{i1} + J_{i2})}{b_{sa}}, \tag{14}$$

$$J_{in} = \frac{2}{n\pi}(J_{i1} + J_{i2}\cos(n\pi))\sin(n\pi d/b_{sa}) \tag{15}$$

and:

$$E_n = \frac{n\pi}{b_{sa}}. \tag{16}$$

where $-\frac{b_{sa}}{2} \leq \alpha \leq \alpha_i + \frac{b_{sa}}{2}$, and α_i and b_{sa} are the slot position and the slot width angle respectively. Thus, Equation (5) becomes:

$$\frac{\partial^2 A_{z1}}{\partial r^2} + \frac{1}{r}\frac{\partial A_{z1}}{\partial r} + \frac{1}{r^2}\frac{\partial^2 A_{z1}}{\partial^2 \alpha^2} = -\frac{\mu_0}{r}\left(\sum_k [(M_{\alpha ck} - kM_{rsk})\cos(k\alpha) + (M_{\alpha sk} + kM_{rck})\sin(k\alpha)]\right). \tag{17}$$

The general solution for Equation (17) is:

$$\begin{aligned}
A_{z1} &= \sum_k \left[A_1\left(\frac{r}{R_m}\right)^k + B_1\left(\frac{r}{R_r}\right)^{-k} + \frac{\mu_0 r}{k^2-1}(M_{\alpha ck} - kM_{rsk})\right]\cos(k\alpha) \\
&+ \sum_k \left[C_1\left(\frac{r}{R_m}\right)^k + D_1\left(\frac{r}{R_r}\right)^{-k} + \frac{\mu_0 r}{k^2-1}(M_{\alpha sk} + kM_{rck})\right]\sin(k\alpha)
\end{aligned} \tag{18}$$

where A_1, B_1, C_1 and D_1 are coefficients will be determined later and R_r and R_m are the radii of rotor back iron and magnet surfaces respectively.

The general solution for (7), the vector field in the air-gap, is:

$$A_{z2} = \sum_k \left[A_2\left(\frac{r}{R_s}\right)^k + B_2\left(\frac{r}{R_m}\right)^{-k}\right]\cos(k\alpha) + \sum_k \left[C_2\left(\frac{r}{R_s}\right)^k + D_2\left(\frac{r}{R_m}\right)^{-k}\right]\sin(k\alpha) \tag{19}$$

where A_2, B_2, C_2 and D_2 are coefficients to be determined and R_s is the radius of inner stator surface. For non-overlapping windings, the general solution of (6), the vector field in the ith slot, can be derived by incorporating the boundary condition along the slot bottom on the surface of infinite permeable lamination when the circumferential component flux density is zero, namely:

$$A_{z3i} = A_0 + \sum_n A_n \cos\left[E_n\left(\alpha + \frac{b_{sa}}{2} - \alpha_i\right)\right] \tag{20}$$

where:

$$A_0 = \frac{\mu_0 J_{i0}}{4}\left(2R_{sb}^2 \ln(r) - r^2\right) + Q_{3i}, \tag{21}$$

$$A_n = D_{3i}\left[G_3\left(\frac{r}{R_{sb}}\right)^{E_n} + \left(\frac{r}{R_t}\right)^{-E_n}\right] + \mu_0 \frac{J_{in}}{E_n^2 - 4}\left[r^2 - \frac{2R_{sb}^2}{E_n}\left(\frac{r}{R_{sb}}\right)^{E_n}\right] \tag{22}$$

And:

$$G_3 = \left(\frac{R_t}{R_{sb}}\right)^{E_n} \tag{23}$$

Q_{3i} and D_{3i} are coefficients to be derived using regions interactions. In the ith slot opening, the vector field is derived by considering the boundary condition on both sides of the slot opening:

$$A_{z4i} = D\ln(r) + Q_{4i} + \sum_m\left[C_{4i}\left(\frac{r}{R_t}\right)^{F_m} + D_{4i}\left(\frac{r}{R_s}\right)^{-F_m}\right]\cos\left(F_m\left(\alpha + \frac{b_{oa}}{2} - \alpha_i\right)\right) \tag{24}$$

where D, Q_{4i}, C_{4i} and D_{4i} are unknown coefficients to be determined, b_{oa} is the slot opening width and F_m is defined as follows:

$$F_m = \frac{m\pi}{b_{oa}}. \tag{25}$$

The general solution for the vector field in the rotor back iron is:

$$A_{z5} = \sum_k\left[A_5\left(\frac{r}{R_r}\right)^k + B_5\left(\frac{r}{R_0}\right)^{-k}\right]\cos(k\alpha) + \sum_k\left[C_5\left(\frac{r}{R_r}\right)^k + D_5\left(\frac{r}{R_0}\right)^{-k}\right]\sin(k\alpha) \tag{26}$$

where A_5, B_5, C_5 and D_5 are coefficients to be determined later, R_0 is the radius of inner rotor back iron surface.

The unknown coefficients in the expressions of vector potentials (18)–(26) are determined by applying the continuations of normal flux density and circumferential vector potential between subdomains. The details derivations are presented in the following section.

2.2. Boundary Conditions

The radial and circumferential components of flux density is calculated from the vector potential distribution as:

$$B_r = \frac{1}{r}\frac{\partial A_z}{\partial \alpha} \text{ and } B_\alpha = -\frac{\partial A_z}{\partial r}. \tag{27}$$

In the magnet region, the magnetic flux density is:

$$B = \mu_0\mu_r H + \mu_0 M. \tag{28}$$

while it is expressed in the rotor back iron, air gap, slot and slot opening as:

$$B = \mu_0 H. \tag{29}$$

2.2.1. Interface Between Air and Rotor Back Iron

Applying (29) to calculate the circumferential magnetic field intensity $H_{5\alpha}$ in rotor back iron, the boundary condition on the surface requires:

$$H_{5\alpha}\big|_{r=R_0} = \frac{1}{\mu_0\mu_r}B_{5\alpha}\bigg|_{r=R_0} = 0. \tag{30}$$

Based on (26) and (27), $B_{5\alpha}$ expression is calculated and then applied in (30) which leads to the following equations:

$$\begin{cases} B_5 = G_5 A_5 \\ D_5 = G_5 C_5 \end{cases} \tag{31}$$

where:

$$G_5 = \left(\frac{R_0}{R_r}\right)^k. \tag{32}$$

2.2.2. Interface Between Rotor Back Iron and Halbach Permanent Magnets

The first boundary condition in this surface requires that the circumferential magnetic field intensity is continuous between the rotor back iron and Halbach Array permanent magnets giving:

$$H_{5\alpha}\big|_{r=R_r} = H_{1\alpha}\big|_{r=R_r}. \tag{33}$$

Referring to (28) $H_{1\alpha}$ is calculated as:

$$H_{1\alpha} = \frac{1}{\mu_r}\left(\frac{B_{1\alpha}}{\mu_0} - M_\alpha\right) \tag{34}$$

substituting (34) into (33) gives the expressions:

$$\begin{cases} A_1G_1 - B_1 + A_5\left(G_5^2 - 1\right) = \frac{\mu_0 R_r}{k^2-1}\left(M_{rsk} - kM_{\alpha ck}\right) \\ C_1G_1 - D_1 + C_5\left(G_5^2 - 1\right) = -\frac{\mu_0 R_r}{k^2-1}\left(M_{rck} + kM_{\alpha sk}\right) \end{cases}. \tag{35}$$

The second boundary condition is that the normal flux density is continuous between rotor back iron and Halbach Array permanent magnets. B_{5r} is derived from Equations (27) and (29) while B_{1r} is calculated from Equations (27) and (28):

$$B_{5r}\big|_{r=R_r} = B_{1r}\big|_{r=R_r} \tag{36}$$

leading to:

$$\begin{cases} A_1G_1 + B_1 - A_5\left(G_5^2 + 1\right) = \frac{\mu_0 R_r}{k^2-1}\left(kM_{rsk} - M_{\alpha ck}\right) \\ C_1G_1 + D_1 - C_5\left(G_5^2 + 1\right) = -\frac{\mu_0 R_r}{k^2-1}\left(kM_{rck} + M_{\alpha sk}\right) \end{cases}. \tag{37}$$

2.2.3. Interface Between Air Gap and Halbach Permanent Magnets

Firstly, the circumferential magnetic field intensity is continuous between the air gap and the Halbach Array permanent magnets:

$$H_{1\alpha}\big|_{r=R_m} = H_{2\alpha}\big|_{r=R_m}. \tag{38}$$

Referring to (29), $H_{2\alpha}$ is calculated as:

$$H_{2\alpha} = \frac{B_{2\alpha}}{\mu_0}. \tag{39}$$

Substituting (39) into (38) and given $H_{1\alpha}$ from (34), we conclude:

$$\begin{cases} -A_1 + B_1 G_1 + \mu_r A_2 G_2 - \mu_r B_2 = \frac{\mu_0 R_m}{k^2-1}(kM_{\alpha ck} - M_{rsk}) \\ -C_1 + D_1 G_1 + \mu_r C_2 G_2 - \mu_r B_2 = \frac{\mu_0 R_m}{k^2-1}(M_{rck} + kM_{\alpha sk}) \end{cases}. \tag{40}$$

Secondly, the normal flux density is continuous between the air gap and the Halbach Array permanent magnets:

$$B_{2r}|_{r=R_m} = B_{1r}|_{r=R_m} \tag{41}$$

from which the following expressions can be derived:

$$\begin{cases} A_1 + B_1 G_1 - B_2 - A_2 G_2 = \frac{\mu_0 R_m}{k^2-1}(kM_{rsk} - M_{\alpha ck}) \\ C_1 + D_1 G_1 - D_2 2 - C_2 G_2 = -\frac{\mu_0 R_m}{k^2-1}(kM_{rck} + M_{\alpha sk}) \end{cases}. \tag{42}$$

2.2.4. Interface Between Air Gap and Slot Opening

Based on (24) and (27), the circumferential component flux density within the slot opening:

$$B_{4i\alpha} = -\frac{D}{r} - \sum_m \frac{F_m}{R_s}\left[C_{4i}\left(\frac{R_s}{R_t}\right)\left(\frac{r}{R_t}\right)^{F_m-1} - D_{4i}\left(\frac{r}{R_s}\right)^{-F_m-1}\right]\cos\left(F_m\left(\alpha + \frac{b_{oa}}{2} - \alpha_i\right)\right). \tag{43}$$

Since the stator core material is infinitely permeable which make the circumferential component flux density $B_{s\alpha}$ along the stator bore outside the slot opening equal to zero. Along the stator bore, the circumferential component flux density is expressed as [23]:

$$B_{s\alpha} = \sum_k [C_s \cos(k\alpha) + D_s \sin(k\alpha)] \tag{44}$$

where:

$$C_s = \sum_i \sum_m \left(-\frac{F_m}{R_s}(C_{4i}G_4 - D_{4i})\right)\eta_i + \sum_i \left(-\frac{D}{R_s}\right)\eta_{i0}, \tag{45}$$

$$D_s = \sum_i \sum_m \left(-\frac{F_m}{R_s}(C_{4i}G_4 - D_{4i})\right)\xi_i + \sum_i \left(-\frac{D}{R_s}\right)\xi_{i0}, \tag{46}$$

$$\eta_i(m,k) = -\frac{k}{\pi(F_m^2 - k^2)}[\cos(m\pi)\sin(k\alpha_i + kb_{oa}/2) - \sin(k\alpha_i - kb_{oa}/2)], \tag{47}$$

$$\xi_i(m,k) = \frac{k}{\pi(F_m^2 - k^2)}[\cos(m\pi)\cos(k\alpha_i + kb_{oa}/2) - \cos(k\alpha_i - kb_{oa}/2)], \tag{48}$$

$$\eta_{i0}(k) = 2\sin(kb_{oa}/2)\frac{\cos(k\alpha_i)}{k\pi} \tag{49}$$

and:

$$\xi_{i0}(k) = 2\sin(kb_{oa}/2)\frac{\sin(k\alpha_i)}{k\pi}. \tag{50}$$

The first boundary condition demands that the circumferential flux density is continuous between the air gap and the slot opening giving:

$$B_{2\alpha}|_{r=R_s} = B_{s\alpha}|_{r=R_s}. \tag{51}$$

Having $B_{s\alpha}$ from (44) and calculating $B_{2\alpha}$ from (19) and (27), condition (51) leads to:

$$\begin{cases} R_s C_s = -kA_2 + kG_2 B_2 \\ R_s D_s = -kC_2 + kG_2 D_2 \end{cases}. \tag{52}$$

The vector potential distribution within air gap along the stator bore is given as

$$A_s = A_{z2}|_{r=R_s} = \sum_k [(A_2 + B_2 G_2)\cos(k\alpha) + (C_2 + D_2 G_2)\sin(k\alpha)]. \tag{53}$$

The vector potential distribution over the slot opening along the stator bore is

$$A_s = \left(\sum_k ((A_2 + B_2 G_2)\sigma_{i0} + (C_2 + D_2 G_2)\tau_{i0}) \right) \\ + \sum_m \left[\left(\sum_k ((A_2 + B_2 G_2)\sigma_i + (C_2 + D_2 G_2)\tau_i) \right) \cos(F_m(\alpha + b_{oa}/2 - \alpha_i)) \right] \tag{54}$$

where

$$\sigma_{i0} = \left(\frac{\pi}{b_{oa}} \right) \eta_{i0}(k), \tag{55}$$

$$\tau_{i0} = \left(\frac{\pi}{b_{oa}} \right) \xi_{i0}(k), \tag{56}$$

$$\sigma_i = \left(\frac{2\pi}{b_{oa}} \right) \eta_i(m,k), \tag{57}$$

and

$$\tau_i = \left(\frac{2\pi}{b_{oa}} \right) \xi_i(m,k). \tag{58}$$

The vector potential distribution within slot opening along the stator bore is

$$A_{z4i}|_{r=R_s} = D\ln(R_s) + Q_{4i} + \sum_m [C_{4i}G_4 + D_{4i}]\cos\left(F_m\left(\alpha + \frac{b_{oa}}{2} - \alpha_i \right) \right) \tag{59}$$

where

$$G_4 = \left(\frac{R_s}{R_t} \right)^{F_m}. \tag{60}$$

The second boundary condition involves the continuation vector potential for $\left(\alpha_i - \frac{b_{sa}}{2} \le \alpha \le \alpha_i + \frac{b_{sa}}{2} \right)$ is

$$A_{z4i}|_{r=R_s} = A_s \tag{61}$$

thus

$$Q_{4i} = \sum_k [(A_2 + B_2 G_2)\sigma_{i0} + (C_2 + D_2 G_2)\tau_{i0}] - D\ln(R_s) \tag{62}$$

and

$$D_{4i} + C_{4i}G_4 = \sum_k [(A_2 + B_2 G_2)\sigma_i + (C_2 + D_2 G_2)\tau_i]. \tag{63}$$

2.2.5. Interface Between Slot Opening and the Slot

Along the interface between the slot and the slot opening, the circumferential component of the flux density is derived from (24) and (27) [1]

$$B_{4i\alpha}|_{r=R_t} = B_{4i\alpha 0} + \sum_m B_{4i\alpha m}\cos(F_m(\alpha + b_{oa}/2 - \alpha_i)) \tag{64}$$

where

$$B_{4i\alpha 0} = -\frac{D}{R_t} \tag{65}$$

and

$$B_{4i\alpha m} = -\frac{F_m}{R_t}(C_{4i} - D_{4i}G_4) \tag{66}$$

The circumferential flux density in the slot along the outer radius of the slot opening is null since the stator core material is infinitely permeable. This component along R_t is expressed into Fourier series over $\left(\alpha_i - \frac{b_{sa}}{2} \leq \alpha \leq \alpha_i + \frac{b_{sa}}{2} \right)$ as

$$B_{4i\alpha}|_{r=R_t} = B_0 + \sum_n B_n \cos(E_n(\alpha + b_{sa}/2 - \alpha_i)) \tag{67}$$

where

$$B_0 = B_{4i\alpha 0}\varphi_{ex}, \tag{68}$$

$$B_n = B_{4i\alpha 0}\varphi_0 + \sum_m B_{4i\alpha m}\varphi, \tag{69}$$

$$\varphi_{ex} = \frac{b_{oa}}{b_{sa}}, \tag{70}$$

$$\varphi_0(n) = \frac{4}{n\pi}\cos(n\pi/2)\sin(E_n b_{oa}/2) \tag{71}$$

and

$$\varphi(m,n) = -\frac{2}{b_{sa}}\frac{E_n}{F_m^2 - E_n^2}\left[\cos(m\pi)\sin\left(E_n\frac{b_{sa} + b_{oa}}{2}\right) - \sin\left(E_n\frac{b_{sa} - b_{oa}}{2}\right)\right]. \tag{72}$$

The circumferential flux density in the slot opening along the interface between the slot opening and the slot is

$$B_{3i\alpha}|_{r=R_t} = B_{3i\alpha 0} + \sum_n B_{3i\alpha n}\cos(E_n(\alpha + b_{sa}/2 - \alpha_i)) \tag{73}$$

where for non-overlapping winding

$$B_{3i\alpha 0} = -\frac{\mu_0 J_{i0}\left(R_{sb}^2 - R_t^2\right)}{2R_t} \tag{74}$$

and

$$B_{3i\alpha n} = -\frac{E_n D_{3i}\left(G_3^2 - 1\right)}{R_t} - \frac{2\mu_0 J_{in}}{R_t(E_n^2 - 4)}\left(R_t^2 - R_{sb}^2 G_3\right). \tag{75}$$

Applying the continuity of the circumferential component of flux density gives

$$B_{3i\alpha 0} = B_{4i\alpha 0}\varphi_a \tag{76}$$

and

$$B_{3i\alpha n} = B_{3i\alpha n}\varphi_0 + \sum_m B_{4i\alpha m}\varphi. \tag{77}$$

From (65), (71), (74) and (76)

$$D = \frac{\mu_0 J_{i0}\left(R_{sb}^2 - R_t^2\right)}{2}\left(\frac{b_{sa}}{b_{oa}}\right). \tag{78}$$

The vector potential distribution in the slot along the radius R_t is

$$A_t = A_{z3i}|_{r=R_t} = A_{3i0} + \sum_n A_{3in}\cos\left[E_n\left(\alpha + \frac{b_{sa}}{2} - \alpha_i\right)\right] \tag{79}$$

where for non-overlapping winding

$$A_{3i0} = \frac{\mu_0 J_{i0}\left(2R_{sb}^2\ln(R_t) - R_t^2\right)}{4} + Q_{3i} \tag{80}$$

and

$$A_{3in} = D_{3i}\left(G_3^2 + 1\right) + \frac{\mu_0 J_{in}}{\left(E_n^2 - 4\right)}\left(R_t^2 - \frac{2R_{sb}^2 G_3}{E_n}\right). \tag{81}$$

The same vector potential is expressed over $\left(\alpha_i - \frac{b_{sa}}{2} \le \alpha \le \alpha_i + \frac{b_{sa}}{2}\right)$

$$A_t = A_{3t0} + \sum_m A_{3tn} \cos\left[F_m\left(\alpha + \frac{b_{oa}}{2} - \alpha_i\right)\right] \tag{82}$$

where

$$A_{3t0} = \sum_n A_{3in}\kappa_0 + \frac{\mu_0}{4}J_{i0}\left(2R_{sb}^2 \ln(R_t) - R_t^2\right) + Q_{3i}, \tag{83}$$

$$A_{3tm} = \sum_n A_{3in}\kappa, \tag{84}$$

$$\kappa_0 = \left(\frac{b_{sa}}{b_{oa}}\right)\varphi_0(n) \tag{85}$$

and

$$\kappa = \left(\frac{b_{sa}}{b_{oa}}\right)\varphi(m, n). \tag{86}$$

From the vector potential within the slot opening

$$A_{z4i}\big|_{r=R_t} = D\ln(R_t) + Q_{4i} + \sum_m (C_{4i} + D_{4i}G_4)\cos\left(F_m\left(\alpha + \frac{b_{oa}}{2} - \alpha_i\right)\right). \tag{87}$$

Applying the continuity of the vector potential along the interface between the slot opening and the slot

$$A_t = A_{z4i}\big|_{r=R_t}. \tag{88}$$

The following equations can now be developed

$$Q_{3i} = Q_{4i} + D\ln(R_t) - \frac{\mu_0 J_{i0}}{4}\left(2R_{sb}^2 \ln(R_t) - R_t^2\right) - \sum_n A_{3in}\kappa_0 \tag{89}$$

and

$$C_{4i} + D_{4i}G_4 = \sum_n A_{3in}\kappa. \tag{90}$$

2.3. Back EMF and Torque Calculations

From the field solutions generated in the previous subsection, relevant electromagnetic characteristics, such as the flux linkage and the back EMF, are obtained based on the subdomain models. Also developed are expressions for the electromagnetic torque and cogging torque.

2.3.1. Flux Linkage

Flux linkage occurs when a magnetic field interacts with a material such as a magnetic field travels through a coil of wire. The flux linkage from one coil side of an arbitrary winding is calculated from the magnetic vector potential in the general equation for flux linkage. The calculation can be simplified by means of the Stokes integral theorem:

$$\psi_{i1} = \frac{LN_c}{A_s}\iint_S \vec{B} \cdot d\vec{s} = \frac{LN_c}{A_s}\iint_S \nabla \times \vec{A} \cdot d\vec{s} = \frac{LN_c}{A_s}\oint_C \vec{A} \cdot d\vec{s} = \frac{LN_c}{A_s}\int_{\alpha_i - b_{sa}/2}^{\alpha_i - b_{sa}/2 + d} \int_{R_t}^{R_{sb}} A_{z3i} \cdot r\,dr\,d\alpha \tag{91}$$

where L is the stack length of the motor, A_s is the area of one coil side and N_c is the number of turns per coil.

Based on A_{z3i} in (20) and (91):

$$\psi_{i1} = \frac{LN_c}{A_s}\left[Q_0 d + \sum_n \frac{Q_n}{E_n}\sin(E_n d)\right] \tag{92}$$

where:

$$Q_0 = \int_{R_t}^{R_{sb}} A_0 \cdot r dr \tag{93}$$

and:

$$Q_n = \int_{R_t}^{R_{sb}} A_n \cdot r dr. \tag{94}$$

Following the same approach, the flux linkage in the other coil side of the same slot is:

$$\psi_{i2} = \frac{LN_c}{A_s}\left[Q_0 d - \sum_n \frac{Q_n}{E_n}\sin(n\pi - E_n d)\right]. \tag{95}$$

The summation of the flux linkages associated with all the coil sides of the corresponding phase results in the total flux linkage of each phase. This calculation includes a connection matrix S_w that represents the winding distribution in the slot:

$$\begin{bmatrix} \psi_a \\ \psi_b \\ \psi_c \end{bmatrix} = S_w \psi_c. \tag{96}$$

Presented is an estimation method for the flux linkage. These results will be used to estimate the phase back-EMF for the Halbach motor.

2.3.2. Back EMF

Based on Lenz's law, the induced EMF always counters the source of its creation. Here, the source of back-EMF is the rotation of armature. The torque produces rotation of armature. Torque is due to armature current and armature current is created by the supply voltage. Therefore, the ultimate cause of production of the back EMF is the supply voltage. The three-phase back-EMF vector is calculated by the derivative of flux linkage calculated in (96) with respect to time when the motor is in open circuit:

$$\begin{bmatrix} E_a \\ E_b \\ E_c \end{bmatrix} = \omega \begin{bmatrix} \frac{d\psi_a}{dt} \\ \frac{d\psi_b}{dt} \\ \frac{d\psi_c}{dt} \end{bmatrix}. \tag{97}$$

2.3.3. Electromagnetic and Cogging Torques

The electromagnetic torque is calculated as:

$$T_{em} = \frac{(E_a I_a + E_b I_b + E_c I_c)}{\omega}. \tag{98}$$

Concerning the cogging torque, many methods have been used such as the lateral force [30], complex permeance [31,32] and energy [33,34]. In PM machines, calculating and minimizing the cogging torque is classically evaluated using either the virtual work or Maxwell stress tensor methods.

In this development, the Maxwell stress tensor is proposed providing accurate prediction of the air-gap field by way of the subdomain method accounting for tooth-tips. The derivation is based on the open circuit field:

$$T_c = \left(\frac{L R_s^2}{\mu_0} \right) \int_0^{2\pi} B_{2r} B_{2\alpha} d\alpha. \tag{99}$$

3. Results and Model Validation

Validation of the proposed electromagnetic analytical model depends on the modeling accuracy to predict motor torque with slot and tooth tip effects. The proposed validation is based on finite element analysis (FEA) simulations, no experimental results are presented. A Halbach Array machine shown in Figure 1 and a conventional one having both a three-phase 12-slot/10-pole embodiment are adopted to validate the analytical subdomain model developed in this manuscript. In the present validation, the simulation on the conventional and the Halbach Array SPM motors is performed by the help of the commercial FEA software MotorSolve v6.01 (Infolytica Corporation, Montréal, QC, Canada). The motor design parameters applied in the present study are provided in Table 1.

Table 1. Main Motor Parameters.

Parameter	Value
Slot Number, N_s	12
Pole Number, p	10
Stator Outer Diameter, D_s	100 mm
Magnet Remanence, B_r	1.12 T
Relative Permeability, μ	1.05
Air gap Length, g	1 mm
Magnet Thickness, h_m	3 mm
Active Length, L	50 mm
Winding Turns / Coil, N_c	35
Magnet Magnetization	Radial
Stator Inner Diameter	55 mm
Tooth Tips Edge	3 mm
Slot-opening Angle	5.5°
Winding Slot Angle	14.5°
Stator Yoke Height	5 mm
Rotor Outer Diameter	53 mm
Peak Rated Current	10 A
Motor Speed	400 RPM

The non-overlapped winding layout is shown in Figure 5. The use of this configuration in SPM motors to meet the optimal flux weakening condition is supported. Therefore, machines made with concentrated windings have shorter end coils, which reduce the total copper ohmic losses, the length and the weight of the machine. In addition, this configuration is chosen to maximize the winding factor of 93.3% with a coil span equal to unity [32]. The results for the Fourier series expansion are computed with a finite number of harmonics for k, m, and n. Generally, the limited number of harmonics is considered in the analytical model to generate suitable results, while the mesh in the FE model must be adjusted before achieving acceptable results. The computation time of the analytical field solutions is dependent from the harmonic numbers (k, m, n), for (100, 50, 50), by way of example, the average calculation time is 14.3 s (with 2.3 GHz CPU) on the MatLab platform. For the Halbach Array magnets configuration, when the magnet ratio R_{mp}, equal to 0.5, the predicted 2-D analytical radial flux density and MotorSolve simulation in the centerline of air gap position are presented in Figure 6. The resulting air gap flux density waveform distributions are distorted at the slot-openings showing that the analytical subdomain model is able to capture the slot-openings and tooth tips effects. The proposed model predicts the flux density within the considered five subdomains. MotorSolve only provides the flux density prediction for only the air-gap. The validation of flux density would promote further analysis in the demagnetization withstand ability for different magnet ratio cases and inductance prediction.

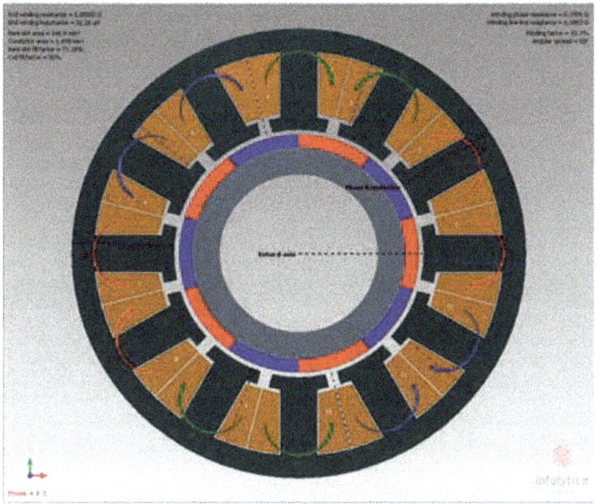

Figure 5. Non-Overlapping Winding Layout in MotorSolve.

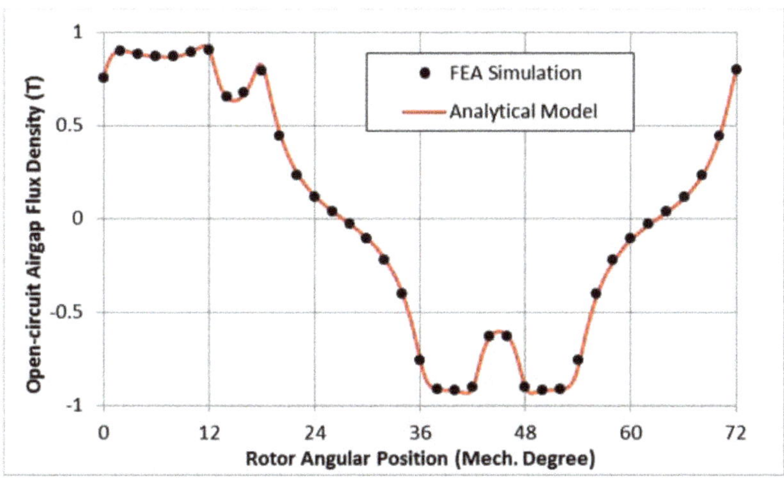

Figure 6. Comparison of FEA and Analytically Predicted Air-gap Flux Density, R_{mp} = 0.5.

By varying the magnet ratio, the proposed model can predict a variety of magnetization patterns as shown in Figure 7. It should be noted that the conventional SPM magnet has a magnet ratio R_{mp} equal to unity, while the magnet ratio should be 0.5 for the model shown in Figure 1, when the side and the central magnet segments have the same size. The highest flux density peak value (FDPV) is given by lowest Halbach Array magnet ratio, which is approximately 10% higher than the conventional magnetization. By increasing the magnet ratio, the FDPV decreases and the shape of the flux becomes wider within each pole range. This is because the central magnet with radial magnetization becomes wider as the magnet ratio increases. The air gap flux density waveform reaches its widest and lowest peak amplitude in the magnet ratio of unity.

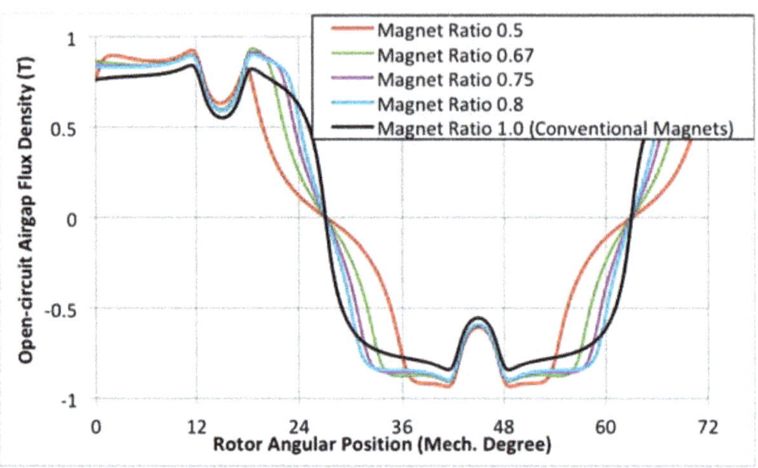

Figure 7. Comparison of Analytically Predicted Air-gap Flux Density with Different Magnet Ratio Values.

The stator windings are excited with sinusoidal currents with a peak value of 10A. Shown in Figures 8–10 are the comparisons between the analytically calculated and FEA simulated waveforms of electromagnetic torque, cogging torque and back-EMF, respectively.

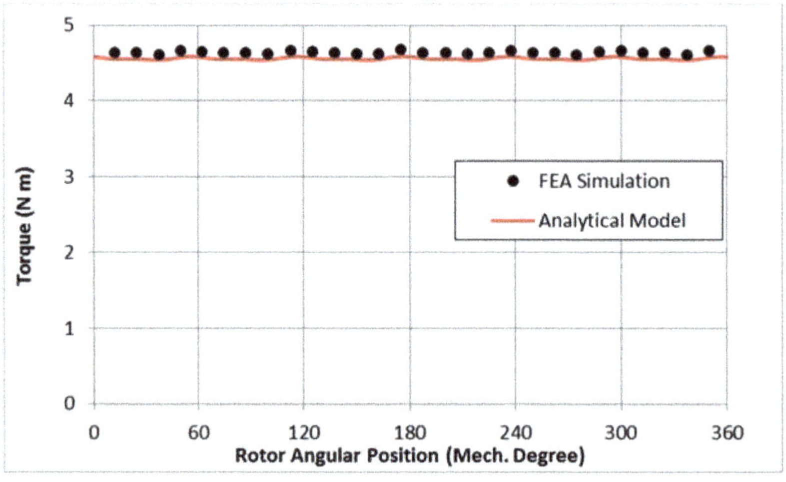

Figure 8. Analytical and MotorSolve FEA Predictions of Electromagnetic in the 12 s/10 p Halbach Machine with $R_{mp} = 0.5$ (average deviation: 1.8%).

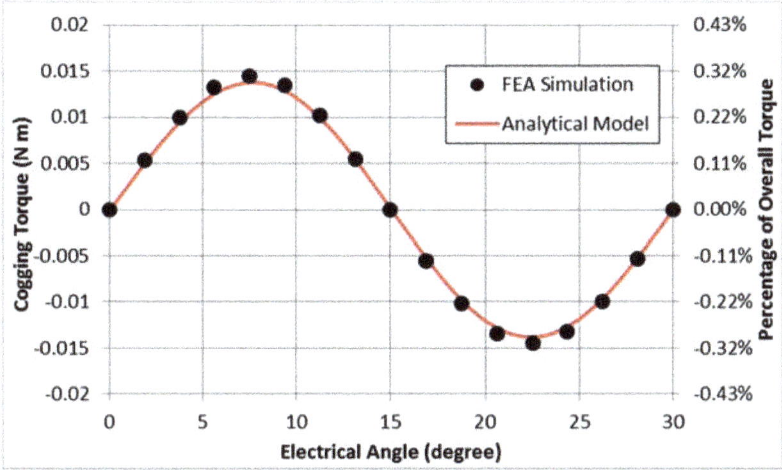

Figure 9. Analytical and MotorSolve Predictions of Cogging Torque in the 12 s/10 p Halbach Machine with R_{mp} = 0.5.

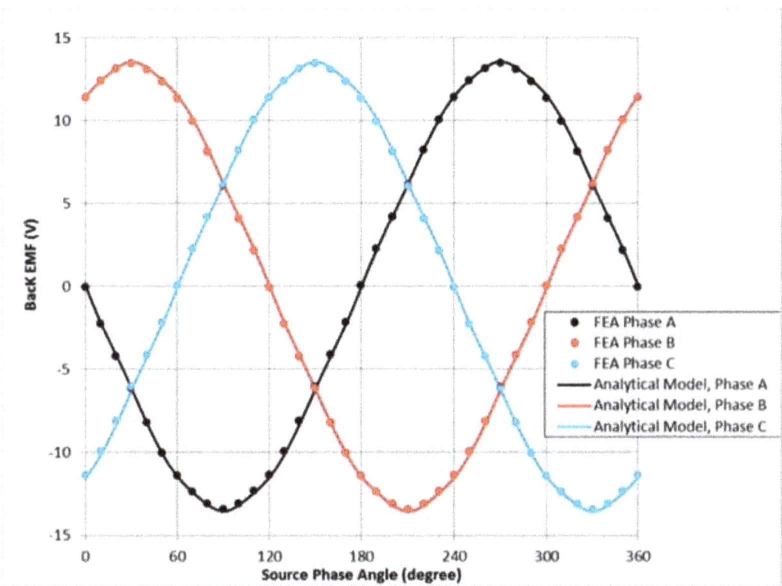

Figure 10. Analytical and MotorSolve Predictions of Back EMF in the 12 s/10 p Halbach Machine with R_{mp} = 0.5.

The analytical results provide good agreements with those obtained from FEA simulation. Shown in Figure 11 are the flux lines with different magnet ratio values. The flux lines are densely low in the passive region (rotor back iron) for the magnet ratio 0.5. When increasing the magnet ratio, the FDPV increases to the value of conventional magnet pattern.

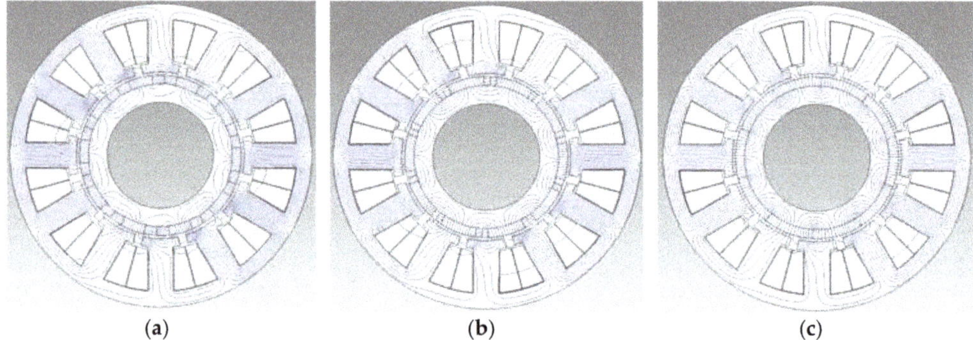

(a)   (b)   (c)

Figure 11. Magnetic Field Distributions across the Motor Area: (**a**) $R_{mp} = 0.5$; (**b**) $R_{mp} = 0.7$; (**c**) $R_{mp} = 1$.

For comparison purposes, the electromagnetic performance for the conventional (non-Halbach) motor design with the same parameters in Table 1 is shown in Figures 12–14. All results for the analytical calculations are in agreement with FEA simulations.

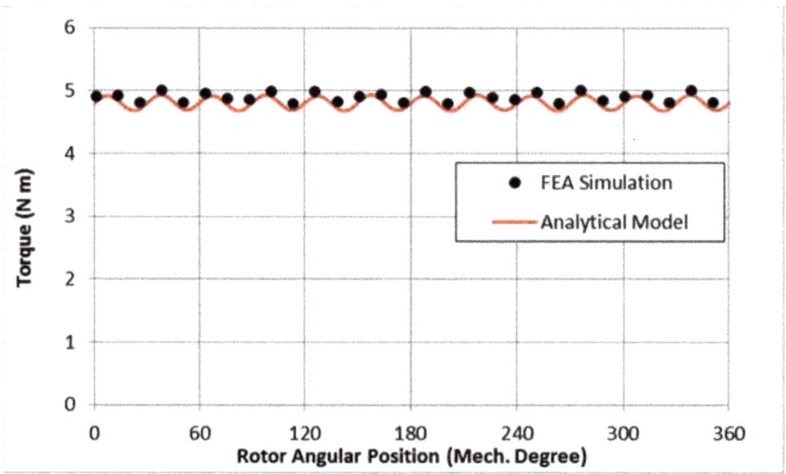

Figure 12. Analytical and MotorSolve FEA Predictions of Electromagnetic in the 12 s/10 p Conventional Machine (average deviation: 1.6%).

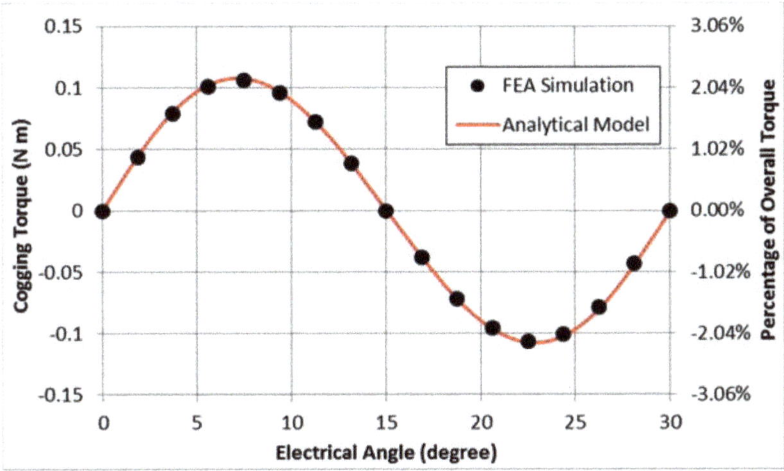

Figure 13. Analytical and MotorSolve Predictions of Cogging Torque in the 12 s/10 p Conventional Machine.

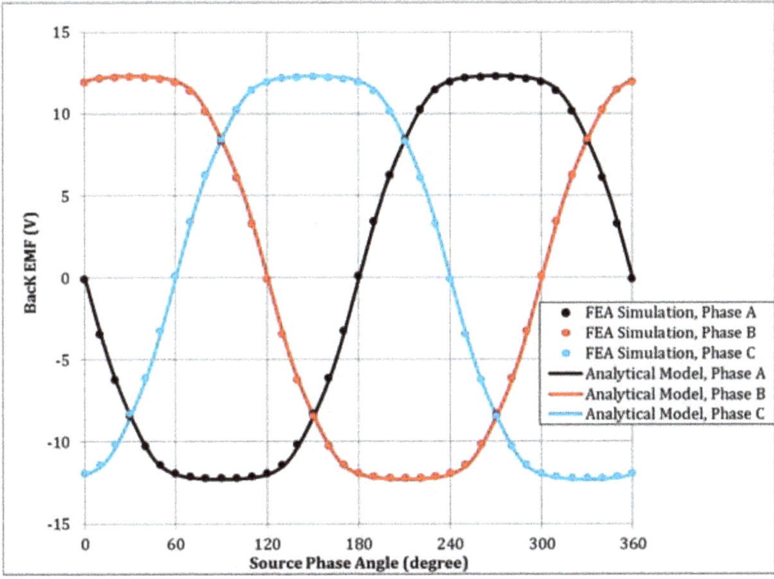

Figure 14. Analytical and MotorSolve Predictions of Back EMF in the 12 s/10 p Conventional Machine.

The back-EMF for Halbach Array motor is more sinusoidal than conventional surface-mounted permanent magnet motor. As a result, the Halbach Array machine, as shown in Figure 9 (R_{mp} = 0.5), provides much lower cogging torque (with a peak magnitude of 0.015 Nm) comparing to conventional motor in Figure 13 (with a peak magnitude of 0.11 Nm).

Comparison motor torques among the different machine designs is presented in Figure 15. Among the magnet ratios of 0.5, 0.667, 0.75, 0.8 and 1, the design for 0.8 provides the highest torque, which is approximately 10% higher than the conventional design (R_{mp} = 1) in this particular configuration while it has reached 25% for other cases. Also note that the design with a magnet ratio of 0.5 actually generates

4.5% less torque comparing to the convention magnet configuration. These results demonstrate that the maximization process of electromagnetic torque is nonlinear with respect to the magnet ratio. Only with an appropriate and optimal magnet ratio can the Halbach Array design provide a higher torque compared to the conventional SPM motor given the same size and supply constraints. From such calculations, it suggests that the PM motor output torque is related to the integration of the half cycle of flux linkage for each coil; and the optimal magnet ratio to maximize the torque requires the maximum integration of this flux linkage for each coil, under the same input conditions [3]:

$$\frac{\partial \int_{\frac{\pi}{2}}^{\frac{3\pi}{2}} \psi_{coil} d\theta}{\partial R_{mp}} = 0. \tag{100}$$

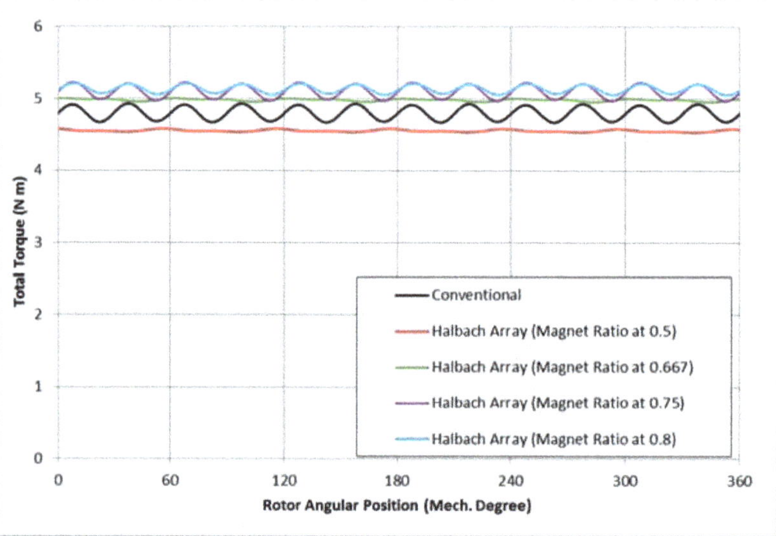

Figure 15. Comparison of Analytically Predicted Torque with Different Magnet Ratio Values.

4. Thermal Analysis

In addition to the electromagnetic modeling, a thermal analysis of the reported 10-pole SPM motor is performed using FEA simulation. Such an analysis is to provide illuminated evaluations for the heat generation and its transfer within the motor embodiment during operation. As a result, the temperature distribution within the motor assembly can be predicted. Overheat spots on copper windings or thermal-sensitive electric components are regarded as constraints to the motor embodiment and will be examined. Unless sufficient heat could be removed from the motor assembly, or limited current magnitude would be allowed for necessary cooling to suffice, over-heat or melting would happen in the motor material and cause malfunction of operation, or even bring about safety issues.

In the present study, the thermal analysis and evaluation are performed under the requirements that the maximum temperature within the motor assembly be lower than 180 °C (any spots with a temperature higher than 180 °C will be considered as motor overheat). There are two main sources for the heat generation while a PM motor machine operates, which are often referred as the copper loss and the iron loss. Copper losses are generally the heat losses generated within the copper winding conductor because of its carried electric current, which are proportional to the winding resistance and the square of the current magnitude. Generally, lower current values or winding resistance (e.g., shorter total length or fewer numbers of turns in stator) will reduce the heat caused by the copper loss. Also, the limitation of the total motor weight would require less use of copper in windings, which could minimize its cross-sectional area, increase the conductor resistance and therefore increase heating.

For the iron loss, two mechanisms occur, namely eddy current loss and hysteresis loss. The eddy current loss is produced by relative movement between conductors (stator iron core) and magnetic flux lines. The hysteresis loss is due to reversal of magnetization of stator iron core whenever it is subjected to changing magnetic forces. The iron loss depends on variables such as the thickness of the laminations in the stator, the magnitude of the flux density, the stator iron material resistivity, and most importantly, the frequency at which the motor operates. To use a laminated stator and decrease the thickness of the laminations with electrical insulation from each other could reduce the iron loss.

The modeling of the iron loss power P_{iron} per unit volume of the stator material in this study was performed as:

$$P_{iron} = k_e B_p^2 f^2 + k_{ex} B_p^{1.5} f^{1.5} + k_h B_p^2 f \tag{101}$$

where k_e, k_{ex} and k_h are the coefficients of classical eddy current loss, excess eddy current loss and hysteresis loss, respectively. The variable B_p is the peak flux density in the three-dimensional (3-D) stator domain and f is the motor supply frequency. The coefficients here for iron losses are related to the given stator material properties from manufacturers, and the integration of the iron heat loss within the entire stator body is calculated in FE model using above equation performed in the stator teeth and yoke (back iron) domains.

In the present study, Hiperco 50A alloy is used as the motor core material and the stator slot encapsulation material (CW 2710/HW 2711, Huntsman, Woodlands, TX, USA) is used for slot potting and thermal conduction enhancement. Further studies regarding different motor materials can be performed. The temperature distribution of the present SPM motor (10p/12s) is modeled to examine whether overheat occurs with present conditions. Shown in Figure 16a is the time-based temperature variation for different motor components, under the operating conditions mentioned in Table 1; Figure 16b shows the axisymmetric temperature distribution in the 3-D motor model, provided by MotorSolve FEA simulation. The ambient temperature for this case study is set to be 40 °C. All three heat transfer mechanisms, conduction (within the motor solid assembly), convection (on motor external surface) and radiation (on motor external surface) are considered. In this case shown in Figure 16, no overheat is observed under the present operating condition. Taking into account the end winding heat effect, the maximum temperature would occur within the copper windings and their ends, where the copper heat loss is generated and has the longest distance (length of thermal transfer path) to the low-temperature ambient.

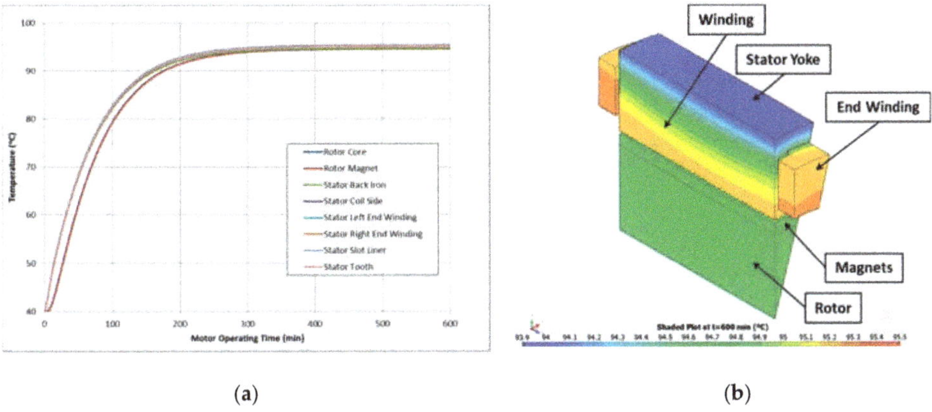

(a)   (b)

Figure 16. Motor Thermal Analysis Results: (a) Time-based Temperature Profiles for Various Motor Components; (b) Temperature Contours for 3D Thermal Modeling at 600 min.

Although the configuration of Halbach Array magnet and its advantages are reported previously, the torque dense capability of a Halbach Array motor is not widely realized. In the present study, Halbach Array SPM motors are verified to be able to produce higher torque than a conventional SPM with the same magnet volume and at the same rotor speed, provided the optimal Halbach Array magnet ratio, R_{mp}, is applied. It is also noted that compared to induction motor (IM) machines, the permanent magnet (PM) motors, including SPM and IPM (interior permanent magnet) machines, are able to achieve greater torque/current density, also with higher efficiency and reliability in low speed operation [35].

On the other hand, the general torque production also implies that in these motor categories, lower supply current may be required to retain the same torque performance using the Halbach Array SPM motor compared to others including the standard IM motors. As a result, lower supply current will present lower copper heat loss. Therefore, for industrial applications such as subsea electrical submersible pumps (ESP) or progressive cavity pumps (PCP), which require of driving motors with high torque density at a relative lower speed [36], the Halbach Array SPM motor is a good choice for its ability of high torque dense performance, with lower heat loss and higher efficiency. Such high-torque, low-speed motors could also eliminate the necessity of using a gear-reduction unit to reduce the high rotational speed driven by an IM machine, and reduce its mechanical loss and possible failures.

5. Conclusions

Developed in this paper is a subdomain model accounting for tooth-tip and slot effects for surface mounted permanent magnet (SPM) Halbach Array motor. The proposed analytical model can accurately predict the electromagnetic flux field distributions in motor applications of conventional and Halbach Array SPM machines. Based on the resultant magnetic field, the electromagnetic performances of the motor, such as the cogging torque, back-EMF, and electromagnetic torque are accurately calculated. The FEA analysis validates the predictions of the proposed analytical model. At this stage, a 2-D model is developed. For future work, there are certain difficulties to perform simulations in 3-D basis; hence an experimental setup is needed to validate the model.

Due to the feature of Halbach Array magnet configuration, the magnetic flux field is augmented on one side (outward of rotor), while cancelled on the other side (inward of rotor). In the present study, much less flux saturation is observed in rotor back iron with Halbach Array configuration compared to the conventional ones. Thus in Halbach Array design, more magnets could be installed in the rotor to achieve higher torque, with lower risk to cause flux saturation within the rotor back iron; on the other hand, less back iron material is needed in the Halbach Array designed rotor.

Also, we conclude that the Halbach Array magnet ratio would be one of the most important design parameters in Halbach Array SPM motors, because it influences on the flux density in both peak value and waveform which will therefore stimulate the resultant torque. With the appropriate magnet ratio in the motor embodiment, the Halbach Array motor is capable to produce higher torque than the conventional SPM motor for the same volume of magnets. Furthermore, potentially lower supply current requirement of the Halbach Array SPM motor would reduce the heat loss and increase the motor operating efficiency. For the industry applications (such as PCP motor) with direct-drive motors operating at low rotor-speed, it suggests that with magnet ratio optimization, the Halbach Array SPM motor is capable to exceed other machines (IM, IPM, or conventional SPM) in both torque performance and efficiency. In the low speed operation, the Halbach Array SPM motor could be able to provide 10-25% more overall torque, with the same motor size and the same input current.

Author Contributions: Methodology, M.M. and J.L.; software, Y.T. and T.W.; validation, M.M., Y.T.; thermal analysis, Y.T.; resources, M.A.F. and J.P.; writing-original draft preparation, M.M., Y.T. and T.W.; writing—review and editing, M.A.F. and J.P.; supervision, M.A.F. and J.P.

Funding: This research received no external funding.

Conflicts of Interest: The authors declare no conflict of interest.

References

1. Zhu, Z.Q.; Howe, D. Halbach permanent magnet machines and applications: A review. *IEE Proc. Electr. Power Appl.* **2001**, *148*, 299–308. [CrossRef]
2. Ng, K.; Zhu, Z.; Howe, D. Open-circuit field distribution in a brushless motor with diametrically magnetised PM rotor, accounting for slotting and eddy current effects. *IEEE Trans. Magn.* **1996**, *32*, 5070–5072. [CrossRef]
3. Shen, Y.; Liu, G.Y.; Xia, Z.P.; Zhu, Z.Q. Determination of maximum electromagnetic torque in PM brushless machines having two-segment Halbach array. *IEEE Trans. Ind. Electr.* **2014**, *61*, 718–729. [CrossRef]
4. Wang, H.; Ye, Y.; Wang, Q.; Dai, Y.; Yu, Y.; Weng, P. Analysis for ring arranged axial field Halbach permanent magnets. *IEEE Trans. Appl. Supercond.* **2006**, *16*, 1562–1565. [CrossRef]
5. Looser, A.; Baumgartner, T.; Kolar, J.W.; Zwyssig, C. Analysis and measurement of three-dimensional torque and forces for slotless permanent-magnet motors. *IEEE Trans. Ind. Appl.* **2012**, *48*, 1258–1266. [CrossRef]
6. Jang, S.-M.; Jeong, S.-S.; Ryu, D.-W.; Choi, S.-K. Design and analysis of high speed slotless PM machine with Halbach array. *IEEE Trans. Magn.* **2001**, *37*, 2827–2830. [CrossRef]
7. Wu, L.; Zhu, Z.Q.; Staton, D.; Popescu, M.; Hawkins, D. An improved subdomain model for predicting magnetic field of surface-mounted permanent magnet machines accounting for tooth-tips. *IEEE Trans. Magn.* **2011**, *47*, 1693–1704. [CrossRef]
8. Zhu, Z. Recent development of Halbach permanent magnet machines and applications. In Proceedings of the Power Conversion Conference-Nagoya, Nagoya, Japan, 2–5 April 2007.
9. Zhu, Z.; Xia, Z.; Howe, D. Comparison of Halbach magnetized brushless machines based on discrete magnet segments or a single ring magnet. *IEEE Trans. Magn.* **2002**, *38*, 2997–2999. [CrossRef]
10. Dwari, S.; Parsa, L. Design of halbach-array-based permanent-magnet motors with high acceleration. *IEEE Trans. Ind. Electron.* **2011**, *58*, 3768–3775. [CrossRef]
11. Wang, J.; Howe, D.; Lin, Z. Design optimization of short-stroke single-phase tubular permanent-magnet motor for refrigeration applications. *IEEE Trans. Ind. Electron.* **2010**, *57*, 327–334. [CrossRef]
12. Halbach, K. Design of permanent multipole magnets with oriented rare earth cobalt material. *Nuclear Instrum. Methods* **1980**, *169*, 1–10. [CrossRef]
13. Dhulipati, H.; Mukundan, S.; Lakshmi Varaha Iyer, K.; Kar, N.C. Skewing of stator windings for reduction of spatial harmonics in concentrated wound PMSM. In Proceedings of the 2017 IEEE 30th Canadian Conference on Electrical and Computer Engineering (CCECE), Windsor, ON, Canada, 30 April–3 May 2017.
14. Hwang, C.; Cheng, S.; Li, P. Design optimization for cogging torque minimization and efficiency maximization of an SPM motor. In Proceedings of the 2007 IEEE International Electric Machines & Drives Conference, Antalya, Turkey, 3–5 May 2007.
15. Islam, J.; Svechkarenko, D.; Chin, R.; Szucs, A.; Mantere, J.; Sakki, R. Cogging torque and vibration analysis of a direct-driven PM wind generator with concentrated and distributed windings. In Proceedings of the 2012 15th International Conference on Electrical Machines and Systems (ICEMS), Sapporo, Japan, 21–24 October 2012.
16. Lee, J.; Nomura, T.; Dede, E.M. Topology optimization of Halbach magnet arrays using isoparametric projection. *J. Magn. Magn. Mate.* **2017**, *432*, 140–153. [CrossRef]
17. Braune, S.; Liu, S.; Mercorelli, P. Design and control of an electromagnetic valve actuator. In Proceedings of the 2006 IEEE Conference on Computer Aided Control System Design, 2006 IEEE International Conference on Control Applications and 2006 IEEE International Symposium on Intelligent Control, Munich, Germany, 4–6 October 2006.
18. Mercorelli, P.; Lehmann, K.; Liu, S. Robust flatness based control of an electromagnetic linear actuator using adaptive PID controller. In Proceedings of the 42nd IEEE International Conference on Decision and Control, Maui, HI, USA, 9–12 December 2003.
19. Furlani, E.P. *Permanent Magnet and Electromechanical Devices: Materials, Analysis, and Applications*, 1st ed.; Academic Press: New York, NY, USA, 2001; ISBN 978-0-12-269951-1.
20. Liu, G.; Shao, M.; Zhao, W.; Ji, J.; Chen, Q.; Feng, Q. Modeling and analysis of halbach magnetized permanent-magnets machine by using lumped parameter magnetic circuit method. *Prog. Electromagn. Res. M* **2015**, *41*, 177–188. [CrossRef]
21. Jin, P.; Yuan, Y.; Lin, H.; Fang, S.; Ho, S.L. General analytical method for magnetic field analysis of Halbach magnet arrays based on magnetic scalar potential. *J. Magn.* **2013**, *18*, 95–104. [CrossRef]

22. Jang, S.-M.; Choi, J.-Y. Analytical prediction for electromagnetic characteristics of tubular linear actuator with Halbach array using transfer relations. *J. Electr. Eng. Technol.* **2007**, *2*, 221–230. [CrossRef]
23. Wu, L.J.; Zhu, Z.Q.; Staton, D.; Popescu, M.; Hawkins, D. Analytical prediction of electromagnetic performance of surface-mounted PM machines based on subdomain model accounting for tooth-tips. *IET Electr. Power Appl.* **2011**, *5*, 597–609. [CrossRef]
24. Wu, L.J.; Zhu, Z.Q.; Staton, D.; Popescu, M.; Hawkins, D. Subdomain model for predicting armature reaction field of surface-mounted permanent-magnet machines accounting for tooth-tips. *IEEE Trans. Magn.* **2011**, *47*, 812–822. [CrossRef]
25. Markovic, M.; Perriard, Y. Optimization design of a segmented Halbach permanent-magnet motor using an analytical model. *IEEE Trans. Magn.* **2009**, *45*, 2955–2960. [CrossRef]
26. Mellor, P.H.; Wrobel, R. Optimization of a multipolar permanent-magnet rotor comprising two arc segments per pole. *IEEE Trans. Ind. Appl.* **2007**, *43*, 942–951. [CrossRef]
27. Kumar, P.; Bauer, P. Improved analytical model of a permanent-magnet brushless DC motor. *IEEE Trans. Magn.* **2008**, *44*, 2299–2309. [CrossRef]
28. Boughrara, K.; Zarko, D.; Ibtiouen, R.; Touhami, O.; Rezzoug, A. Magnetic field analysis of inset and surface-mounted permanent-magnet synchronous motors using Schwarz–Christoffel transformation. *IEEE Trans. Magn.* **2009**, *45*, 3166–3178. [CrossRef]
29. Wu, L.; Zhu, Z.Q.; Staton, D.; Popescu, M.; Hawkins, D. Combined complex permeance and sub-domain model for analytical predicting electromagnetic performance of surface-mounted PM machines. In Proceedings of the 5th IET International Conference on Power Electronics, Machines and Drives, Brighton, UK, 19–21 April 2010.
30. Zhu, Z.; Howe, D. Analytical prediction of the cogging torque in radial-field permanent magnet brushless motors. *IEEE Trans. Magn.* **1992**, *28*, 1371–1374. [CrossRef]
31. Zarko, D.; Ban, D.; Lipo, T.A. Analytical calculation of magnetic field distribution in the slotted air gap of a surface permanent-magnet motor using complex relative air-gap permeance. *IEEE Trans. Magn.* **2006**, *42*, 1828–1837. [CrossRef]
32. Zarko, D.; Ban, D.; Lipo, T.A. Analytical solution for cogging torque in surface permanent-magnet motors using conformal mapping. *IEEE Trans. Magn.* **2008**, *44*, 52–65. [CrossRef]
33. Zhu, L.; Jiang, S.Z.; Zhu, Z.Q.; Chan, C.C. Analytical methods for minimizing cogging torque in permanent-magnet machines. *IEEE Trans. Magn.* **2009**, *45*, 2023–2031. [CrossRef]
34. De La Ree, J.; Boules, N. Torque production in permanent-magnet synchronous motors. *IEEE Trans. Ind. Appl.* **1989**, *25*, 107–112. [CrossRef]
35. Pellegrino, G.; Vagati, A.; Boazzo, B.; Guglielmi, P. Comparison of induction and PM synchronous motor drives for EV application including design examples. *IEEE Trans. Ind. Appl.* **2012**, *48*, 2322–2332. [CrossRef]
36. Cui, J.; Zhang, X.J.; Zhang, X.; Zhang, P. Electric Submersible Progressing Cavity Pump Driven by Low-Speed Permanent Magnet Synchronous Motor. In Proceedings of the SPE Middle East Artificial Lift Conference and Exhibition, Manama, Kingdom of Bahrain, 30 November–1 December 2016.

Article

Magnetically Nonlinear Dynamic Models of Synchronous Machines and Experimental Methods for Determining Their Parameters

Gorazd Štumberger [1,*], Bojan Štumberger [2] and Tine Marčič [3,*]

[1]   Faculty of Electrical Engineering and Computer Science, University of Maribor, 2000 Maribor, Slovenia
[2]   Faculty of Energy Technology, University of Maribor, 8720 Krško, Slovenia; bojan.stumberger@um.si
[3]   Energy Agency of the Republic of Slovenia, 2000 Maribor, Slovenia
*   Correspondence: gorazd.stumberger@um.si (G.Š.); tine.marcic@agen-rs.si (T.M.); Tel.: +386-2220-7075 (G.Š.)

Received: 31 July 2019; Accepted: 10 September 2019; Published: 12 September 2019

Abstract: This paper deals with rotary and linear synchronous reluctance machines and synchronous permanent magnet machines. It proposes a general method appropriate for determining the two-axis dynamic models of these machines, where the effects of slotting, mutual interaction between the slots and permanent magnets, saturation, cross-saturation, and—in the case of linear machines—the end effects, are considered. The iron core is considered to be conservative, without any losses. The proposed method contains two steps. In the first step, the dynamic model state variables are selected. They are required to determine the model structure in an arbitrarily chosen reference frame. In the second step, the model parameters, described as state variable dependent functions, are determined. In this way, the magnetically nonlinear behavior of the machine is accounted for. The relations among the Fourier coefficients of flux linkages and electromagnetic torque/thrust are presented for the models written in dq reference frame. The paper presents some of the experimental methods appropriate for determining parameters of the discussed dynamic models, which is supported by experimental results.

Keywords: synchronous machines; dynamic models; nonlinear magnetics; parameter estimation

---

1. Introduction

Rotary electric machines perform rotary motion or rotation, the origin of which is the electromagnetic torque produced in the machine. Linear electric machines perform linear motion or translation. It is caused by the thrust produced in the machine. In the linear and rotary permanent magnet synchronous machines (PMSMs), the torque or thrust that causes motion appears due to a magnetic field which results from the interaction between magnetic excitations caused by the winding currents and the permanent magnets. In synchronous reluctance machines (SRMs) the origin of motion is a force or torque caused by the magnetic field resulting from the interaction of winding currents and variable reluctance. Modern PMSMs make use of both phenomena in order to increase the thrust or torque.

In the modern modeling of electric machines, the contribution of G. Kron [1–3], where he set a solid and general theoretical background for modeling of electric machines, is often neglected. The generality of Kron's tensor-based approach was reduced by introducing matrices [4–6], where electric machines are mostly treated as magnetically linear systems with neglected magnetically nonlinear properties. However, these properties have to be included in the dynamic models of electric machines when a good agreement between the measured and calculated results is required, or when these models are applied in nonlinear control design for demanding applications.

A constant saliency ratio and a variable inductance are applied in [7–9] to describe the effects of saturation while the authors in [10] use two parameters to describe the magnetically nonlinear properties of a synchronous machine. The authors in [11] show that only using one parameter seems to be insufficient for a proper description of magnetically nonlinear properties of a PMSM. The characteristics of flux linkages are applied in [12] to describe the magnetically nonlinear properties of a linear SRM. The model proposed in [12] is used in the realization of nonlinear input-output linearizing control in [13], where the magnetically nonlinear properties of the machine are considered. To represent a conservative or loss-less system, the characteristics of flux linkages must fulfill the conditions described in [14,15]. An energy function-based magnetically nonlinear model of an SRM is described in [16]. It is applied in the sensorless control realizations reported in [17,18]. The magnetically nonlinear dynamic models of PMSMs based on the characteristics of flux linkages are presented in [19–21]. The authors in [22–24] demonstrate the use of magnetically nonlinear PMSM models in different control applications. On the contrary, the authors in [25–32] use mainly magnetically linear models of PMSMs for similar purposes. Different methods that can be used to determine the parameters of magnetically nonlinear PMSM and SRM models are presented in [11,12], [16], [19–21] and [33–40].

The authors of some of the aforementioned papers often use the magnetically nonlinear dynamic models of synchronous machines without explaining how these models were derived and how their parameters were determined. Therefore, this paper proposes a general procedure for determining the magnetically nonlinear dynamic models of rotary and linear PMSMs and SRMs, which is presented in Section 2. The proposed procedure is completed by the descriptions of experimental methods that can be applied for determining parameters of the obtained models. They are presented in Section 3. The paper ends with the presentation of experimental results given in Section 4, and the Conclusion given in Section 5. The novelties in this paper are related to the proposed straightforward method for determining the structure of the magnetically nonlinear two-axis dynamic models of rotary and linear PMSMs and SRMs written in dq references frame, to the presented relations among Fourier coefficients of flux linkages and electromagnetic force/thrust, and partially to the experimental methods applied for determining required parameters of the obtained dynamic models.

2. Dynamic Models

This section first explains the differences between the magnetically nonlinear properties of ferromagnetic materials, and the magnetically nonlinear properties of an entire electric machine observed from its terminals. Then the three-phase rotary PMSM is described with its three-phase magnetically nonlinear dynamic model in a general form, where the effects of slotting, interaction between the slots and permanent magnets, saturation, and cross-saturation are considered. The iron core is considered to be a conservative, or loss-less, system [14,15], which means that the effects of iron core losses are neglected. The three-phase model is then transformed into a model written in the $dq0$ reference frame, where the d-axis is aligned with the flux linkage vector due to the permanent magnets. In this way, only the structure of the magnetically nonlinear model written in the $dq0$ reference frame is obtained, whereas the model parameters must be determined either experimentally or, e.g., by finite element analysis (FEA). Some of the suitable experimental methods are presented in Section 3. The model flux linkages and the electromagnetic torque are expressed in the form of Fourier series. The relations between the Fourier coefficients of flux linkages and the electromagnetic torque are presented. The obtained magnetically nonlinear PMSM dynamic model written in the $dq0$ reference frame is modified in order to be suitable for a proper description of PMSM performing linear motion, where the end effects are an integral part of this model. By neglecting the effects of permanent magnets in the obtained dynamic models of rotary and linear PMSMs, the magnetically nonlinear dynamic models of rotary and linear SRMs are determined.

2.1. Magnetically Nonlinear Behavior of an Electric Machine

When dealing with a ferromagnetic material, its magnetically nonlinear behavior can be described by the *B(H)* characteristics, where *B* is the magnetic flux density while *H* is the magnetic field strength [41]. The *B(H)* characteristics are normally determined experimentally for a specimen of the material [39]. The differential permeability, given in the form of the partial derivative $\partial B/\partial H$, can be used to describe the local changes in the properties of an isotropic material along the *B(H)* characteristic. Similarly, in the case of an anisotropic material, the changes in material properties influence the relation between the flux density vector **B** and the magnetic field strength vector **H**, which can be described with the differential permeability tensor $\partial \mathbf{B}/\partial \mathbf{H}$. Unfortunately, the approach appropriate to describe the magnetically nonlinear behavior of material is not appropriate to describe the resultant magnetically nonlinear behavior of an entire electric machine, especially in cases where only those variables available on the machine's terminals can be used to describe the machine's magnetically nonlinear behavior.

In an electric machine, different kinds of material can be combined [42–47] in order to reach different goals. Different kinds of material, together with the machine's geometry, influence the resultant magnetically nonlinear behavior of the entire machine as it can be observed through the variables measured on the machine's terminals. These variables are the currents and voltages, while the flux linkages can be determined by the integration of corresponding voltages. If the machine contains only one winding, its magnetically nonlinear behavior can be described by the relation between the flux linkage ψ and the current i, given in the form of $\psi(i)$ characteristic. The local magnetically nonlinear behavior of the machine along the $\psi(i)$ characteristic is described with the partial derivative $\partial\psi/\partial i$. In the case of an electric machine that contains more windings, the currents and corresponding flux linkages of all windings can be arranged in the current vector **i** and flux linkage vector **ψ**. The use of linearly independent variables in both vectors is recommendable. The magnetically nonlinear behavior of the machine can be described by the **ψ**(**i**) characteristics, while the partial derivative $\partial\mathbf{\psi}/\partial\mathbf{i}$ can be used to describe the local magnetically nonlinear behavior along the **ψ**(**i**) characteristics.

2.2. Dynamic Model of a Rotary Three-Phase PMSM

A schematic presentation of the two-pole, three-phase PMSM is given in Figure 1 where the model phase windings are placed in the magnetic axes of actual phase a, b and c stator windings.

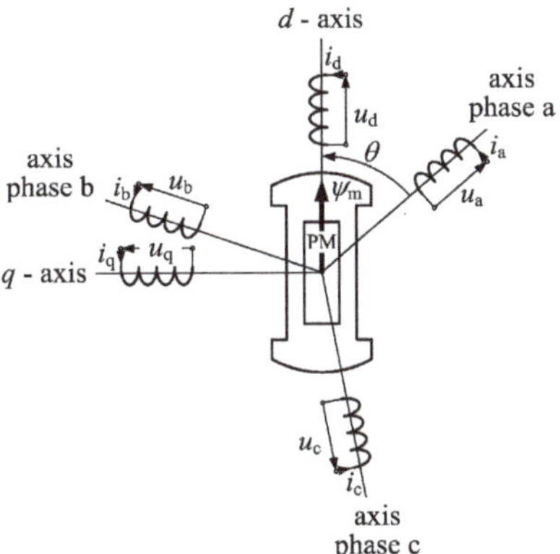

Figure 1. Schematic presentation of a two-pole, three-phase PMSM.

The *d*-axis is aligned with the magnetic axis of the flux linkage vector **ψ**$_m$ due to the permanent magnets placed on the rotor, while ψ_m denotes the length of **ψ**$_m$. The *q*-axis leads the *d*-axis for the electric angle of $\pi/2$. The *d*-axis is displaced with respect to the phase a axis for an electric angle θ which represents the (angular) rotor position.

The mathematical description of the PMSM shown in Figure 1, written in its most general form in the abc reference frame, defined with the magnetic axes of the phase a, b and c windings, is given by (1) to (4):

$$\mathbf{u}_{abc} = \mathbf{R}\mathbf{i}_{abc} + \frac{d}{dt}\boldsymbol{\psi}_{abc} + \frac{d}{dt}\boldsymbol{\psi}_{mabc}, \tag{1}$$

$$J\frac{d^2\theta}{dt^2} = t_e(i_a, i_b, i_c, \psi_m, \theta) - t_l - b\frac{d\theta}{dt}, \tag{2}$$

$$\mathbf{u}_{abc} = \begin{bmatrix} u_a \\ u_b \\ u_c \end{bmatrix}, \qquad \mathbf{i}_{abc} = \begin{bmatrix} i_a \\ i_b \\ i_c \end{bmatrix}, \qquad \mathbf{R} = \begin{bmatrix} R_a & & \\ & R_b & \\ & & R_c \end{bmatrix}, \tag{3}$$

$$\boldsymbol{\psi}_{abc} = \begin{bmatrix} \psi_a(i_a, i_b, i_c, \theta) \\ \psi_b(i_a, i_b, i_c, \theta) \\ \psi_c(i_a, i_b, i_c, \theta) \end{bmatrix}, \boldsymbol{\psi}_{mabc} = \begin{bmatrix} \psi_{ma}(\theta) \\ \psi_{mb}(\theta) \\ \psi_{mc}(\theta) \end{bmatrix}, \tag{4}$$

where u_a, u_b and u_c are the phase a, b and c voltages, i_a, i_b and i_c are the phase a, b and c currents, R_a, R_b and R_c are the phase a, b and c resistances, J is the moment of inertia, t_e is the electromagnetic torque, t_l is the load torque while b is the coefficient of friction. The phase a, b and c flux linkages caused by the permanent magnets are marked with ψ_{ma}, ψ_{mb} and ψ_{mc}, respectively. Similarly, ψ_a, ψ_b and ψ_c are the phase a, b and c flux linkages caused by the magnetic excitation due to the stator currents.

The three-phase magnetically nonlinear PMSM dynamic model, given in a general form by (1) to (4), is not intended to be used in the design of electric machines, but in dynamic simulations and control design. It is given for a two-pole machine or two poles of a multi-pole machine. For PMSMs with a higher number of the pole pairs ($p > 1$), the electromagnetic torque in (2) must be multiplied by p and the relation between the mechanical angle θ_m and the electric angle $\theta = p\theta_m$ must be considered. The electromagnetic torque t_e in (2) depends on the stator currents i_a, i_b, i_c, flux linkages due to the permanent magnets ψ_m and the position θ. The presumption that the flux linkages due to the permanent magnets are position-dependent, while the flux linkages due to the magnetic excitation with the stator currents are current- and position-dependent, seems to be reasonable as a first approximation. Both the position-dependent flux linkages due to the permanent magnets as well as the current- and position-dependent flux linkages due to the stator current excitation are used in model (1)–(4) to consider the effects of slotting, interactions between the slots and permanent magnets, saturation, and cross-saturation. The characteristics of the current- and position-dependent flux linkages can be determined either experimentally or by the FEA.

Most of the PMSMs used in electric drives are wye connected, which means that the currents i_a, i_b and i_c are linearly dependent. Since only the models with independent state variables can be used in the control design, the model (1)–(4), written in the abc reference frame, is transformed into another reference frame, where the state variables are independent while the obtained model is appropriate for the control design. A usual choice is the $dq0$ reference frame, where the d- and q-axes shown in Figure 1 are orthogonal, while the 0-axis is orthogonal to both of them. Since the flux linkages (4) are nonlinearly dependent on currents and position in order to consider the magnetically nonlinear behavior of the PMSM, the well-known methods for derivation of magnetically linear PMSM models [4,5] cannot be applied.

The procedure described in this work makes it possible only to determine the structure of the magnetically nonlinear PMSM model written in the $dq0$ reference frame. The corresponding model parameters in the form of current- and position-dependent characteristics of flux linkages have to be determined separately using experimental or FEA-based methods.

Independently of (1) and (2), which describe the voltage balances and motion of the PMSM, the voltage and current vectors $\mathbf{u}_{abc}$ and $\mathbf{i}_{abc}$, as well as the position-dependent vector of flux linkages due to the permanent magnets $\boldsymbol{\psi}_{mabc}$, can be always written in a new reference frame. The relation between the vector written in the reference frame abc ($\bullet_{abc}$) and the one written in the reference frame $dq0$ ($\bullet_{dq0}$) is given by the transformation matrix $\mathbf{T}$ (5) and Equations (6) and (7):

$$\mathbf{T} = \sqrt{\frac{2}{3}}\begin{bmatrix} \cos(\theta) & -\sin(\theta) & \frac{\sqrt{2}}{2} \\ \cos\left(\theta + \frac{4}{3}\pi\right) & -\sin\left(\theta + \frac{4}{3}\pi\right) & \frac{\sqrt{2}}{2} \\ \cos\left(\theta + \frac{2}{3}\pi\right) & -\sin\left(\theta + \frac{2}{3}\pi\right) & \frac{\sqrt{2}}{2} \end{bmatrix} \tag{5}$$

$$\mathbf{u}_{abc} = \mathbf{T}\mathbf{u}_{dq0}; \quad \mathbf{i}_{abc} = \mathbf{T}\mathbf{i}_{dq0}; \quad \boldsymbol{\psi}_{mabc} = \mathbf{T}\boldsymbol{\psi}_{mdq0}; \quad \boldsymbol{\psi}_{abc} = \mathbf{T}\boldsymbol{\psi}_{dq0}; \tag{6}$$

$$\mathbf{u}_{dq0} = \begin{bmatrix} u_d \\ u_q \\ u_0 \end{bmatrix}, \quad \mathbf{i}_{dq0} = \begin{bmatrix} i_d \\ i_q \\ i_0 \end{bmatrix}, \quad \boldsymbol{\psi}_{mdq0} = \begin{bmatrix} \psi_{md}(\theta) \\ \psi_{mq}(\theta) \\ \psi_{m0}(\theta) \end{bmatrix}, \quad \boldsymbol{\psi}_{dq0} = \begin{bmatrix} \psi_d(i_d, i_q, i_0, \theta) \\ \psi_q(i_d, i_q, i_0, \theta) \\ \psi_0(i_d, i_q, i_0, \theta) \end{bmatrix} \tag{7}$$

where indices d, q and 0 denote the currents, voltages and flux linkages written in the $dq0$ reference frame. Again, $\boldsymbol{\psi}_{mdq0}$ is the position-dependent flux linkage vector due to the permanent magnets, while $\boldsymbol{\psi}_{dq0}$ is the current and position vector dependent flux linkage vector due to magnetic excitation of the stator currents. Before the relation $\boldsymbol{\psi}_{abc} = \mathbf{T}\,\boldsymbol{\psi}_{dq0}$ can be written, the currents i_a, i_b and i_c in $\boldsymbol{\psi}_{abc}(i_a, i_b, i_c, \theta)$ (4) must be replaced with the currents i_d, i_q and i_0. This means that the flux linkage vector $\boldsymbol{\psi}_{abc}$, that describes magnetically nonlinear behavior of the PMSM, is expressed as a nonlinear function $\boldsymbol{\psi}_{abc}(i_d, i_q, i_0, \theta)$.

Considering (6) in (1) yields (8), while (11) is obtained after simple mathematical manipulations in (9) and (10).

$$\mathbf{T}\mathbf{u}_{dq0} = \mathbf{R}\mathbf{T}\mathbf{i}_{dq0} + \frac{d}{dt}\left\{\mathbf{T}\boldsymbol{\psi}_{dq0}\right\} + \frac{d}{dt}\left\{\mathbf{T}\boldsymbol{\psi}_{mdq0}\right\} \tag{8}$$

$$\mathbf{u}_{dq0} = \mathbf{T}^{-1}\mathbf{R}\mathbf{T}\mathbf{i}_{dq0} + \mathbf{T}^{-1}\frac{d}{dt}\left\{\mathbf{T}\boldsymbol{\psi}_{dq0}\right\} + \mathbf{T}^{-1}\frac{d}{dt}\left\{\mathbf{T}\boldsymbol{\psi}_{mdq0}\right\} \tag{9}$$

$$\mathbf{u}_{dq0} = \mathbf{T}^{-1}\mathbf{R}\mathbf{T}\mathbf{i}_{dq0} + \mathbf{T}^{-1}\frac{d}{dt}\{\mathbf{T}\}\boldsymbol{\psi}_{dq0} + \mathbf{T}^{-1}\mathbf{T}\frac{d}{dt}\left\{\boldsymbol{\psi}_{dq0}\right\} + \mathbf{T}^{-1}\frac{d}{dt}\{\mathbf{T}\}\boldsymbol{\psi}_{mdq0} + \mathbf{T}^{-1}\mathbf{T}\frac{d}{dt}\left\{\boldsymbol{\psi}_{mdq0}\right\} \tag{10}$$

$$\mathbf{u}_{dq0} = \mathbf{T}^{-1}\mathbf{R}\mathbf{T}\mathbf{i}_{dq0} + \mathbf{T}^{-1}\frac{d}{dt}\{\mathbf{T}\}\boldsymbol{\psi}_{dq0} + \frac{d}{dt}\left\{\boldsymbol{\psi}_{dq0}\right\} + \mathbf{T}^{-1}\frac{d}{dt}\{\mathbf{T}\}\boldsymbol{\psi}_{mdq0} + \frac{d}{dt}\left\{\boldsymbol{\psi}_{mdq0}\right\} \tag{11}$$

The form of the matrix equation that describes voltage balances in the magnetically nonlinear PMSM dynamic model, written in the $dq0$ reference frame, is given by (11). Considering the balanced stator resistances $R = R_a = R_b = R_c$ in (3) and (5), (7), and (9) in (11) gives (12):

$$\begin{bmatrix} u_d \\ u_q \\ u_0 \end{bmatrix} = R\begin{bmatrix} i_d \\ i_q \\ i_0 \end{bmatrix} + \begin{bmatrix} \frac{\partial \psi_d}{\partial i_d} & \frac{\partial \psi_d}{\partial i_q} & \frac{\partial \psi_d}{\partial i_0} \\ \frac{\partial \psi_q}{\partial i_d} & \frac{\partial \psi_q}{\partial i_q} & \frac{\partial \psi_q}{\partial i_0} \\ \frac{\partial \psi_0}{\partial i_d} & \frac{\partial \psi_0}{\partial i_q} & \frac{\partial \psi_0}{\partial i_0} \end{bmatrix}\frac{d}{dt}\begin{bmatrix} i_d \\ i_q \\ i_0 \end{bmatrix} + \frac{d\theta}{dt}\cdot\left\{\begin{bmatrix} \frac{\partial \psi_d}{\partial \theta} \\ \frac{\partial \psi_q}{\partial \theta} \\ \frac{\partial \psi_0}{\partial \theta} \end{bmatrix} + \begin{bmatrix} -\psi_q \\ \psi_d \\ 0 \end{bmatrix} + \begin{bmatrix} \frac{\partial \psi_{md}}{\partial \theta} \\ \frac{\partial \psi_{mq}}{\partial \theta} \\ \frac{\partial \psi_{m0}}{\partial \theta} \end{bmatrix} + \begin{bmatrix} -\psi_{mq} \\ \psi_{md} \\ 0 \end{bmatrix}\right\} \tag{12}$$

where the time derivatives are expressed by the partial derivatives using the chain rule. For the wye connected PMSM, the current i_0 from (7) can be expressed using (6) and the inverse $\mathbf{T}^{-1}$ of the transformation matrix $\mathbf{T}$ (5), which gives (13).

$$i_a + i_b + i_c = 0 \Rightarrow i_0 = \sqrt{\frac{2}{3}}\frac{1}{\sqrt{2}}(i_a + i_b + i_c) = 0 \tag{13}$$

Since i_0 equals 0, all terms containing i_0 in (12) can be eliminated. The machine's neutral point voltage u_0 is caused by the locally changing saturation level inside the machine. Considering (12), u_0 can be expressed by (14) as:

$$u_0 = \frac{\partial \psi_0}{\partial i_d}\frac{di_d}{dt} + \frac{\partial \psi_0}{\partial i_q}\frac{di_q}{dt} + \frac{\partial \psi_0}{\partial \theta}\frac{d\theta}{dt} + \frac{\partial \psi_{m0}}{\partial \theta}\frac{d\theta}{dt} \tag{14}$$

where on the right-hand side of (14) the dominant last term represents the changes of the flux linkage vector component ψ_{m0}, caused by the changing position of the permanent magnets with respect to the stator windings. The first three normally recessive terms appear due to the changes in ψ_0 caused by the changing i_d, i_q and θ. In the cases when iron core losses are important, the changing flux linkages ψ_0 and ψ_{m0} can contribute to the increase of iron core losses. The effects of changing neutral point voltage u_0 are often neglected, since they are implicitly compensated by the closed-loop current controllers, which force the currents i_d and i_q to follow their references. Moreover, according to (12), u_0 cannot directly influence the d- and q-axis currents, voltages and flux linkages. After neglecting all terms in (12) that contribute to u_0, (12) changes to (15). Equation (15) describes the voltage balances in the magnetically nonlinear two-axis dynamic model of PMSM written in the dq reference frame. The model is completed by (16), describing motion while the torque equation is still missing.

$$\begin{bmatrix} u_d \\ u_q \end{bmatrix} = R\begin{bmatrix} i_d \\ i_q \end{bmatrix} + \begin{bmatrix} \frac{\partial \psi_d}{\partial i_d} & \frac{\partial \psi_d}{\partial i_q} \\ \frac{\partial \psi_q}{\partial i_d} & \frac{\partial \psi_q}{\partial i_q} \end{bmatrix}\frac{d}{dt}\begin{bmatrix} i_d \\ i_q \end{bmatrix} + \frac{d\theta}{dt}\left\{ \begin{bmatrix} \frac{\partial \psi_d}{\partial \theta} \\ \frac{\partial \psi_q}{\partial \theta} \end{bmatrix} + \begin{bmatrix} -\psi_q \\ \psi_d \end{bmatrix} + \begin{bmatrix} \frac{\partial \psi_{md}}{\partial \theta} \\ \frac{\partial \psi_{mq}}{\partial \theta} \end{bmatrix} + \begin{bmatrix} -\psi_{mq} \\ \psi_{md} \end{bmatrix} \right\} \tag{15}$$

$$J\frac{d^2\theta}{dt^2} = t_e - t_l - b\frac{d\theta}{dt} \tag{16}$$

The electromagnetic torque t_e can be determined as a partial derivative of coenergy W_c [5] (17).

$$t_e(i_d, i_q, \theta) = \frac{\partial W_c(i_d, i_q, \theta)}{\partial \theta} \tag{17}$$

To consider the effects of cogging torque in the cases when the stator currents i_d, i_q are not flowing, it is wise to describe the flux linkages due to the permanent magnets ψ_{md} and ψ_{mq} as position-dependent functions of a fictive excitation current. In this way, the coenergy can change with the position even when i_d, i_q are not flowing, while its partial derivative gives a position-dependent term in the electromagnetic torque which represents the cogging torque. The calculation of coenergy could be quite a demanding task especially in the cases when the flux linkages due to the permanent magnets are position-dependent while the flux linkages due to the stator current excitation are current- and position-dependent. Therefore, the expression for electromagnetic torque is determined from the power balance (18), where the products of induced voltages e_d, e_q and currents i_d, i_q in d- and q-axis are given by (19).

$$\frac{d\theta}{dt}t_e = e_d i_d + e_q i_q \tag{18}$$

$$\begin{aligned} e_d i_d &= \frac{d\theta}{dt}\left(\frac{\partial \psi_d}{\partial \theta} + \frac{\partial \psi_{md}}{\partial \theta} - \psi_q - \psi_{mq}\right)i_d \\ e_q i_q &= \frac{d\theta}{dt}\left(\frac{\partial \psi_q}{\partial \theta} + \frac{\partial \psi_{mq}}{\partial \theta} + \psi_d + \psi_{md}\right)i_q \end{aligned} \tag{19}$$

The comparison of the left-hand side and the right-hand side of (18), considering products (19), gives (20).

$$t_e = \left(\frac{\partial \psi_d}{\partial \theta} + \frac{\partial \psi_{md}}{\partial \theta} - \psi_q - \psi_{mq}\right)i_d + \left(\frac{\partial \psi_q}{\partial \theta} + \frac{\partial \psi_{mq}}{\partial \theta} + \psi_d + \psi_{md}\right)i_q \tag{20}$$

Equation (20) represents a very good approximation for electromagnetic torque calculation. It gives acceptable results with the exception of no current condition, where the cogging torque cannot be calculated properly.

The magnetically nonlinear two-axis PMSM model is given by (15), (16) and (20). To consider the magnetically nonlinear behavior of the PMSM, which means the effects of slotting, the interaction between the slots and permanent magnets as well as the saturation and cross-saturation, the characteristics of flux linkages must be determined properly, which is discussed in the next section. The next subsection describes the relations between individual harmonic components of the flux linkages and electromagnetic torque.

2.3. Fourier Analysis of Flux Linkages and Torque

The position- and current-dependent flux linkages ψ_d and ψ_q, caused by the stator currents, are given in the form of Fourier series (21) and (22) for different constant values of i_d and i_q:

$$\psi_d = \psi_{d0} + \sum_{h=1}^{N} \left(\psi_{dch} \cos(h\theta) + \psi_{dsh} \sin(h\theta) \right) \tag{21}$$

$$\psi_q = \psi_{q0} + \sum_{h=1}^{N} \left(\psi_{qch} \cos(h\theta) + \psi_{qsh} \sin(h\theta) \right) \tag{22}$$

where h is the harmonic order, N is the highest harmonic order while ψ_{d0}, ψ_{q0}, ψ_{dch}, ψ_{qch} and ψ_{dsh}, ψ_{qsh} are the Fourier coefficients. Similarly, ψ_{md0}, ψ_{mq0}, ψ_{mdch}, ψ_{mqch} and ψ_{mdsh}, ψ_{mqsh} are the Fourier coefficients of the position-dependent flux linkages ψ_{md} and ψ_{mq}, caused by the permanent magnets and given in the form of Fourier series (23) and (24).

$$\psi_{md} = \psi_{md0} + \sum_{h=1}^{N} \left(\psi_{mdch} \cos(h\theta) + \psi_{mdsh} \sin(h\theta) \right) \tag{23}$$

$$\psi_{mq} = \psi_{mq0} + \sum_{h=1}^{N} \left(\psi_{mqch} \cos(h\theta) + \psi_{mqsh} \sin(h\theta) \right) \tag{24}$$

The electromagnetic torque t_e can be expressed in the form of Fourier series (25), with the Fourier coefficients T_{d0}, T_{ch} and T_{sh}.

$$t_e = T_{e0} + \sum_{h=1}^{N} \left(T_{ch} \cos(h\theta) + T_{sh} \sin(h\theta) \right) \tag{25}$$

Considering (21) to (24) in (20) leads to the relations between the Fourier coefficients of electromagnetic torque and flux linkages described by (26) to (28), where h is the harmonic order.

$$T_{e0} = \left((\psi_{d0} + \psi_{md0}) i_q - (\psi_{q0} + \psi_{mq0}) i_d \right) \tag{26}$$

$$T_{ech} = (\psi_{dch} + \psi_{mdch}) i_q - (\psi_{qch} + \psi_{mqch}) i_d + h(\psi_{dsh} + \psi_{mdsh}) i_d + h(\psi_{qsh} + \psi_{mqsh}) i_q \tag{27}$$

$$T_{esh} = (\psi_{dsh} + \psi_{mdsh}) i_q - (\psi_{qsh} + \psi_{mqsh}) i_d - h(\psi_{dch} + \psi_{mdch}) i_d - h(\psi_{qch} + \psi_{mqch}) i_q \tag{28}$$

The well-known equation for electromagnetic torque of a PMSM is obtained when only the DC torque component (26) is applied. Such a description of electromagnetic torque is normally sufficient for application at higher speeds, where the losses are not of primary interest, while the higher harmonic order torque pulsation is filtered out through the mechanical subsystem of the PMSM and does not influence the speed and position trajectories. However, in low-speed applications, the higher harmonic order torque pulsation cannot be filtered out through the mechanical subsystem of the PMSM and directly influence the speed and position trajectories. In such cases, the terms (27) and (28), containing only a few dominant torque harmonics, can be substantial for a proper, often nonlinear, control design and smooth tracking of position and speed references.

2.4. Dynamic Model of Linear PMSM

In the case of linear PMSM, the angular position θ is expressed with the pole pitch τ_p and the position x (29), while the electromagnetic torque t_e (20) and the moment of inertia J in (2) are replaced with the thrust f_e (31) and the mass m in (32). Considering these changes, the magnetically nonlinear two-axis dynamic model of a linear PMSM is given by (29) to (32):

$$\theta = \frac{\pi}{\tau_p}x \tag{29}$$

$$\begin{bmatrix} u_d \\ u_q \end{bmatrix} = R\begin{bmatrix} i_d \\ i_q \end{bmatrix} + \begin{bmatrix} \frac{\partial \psi_d}{\partial i_d} & \frac{\partial \psi_d}{\partial i_q} \\ \frac{\partial \psi_q}{\partial i_d} & \frac{\partial \psi_q}{\partial i_q} \end{bmatrix}\frac{d}{dt}\begin{bmatrix} i_d \\ i_q \end{bmatrix} + \frac{\pi}{\tau_p}\frac{dx}{dt}\left\{\begin{bmatrix} \frac{\partial \psi_d}{\partial x} \\ \frac{\partial \psi_q}{\partial x} \end{bmatrix} + \begin{bmatrix} -\psi_q \\ \psi_d \end{bmatrix} + \begin{bmatrix} \frac{\partial \psi_{md}}{\partial x} \\ \frac{\partial \psi_{mq}}{\partial x} \end{bmatrix} + \begin{bmatrix} -\psi_{mq} \\ \psi_{md} \end{bmatrix}\right\} \tag{30}$$

$$f_e = \frac{\pi}{\tau_p}\left(\frac{\partial \psi_d}{\partial x} + \frac{\partial \psi_{md}}{\partial x} - \psi_q - \psi_{mq}\right)i_d + \frac{\pi}{\tau_p}\left(\frac{\partial \psi_q}{\partial x} + \frac{\partial \psi_{mq}}{\partial x} + \psi_d + \psi_{md}\right)i_q \tag{31}$$

$$m\frac{d^2x}{dt^2} = f_e - f_l - f_f \tag{32}$$

where f_l and f_f are the load force and friction force, respectively. Equation (30) describes voltage balances, (31) is the thrust expression, while (32) describes linear motion (translation). The expressions for the flux linkages (21) to (24) preserve their forms while θ is replaced with $x\,\pi/\tau_p$ (29). Similarly, the expressions for electromagnetic torque (25) to (28) change to (33) to (36), describing the position-dependent thrust produced by the linear PMSM, where F_0, F_{ch} and F_{sh} are the Fourier coefficients of the thrust f_e.

$$f_e = F_0 + \sum_{h=1}^{N}\left(F_{ch}\cos\left(h\frac{\pi}{\tau_p}x\right) + F_{sh}\sin\left(h\frac{\pi}{\tau_p}x\right)\right) \tag{33}$$

$$F_0 = \frac{\pi}{\tau_p}\left((\psi_{d0} + \psi_{md0})i_q - (\psi_{q0} + \psi_{mq0})i_d\right) \tag{34}$$

$$F_{ch} = \frac{\pi}{\tau_p}(\psi_{dch} + \psi_{mdch})i_q - \frac{\pi}{\tau_p}(\psi_{qch} + \psi_{mqch})i_d + \frac{\pi}{\tau_p}h(\psi_{dsh} + \psi_{mdsh})i_d + \frac{\pi}{\tau_p}h(\psi_{qsh} + \psi_{mqsh})i_q \tag{35}$$

$$F_{sh} = \frac{\pi}{\tau_p}(\psi_{dsh} + \psi_{mdsh})i_q - \frac{\pi}{\tau_p}(\psi_{qsh} + \psi_{mqsh})i_d - \frac{\pi}{\tau_p}h(\psi_{dch} + \psi_{mdch})i_d - \frac{\pi}{\tau_p}h(\psi_{qch} + \psi_{mqch})i_q \tag{36}$$

The magnetically nonlinear dynamic model of the linear PMSM is given by (29) to (31). It is completed by (33) to (36), required for the thrust calculation. In addition to the effects of slotting, interaction between the slots and permanent magnets, saturation and cross-saturation, this model implicitly also considers the end effects which are specific for the linear machines.

2.5. Dynamic Models of Rotary and Linear Synchronous Reluctance Machines

When the schematic presentations of the PMSM shown in Figure 1 and the one showing synchronous reluctance machine (SRM) in Figure 2 are compared, and neglecting the actual designs of the machines, the only substantial difference that can be pointed out is the missing permanent magnets in the case of SRMs. Thus, the d-axis is defined with the axis of the lowest reluctance.

The magnetically nonlinear two-axis dynamic models of the rotary and linear SRMs can be described with the same sets of equations as the models of corresponding PMSMs, omitting the flux linkages due to the permanent magnets and their partial derivatives. Considering $\psi_{md} = 0$, $\psi_{mq} = 0$, $\partial\psi_{md}/\partial\theta = 0$, and $\partial\psi_{mq}/\partial\theta = 0$ in (15) and (20) leads to the magnetically nonlinear two-axis dynamic model of a rotary SRM given by (37) describing voltage balances, (38) describing electromagnetic torque, and (16) describing motion.

$$\begin{bmatrix} u_d \\ u_q \end{bmatrix} = R \begin{bmatrix} i_d \\ i_q \end{bmatrix} + \begin{bmatrix} \frac{\partial \psi_d}{\partial i_d} & \frac{\partial \psi_d}{\partial i_q} \\ \frac{\partial \psi_q}{\partial i_d} & \frac{\partial \psi_q}{\partial i_q} \end{bmatrix} \frac{d}{dt} \begin{bmatrix} i_d \\ i_q \end{bmatrix} + \frac{d\theta}{dt} \left\{ \begin{bmatrix} \frac{\partial \psi_d}{\partial \theta} \\ \frac{\partial \psi_q}{\partial \theta} \end{bmatrix} + \begin{bmatrix} -\psi_q \\ \psi_d \end{bmatrix} \right\} \tag{37}$$

$$t_e = \left(\frac{\partial \psi_d}{\partial \theta} - \psi_q \right) i_d + \left(\frac{\partial \psi_q}{\partial \theta} + \psi_d \right) i_q \tag{38}$$

Similarly, if $\psi_{md} = 0$, $\psi_{mq} = 0$, $\partial \psi_{md}/\partial x = 0$, and $\partial \psi_{mq}/\partial x = 0$ are considered in (30) and (31), the magnetically nonlinear two-axis dynamic model of a linear SRM can be given by (39) describing voltage balances, (40) describing thrust, and (32) describing motion.

$$\begin{bmatrix} u_d \\ u_q \end{bmatrix} = R \begin{bmatrix} i_d \\ i_q \end{bmatrix} + \begin{bmatrix} \frac{\partial \psi_d}{\partial i_d} & \frac{\partial \psi_d}{\partial i_q} \\ \frac{\partial \psi_q}{\partial i_d} & \frac{\partial \psi_q}{\partial i_q} \end{bmatrix} \frac{d}{dt} \begin{bmatrix} i_d \\ i_q \end{bmatrix} + \frac{\pi}{\tau_p} \frac{dx}{dt} \left\{ \begin{bmatrix} \frac{\partial \psi_d}{\partial x} \\ \frac{\partial \psi_q}{\partial x} \end{bmatrix} + \begin{bmatrix} -\psi_q \\ \psi_d \end{bmatrix} \right\} \tag{39}$$

$$f_e = \frac{\pi}{\tau_p} \left(\frac{\partial \psi_d}{\partial x} - \psi_q \right) i_d + \frac{\pi}{\tau_p} \left(\frac{\partial \psi_q}{\partial x} + \psi_d \right) i_q \tag{40}$$

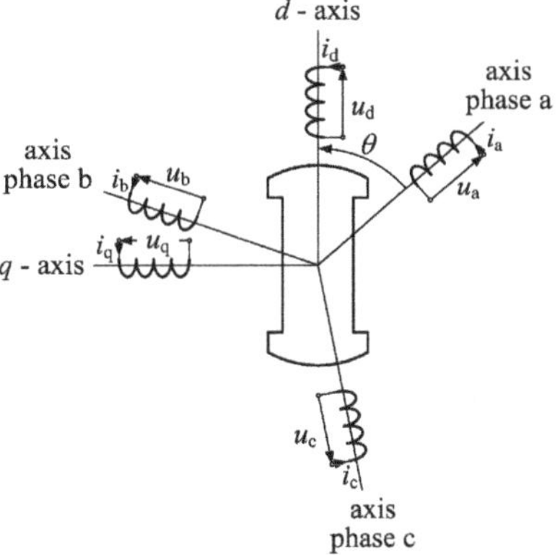

Figure 2. Schematic presentation of a two-pole, three-phase SRM.

3. Procedures for Determining Parameters of Dynamic Models

The magnetically nonlinear two-axis dynamic models of rotary and linear PMSMs and SRMs, written in the *dq* reference frame, are given in Section 2. The magnetically nonlinear properties of the discussed machines are described in the form of position-dependent characteristics of flux linkages due to the permanent magnets and in the form of current- and position-dependent characteristics of flux linkages due to the magnetic excitation of stator currents. These characteristics complete the models. They can be determined using FEA or experimental methods. Some of the experimental methods that can be applied for the rotary and linear machines are described in the next subsections. The other parameters required in the model, like the stator resistance R, the moment of inertia J, the mass m, and the coefficient of viscous friction b, are not discussed in this paper since they can be measured directly or determined by some of the well-known methods.

3.1. Flux Linkages Caused by the Permanent Magnets

The tested three-phase PMSM is driven by another speed-controlled machine at the constant speed. The position θ and the speed $\omega = d\theta/dt$ are measured together with the waveforms of the three-phase back electromotive forces (EMFs) e_a, e_b and e_c available on the open PMSM terminals. Considering (5) to (7), the waveforms of e_d, e_q and e_0 are calculated. Neglecting e_0 and ψ_{m0}, and considering $i_d = 0$, $i_q = 0$, and $\omega = d\theta/dt$, (15) reduces to (41) and (42).

$$\frac{e_d}{\omega} = -\psi_{mq} + \frac{\partial\psi_{md}}{\partial\theta} \tag{41}$$

$$\frac{e_q}{\omega} = \psi_{md} + \frac{\partial\psi_{mq}}{\partial\theta} \tag{42}$$

After performing partial derivation of (41) and (42) with respect to θ, (43) and (44) are obtained. The partial derivatives $\partial\psi_{md}/\partial\theta$ and $\partial\psi_{mq}/\partial\theta$ expressed from (41) and (42) are considered in (43) and (44), which gives (45) and (46) [19–21].

$$\frac{\partial}{\partial\theta}\left(\frac{e_d}{\omega}\right) = -\frac{\partial\psi_{mq}}{\partial\theta} + \frac{\partial^2\psi_{md}}{\partial\theta^2} \tag{43}$$

$$\frac{\partial}{\partial\theta}\left(\frac{e_q}{\omega}\right) = \frac{\partial\psi_{md}}{\partial\theta} + \frac{\partial^2\psi_{mq}}{\partial\theta^2} \tag{44}$$

$$\frac{\partial^2\psi_{md}}{\partial\theta^2} + \psi_{md} = \frac{\partial}{\partial\theta}\left(\frac{e_d}{\omega}\right) + \frac{e_q}{\omega} \tag{45}$$

$$\frac{\partial^2\psi_{mq}}{\partial\theta^2} + \psi_{mq} = \frac{\partial}{\partial\theta}\left(\frac{e_q}{\omega}\right) - \frac{e_d}{\omega} \tag{46}$$

The back EMFs e_d and e_q, as well as the flux linkages ψ_{md} and ψ_{mq}, are expressed in the form of Fourier series (47) to (50):

$$e_d = e_{d0} + \sum_{h=1}^{N}\left(e_{dch}\cos(h\theta) + e_{dsh}\sin(h\theta)\right) \tag{47}$$

$$e_q = e_{q0} + \sum_{h=1}^{N}\left(e_{qch}\cos(h\theta) + e_{qsh}\sin(h\theta)\right) \tag{48}$$

$$\psi_{md} = \psi_{md0} + \sum_{h=1}^{N}\left(\psi_{mdch}\cos(h\theta) + \psi_{mdsh}\sin(h\theta)\right) \tag{49}$$

$$\psi_{mq} = \psi_{mq0} + \sum_{h=1}^{N}\left(\psi_{mqch}\cos(h\theta) + \psi_{mqsh}\sin(h\theta)\right) \tag{50}$$

where h is the harmonic order and N is the highest harmonic order. The Fourier coefficients of the back EMFs are denoted with e_{d0}, e_{dch}, e_{dsh} and e_{q0}, e_{qch}, e_{qsh}, while the ones describing the flux linkages are denoted with ψ_{md0}, ψ_{mdch}, ψ_{mdsh} and ψ_{mq0}, ψ_{mqch}, ψ_{mqsh}. After inserting (47) to (50) into (45) and (46), and performing the required partial derivations of (47) to (50), (51) and (52) are obtained. They describe the relations among individual Fourier coefficients of the flux linkages and back EMFs.

$$\left(1 - h^2\right)\left(\psi_{mdch}\cos(h\theta) + \psi_{mdsh}\sin(h\theta)\right) = \frac{he_{dsh} + e_{qch}}{\omega}\cos(h\theta) + \frac{-he_{dch} + e_{qsh}}{\omega}\sin(h\theta) \tag{51}$$

$$\left(1 - h^2\right)\left(\psi_{mqch}\cos(h\theta) + \psi_{mqsh}\sin(h\theta)\right) = \frac{he_{qsh} - e_{dch}}{\omega}\cos(h\theta) + \frac{-he_{qch} - e_{dsh}}{\omega}\sin(h\theta) \tag{52}$$

The comparison of the terms that multiply the same harmonic function on the left- and right-hand side of (51) and (52) gives (53) and (54), where the Fourier coefficients of flux linkages are expressed with the Fourier coefficients of back EMFs.

$$\psi_{mdch} = \frac{he_{dsh} + e_{qch}}{(1 - h^2)\omega}, \quad \psi_{mdsh} = \frac{-he_{dch} + e_{qsh}}{(1 - h^2)\omega} \tag{53}$$

$$\psi_{mqch} = \frac{he_{qsh} - e_{dch}}{(1 - h^2)\omega}, \quad \psi_{mqsh} = \frac{-he_{qch} - e_{dsh}}{(1 - h^2)\omega} \tag{54}$$

Thus, the position-dependent characteristics of flux linkages ψ_{md} and ψ_{mq} are completely defined with the Fourier coefficients of the back EMFs measured on the open terminals of the tested PMSM. To achieve acceptable results, it is often sufficient to consider the DC component supplemented by a few dominant higher order harmonics.

3.2. Flux Linkages Caused by the Stator Currents

The characteristics of current- and position-dependent flux linkages ψ_d and ψ_q, caused by the magnetic excitation of the stator currents, can be determined with tests performed at the locked rotor of the PMSM, where $d\theta/dt = 0$. The tested PMSM is supplied with a controlled voltage source inverter (VSI). The current in the one axis is closed loop controlled in order to keep the constant value, while the voltage in the orthogonal axis is changing in a stepwise manner. Considering the described conditions and a constant value of the current i_q, the first row of (15) is reduced to (55).

$$u_d = Ri_d + \frac{\partial \psi_d}{\partial i_d}\frac{di_d}{dt} = Ri_d + \frac{d\psi_d}{dt} \tag{55}$$

The stepwise changing voltage u_d and the responding current i_d, measured during the test, are applied to determine the time behavior of the flux linkage $\psi_d(t)$ by numerical integration (56):

$$\psi_d(t) = \psi_d(0) + \int_0^t \left(u_d(\tau) - Ri_d(\tau)\right) d\tau \tag{56}$$

where $\psi_d(0)$ is the initial d-axis flux linkage due to the permanent magnets and remanent flux. If $\psi_d(0)$ is considered to have a value of 0, the flux linkage $\psi_d(t)$ (56) contains only changes around pre-magnetization $\psi_d(0) = \psi_{md}$ caused by the permanent magnets. $\psi_d(t)$ can be presented as a current $i_d(t)$ dependent function given in the form of a hysteresis loop $\psi_d(i_d)$ at the constant values of i_q and θ. A family of hysteresis loops $\psi_d(i_d)$ is obtained by repeating the described procedure for different amplitudes of the stepwise changing voltages u_d. The end points of individual hysteresis loops are used to define a unique characteristic $\psi_d(i_d)$. If the described procedure is repeated first for different constant values of the closed-loop controlled current i_q, and after that for different positions of the locked rotor θ, the current- and position-dependent characteristics $\psi_d(i_d, i_q, \theta)$, required in (15) and (20), can be determined over the entire range of operation. Similar procedure is applied to determine the characteristics $\psi_q(i_d, i_q, \theta)$. It must be pointed out that $\psi_d(i_d, i_q, \theta)$ and $\psi_q(i_d, i_q, \theta)$ are the characteristics of flux linkages caused by the magnetic excitation of the stator current. Implicitly, these characteristics also contain the effects of pre-magnetization caused by the permanent magnets, considered as $\psi_d(0) = \psi_{md}$ (56) and similarly $\psi_q(0) = \psi_{mq}$.

The proposed method for determining the characteristics of flux linkages is robust, if the resistance R in (56) is determined and updated from the steady state current and voltage after each voltage step change

4. Experiments and Results

In this section, the descriptions of the applied experimental set-up, tested objects and presented results are given.

4.1. Experimental Set-Up and Tested Machines

The applied experimental set-up is schematically presented in Figure 3. It consists of the tested rotary (or linear) PMSM or SRM, controlled VSI, current measurement chains based on LEM current sensors, voltage measurement chains based on differential probes, torque sensor and measurement chain, external source of torque in the form of speed or torque controlled machine acting as an active load, and control system dSPACE 1103 PPC, which is used for the open-loop voltage control and closed-loop current control of the tested machine performed in the dq reference frame.

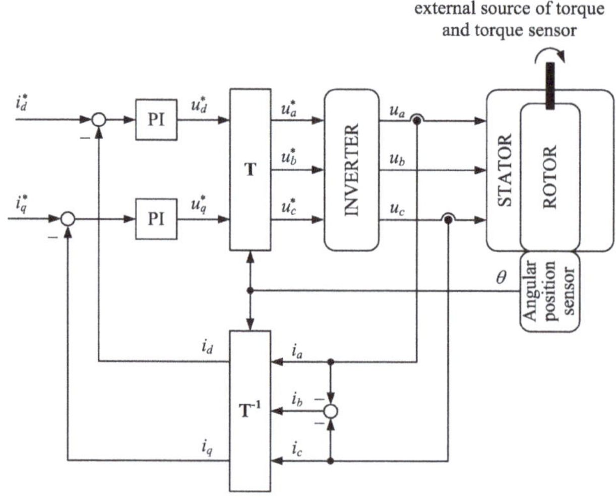

Figure 3. Schematic presentation of applied experimental set–up.

A photograph of the experimental set-up applied to determine the magnetically nonlinear characteristics of the tested rotary SRM and PMSM at locked rotor is shown in Figure 4. The tested synchronous machine shaft is connected to the mechanical brake with a clutch as shown in the photograph on the right-hand side in Figure 4. However, when the back-EMFs are measured, as is described in the next subsection, the mechanical brake is replaced by a speed-controlled driving motor, providing a constant angular speed. The stator terminals with connected differential probes are open which enables the measurement of the back-EMFs with differential probes.

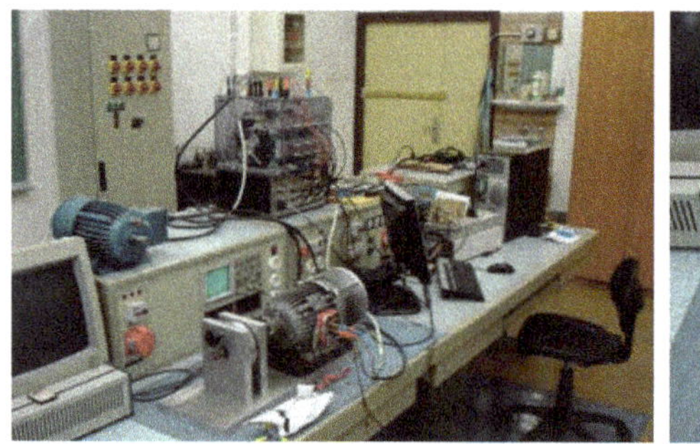
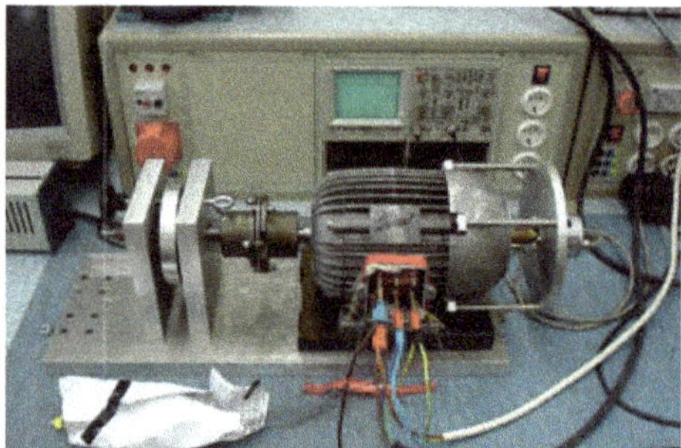

Figure 4. Experimental set-up for testing rotary synchronous machines at locked rotor.

Similarly, Figure 5 shows the experimental set-up applied for experimental work on linear synchronous machines [12,48].

Figure 5. Experimental set-up for testing linear synchronous machines.

Two prototypes of rotary machines were built, with the aim of confirming the here-proposed methods experimentally. The stator of a standard 1.1 kW induction motor (frame size 90) with a four-pole three-phase winding in wye connection was employed. Two equal reluctance rotors with flux barriers were constructed. Permanent magnets of Nd-Fe-B type were inserted in the flux barriers of one of the rotors, thus producing a pure SRM and an interior PMSM; these were employed in the experimental study. The cross-sections of the aforementioned rotors are schematically represented in Figure 6.

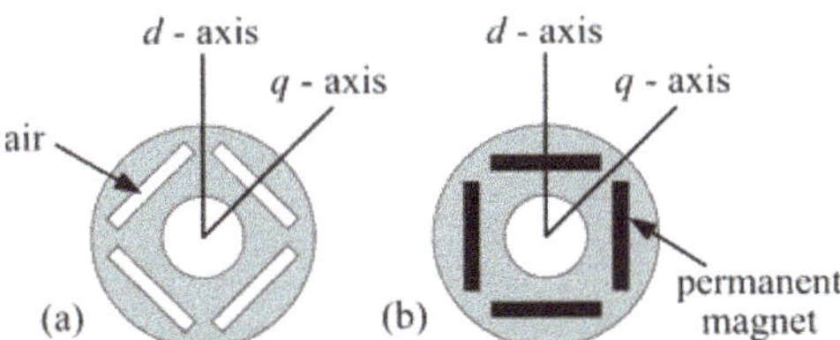

Figure 6. Schematic representation of rotor cross-sections of the SRM (**a**) and PMSM (**b**).

The third tested object is a prototype of a linear SRM with a short moving primary and a long secondary as shown in Figure 7 [48].

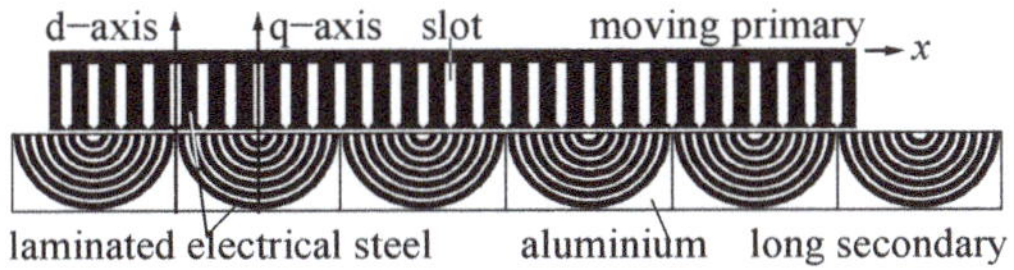

Figure 7. Schematic presentation of linear SRM prototype.

4.2. Determining Flux Linkages Caused by the Permanent Magnets

The procedure for determining these flux linkages is described in Section 3.1. Figure 8 shows the position-dependent three-phase back EMFs e_a, e_b and e_c measured on the open terminals of the tested PMSM driven by another speed-controlled machine. Figure 9 shows the back EMFs e_d, e_q and e_0 determined from e_a, e_b and e_c shown in Figure 8, by (6) considering the inverse of the transformation matrix (5). After performing the procedure described in Section 3.1, considering only the dominant harmonics up to the harmonic order 30, the position-dependent characteristics of flux linkages shown in Figure 10 are determined.

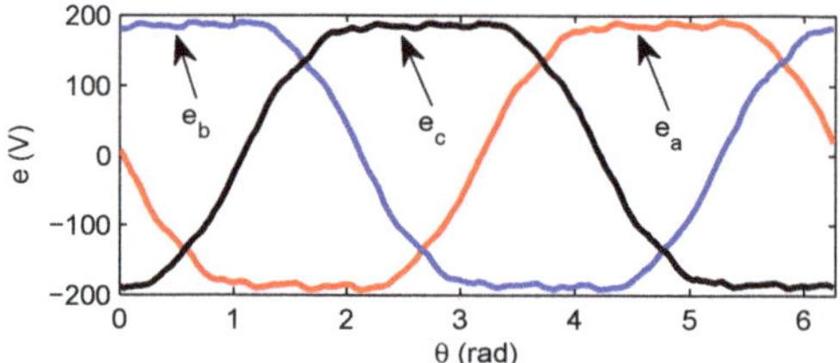

Figure 8. Back EMFs e_a, e_b and e_c measured on the open terminals of the tested PMSM.

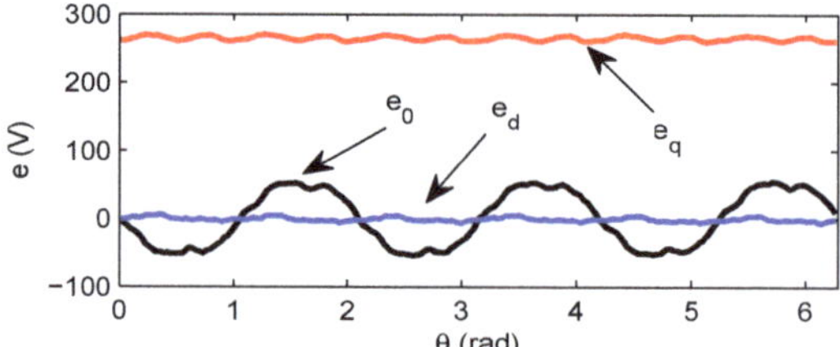

Figure 9. Back EMFs e_d, e_q and e_0 calculated from the measured back EMFs e_a, e_b and e_c.

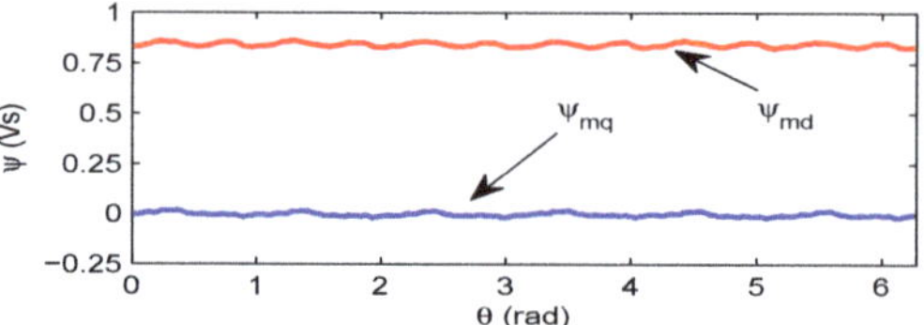

Figure 10. Position-dependent characteristics of flux linkages $\psi_{md}(\theta)$ and $\psi_{mq}(\theta)$.

Figure 9 clearly shows that the back EMFs e_a, e_b and e_c are transformed into e_d, e_q and e_0, where $e_0 \neq 0$. In the given case the machine terminals are open, and no currents are flowing. Thus, e_0 equals the last term on the right-hand side of (14) and changes the neutral point voltage ($u_0 = e_0$). In the wye connected machines i_0 (13) cannot flow and cannot influence the voltages u_d and u_q in (12), although $u_0 \neq 0$. Therefore, (12) can be reduced to (15), which means that u_0 is neglected, although it exists. The model obtained in this way is correct. However, in the case of control scheme realization (Figure 3), the neutral point voltage u_0 changes all the time, which means that the current controllers compensate the influence of u_0 implicitly when minimizing the error between reference and measured d- and q-axis current trajectories.

From the results shown in Figure 10, it is evident that both $\psi_{md}(\theta)$ and $\psi_{mq}(\theta)$, change with the position θ. The flux linkage $\psi_{md}(\theta)$ changes around its DC component, while $\psi_{mq}(\theta)$ changes around the value 0.

Figure 11 show the comparison of measured and finite element analysis (FEA) determined phase a back EMF e_a at a constant angular speed of the rotor. The calculated results were obtained by an in house developed program solution for 2D FEA, which is a further development of the one applied in [11].

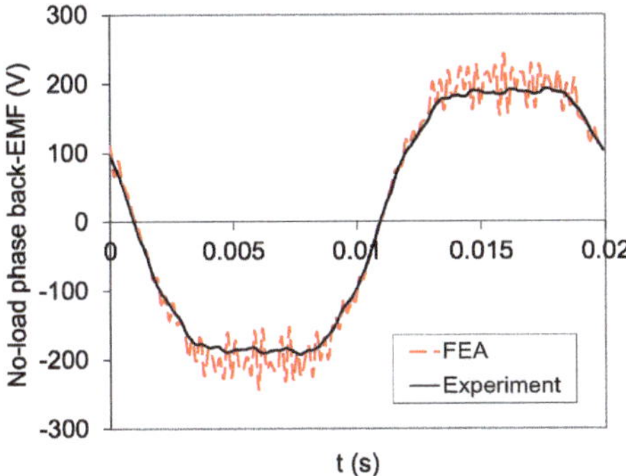

Figure 11. PMSM: measured and FEA calculated phase a back EMF e_a.

4.3. Determining Flux Linkages Caused by the Stator Currents

This procedure, which can be applied to determine the current- and position-dependent characteristics of flux linkages, caused by the stator currents, is described in Section 3.2.

The rotor is locked at a given position, while the current i_d is closed-loop controlled to keep a constant value. Figure 12 shows the applied voltage u_q and the responding current i_q measured during the experiment. Figure 13 shows the flux linkage $\psi_q(t)$ determined by numerical integration (56). The corresponding characteristic $\psi_q(i_q)$ in the form of a hysteresis loop is shown in Figure 14. The hysteresis loops determined for different amplitudes of the stepwise changing voltages u_q are shown in Figure 15, while Figure 16 shows a unique $\psi_q(i_q)$ characteristic for the constant value of $i_d = 0$ A. Figures 17 and 18 show the characteristics $\psi_d(i_d)$ at $i_q = 0$ A in the form of a hysteresis loop and in the form of a unique characteristic, respectively. The comparison of characteristics $\psi_d(i_d)$ and $\psi_q(i_q)$ determined for the SRM and PMSM is shown in Figures 19 and 20.

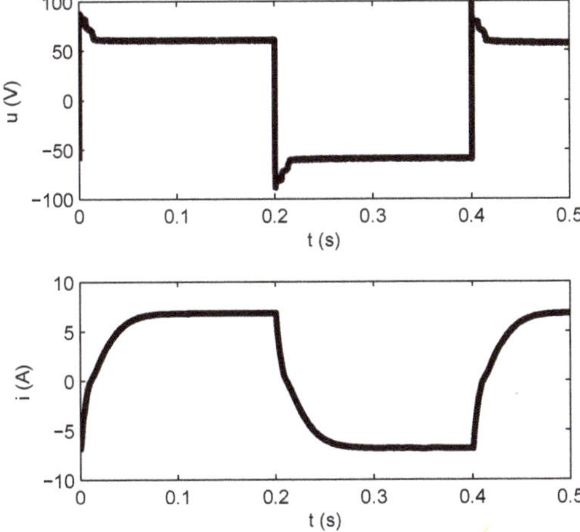

Figure 12. Stepwise changing voltage $u_q(t)$ and responding current $i_q(t)$ measured on the tested SRM for $i_d = 0$ A.

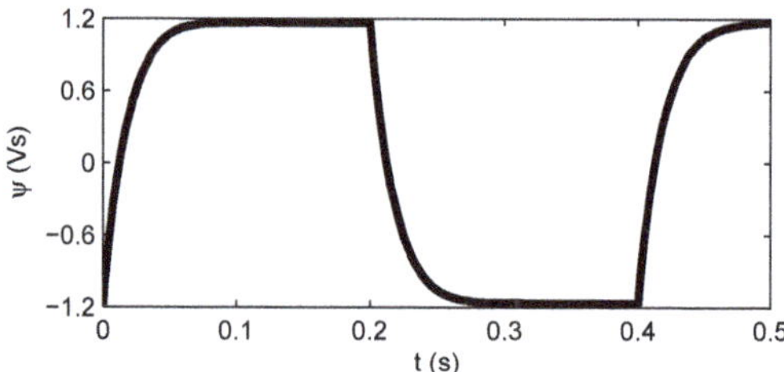

Figure 13. SRM: flux linkage $\psi_q(t)$ calculated using $u_q(t)$ and $i_q(t)$ shown in Figure 9.

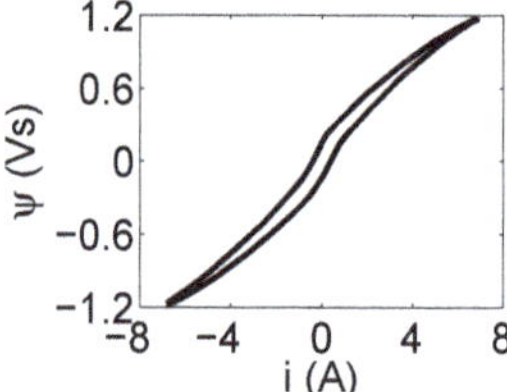

Figure 14. SRM: flux linkage characteristic $\psi_q(i_q)$ at $i_d = 0$ A given in the form of a hysteresis loop.

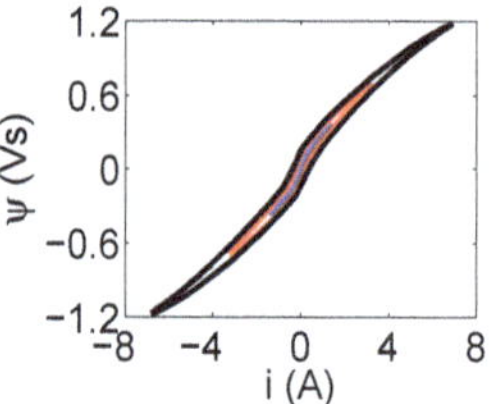

Figure 15. SRM: flux linkage characteristics $\psi_q(i_q)$ at $i_d = 0$ A given in the form of hysteresis loops determined for different amplitudes of applied voltage u_q.

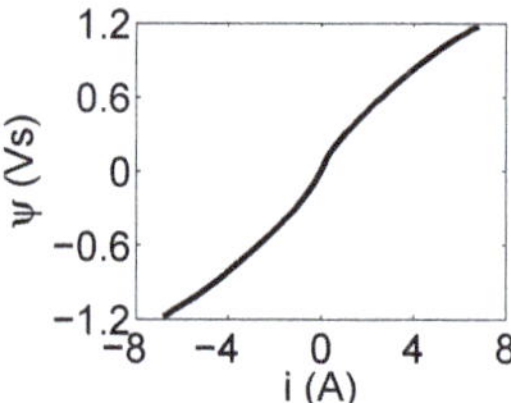

Figure 16. SRM: unique flux linkage characteristics $\psi_q(i_q)$ at $i_d = 0$ A.

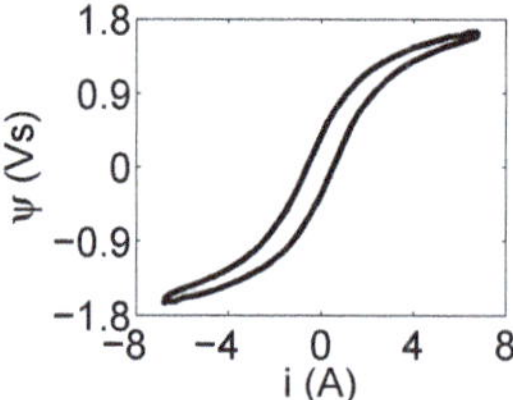

Figure 17. SRM: flux linkage characteristics $\psi_d(i_d)$ at $i_q = 0$ A given in the form of a hysteresis loop.

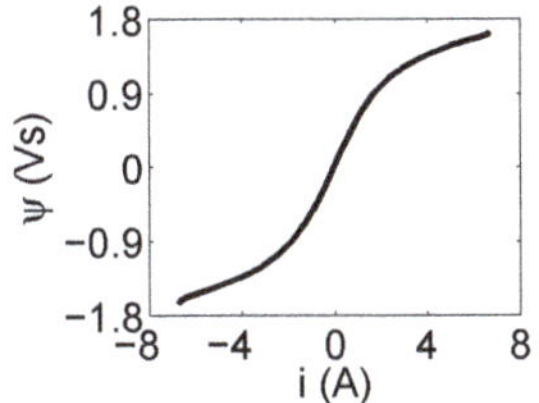

Figure 18. SRM: unique flux linkage characteristics $\psi_d(i_d)$ at $i_q =0$ A.

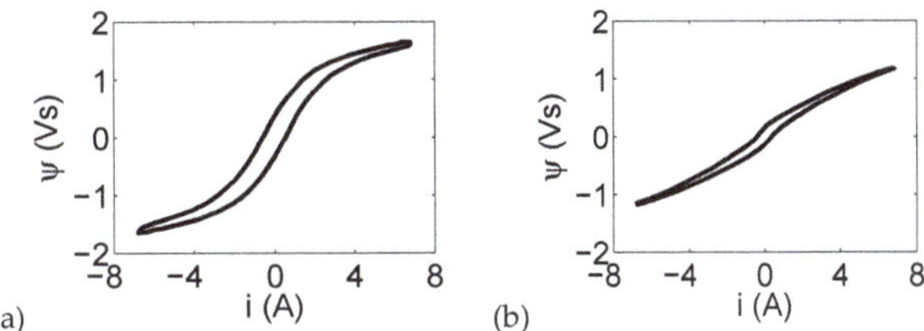

Figure 19. SRM: hysteresis loops. $\psi_d(i_d)$ at $i_q = 0$ A (**a**) and $\psi_q(i_q)$ at $i_d = 0$ A (**b**).

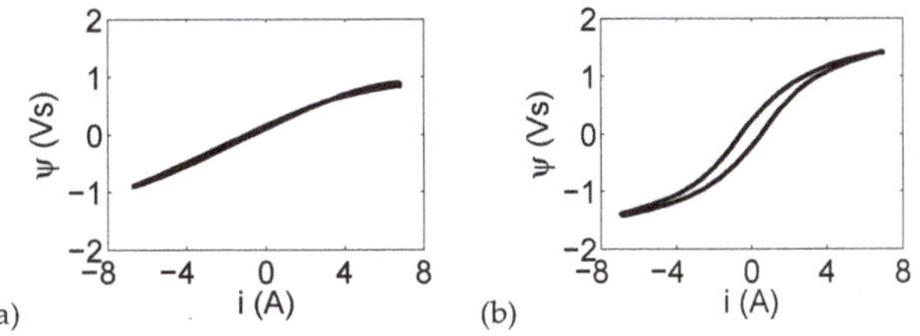

Figure 20. PMSM: hysteresis loops. $\psi_d(i_d)$ at $i_q = 0$ A (**a**) and $\psi_q(i_q)$ at $i_d = 0$ A (**b**).

The characteristics $\psi_d(i_d)$ and $\psi_q(i_q)$ shown in Figure 16 are centered. They are given for the SRM. The same characteristics are shown in Figure 17 for the PMSM. They are not centered, due to the pre-magnetization in the form of $\psi_{md}(\theta)$ and $\psi_{mq}(\theta)$ shown in Figure 8. However, this pre-magnetization is not shown explicitly in Figure 17. As mentioned before, the initial flux linkage due to the remanent flux and permanent magnets is considered with the value 0. Thus, the pre-magnetization is shown in Figure 17 rather indirectly through the increased saturation level in comparison to the SRM, shown in Figure 16. The increased saturation level decreases the slope of the $\psi_d(i_d)$ and $\psi_q(i_q)$ characteristics shown in Figure 17.

The flux linkages from Figure 20 are shown in Figure 21 as unique magnetically nonlinear characteristics determined in the same way as in [12]. Please note, that the flux linkage due to the permanent magnets $\psi_d(0) = \psi_{md}$, which appears in (56), is in this case considered with the value 0.

Based on the unique flux linkage characteristics $\psi_d(i_d)$ and $\psi_q(i_q)$ from Figure 21, the dynamic inductances $L_{sd} = \partial\psi_d/\partial i_d$ and $L_{sq} = \partial\psi_q/\partial i_q$ are calculated numerically. The comparison of experimentally and FEA determined dynamic inductances presented in Figure 22 shows an acceptable agreement.

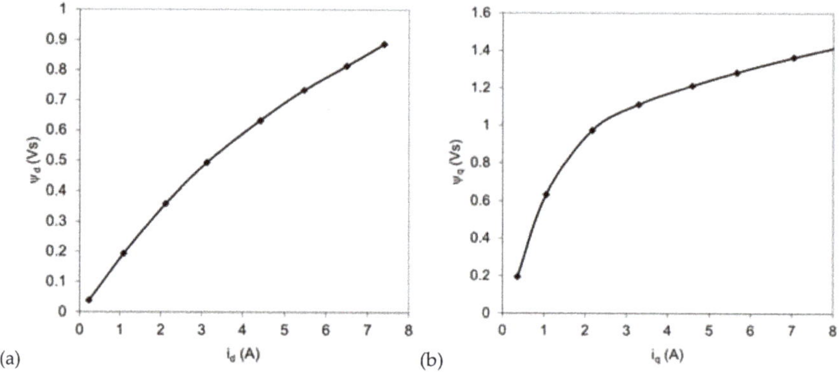

Figure 21. PMSM: unique magnetically nonlinear characteristics $\psi_d(i_d)$ at $i_q = 0$ A (**a**) and $\psi_q(i_q)$ at $i_d = 0$ A (**b**).

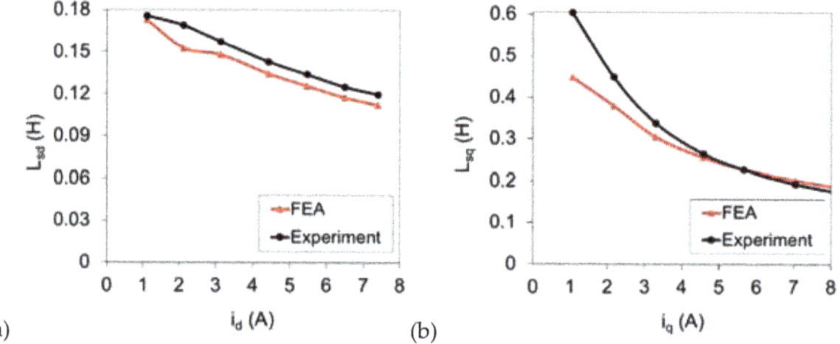

Figure 22. PMSM: FEA and experimentally determined dynamic inductances L_{sd} (**a**) and L_{sq} (**b**).

4.4. Low-Speed Kinematic Control of Linear SRM

The proposed models, as well as the experimental methods applied to determine model parameters in the form of position- and current-dependent flux linkages [12], are confirmed by the comparison of measured (Figure 23) and dynamic model calculated (Figures 24 and 25) trajectories of individual variables in the case of linear SRM kinematic control. The next set of results demonstrates the importance of higher order harmonic components in the thrust equation (33) in the case of low speed kinematic control. The tested linear SRM is shown in Figure 7. Its position x is changed for one pole pitch τ_p. The position x, current i_d, speed $v = dx/dt$, and current i_q, measured during the experiment, are shown in Figure 23. The calculations are performed with the proposed dynamic model of linear SRM given by (39), (32), (33) under the same conditions as the experiment shown in Figure 23. Figure 24 shows the calculated speed v and current i_q for the case when 24 higher order harmonic components are considered in (33). The same calculated variables are given in Figure 25 for the case when only DC components are considered in (33). The results presented clearly show that the higher order harmonic components of the thrust can substantially influence the calculated trajectories of the speed v and current i_q, which can be important in the case of low-speed kinematic control.

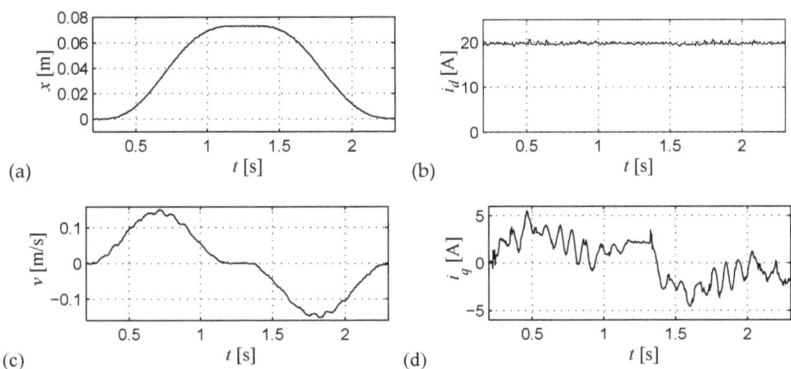

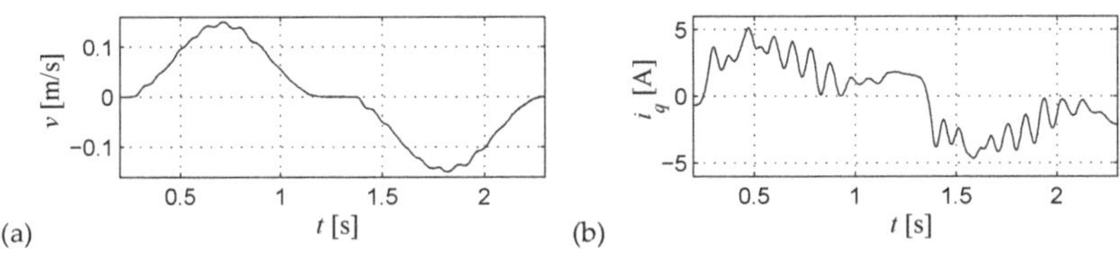

Figure 23. Kinematic control of linear SRM: measured position x (**a**), current i_d (**b**), speed $v = dx/dt$ (**c**) and current i_q. (**d**).

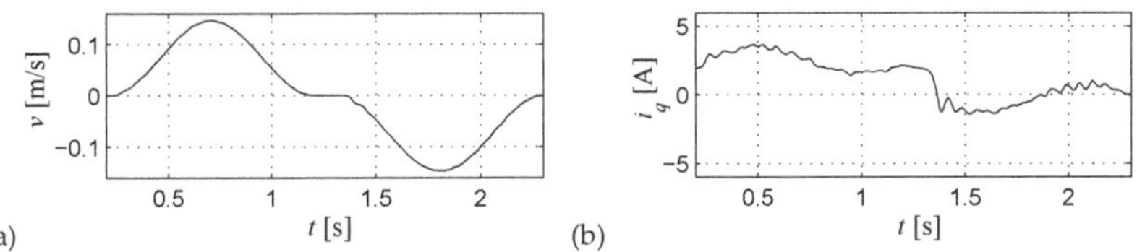

Figure 24. Kinematic control of linear SRM: calculated speed $v = dx/dt$ (**a**) and current i_q (**b**) considering N=24 higher order harmonics in thrust calculation (33).

Figure 25. Kinematic control of linear SRM: calculated speed $v = dx/dt$ (**a**) and current i_q (**b**) considering only DC component in thrust calculation (33).

5. Conclusion

The main contribution of this paper is a proposed general method for deriving magnetically nonlinear two-axis dynamic models of synchronous machines, which is applied to derive the dynamic models of rotary and linear PMSMs and SRMs.

The proposed method consists of two steps. In the first step, only the structure of the magnetically nonlinear dynamic model is determined, while in the second step, the current- and position-dependent characteristics of flux linkages are determined experimentally by measurements on the machine's terminals. With these characteristics, the effects of slotting, interactions between the slots and permanent magnets, saturation, cross-saturation as well as the linear machine specific end effects are considered; and this without any prior knowledge of the machine properties required in FEA-based methods. Some of the experimental methods suitable for determining the aforementioned characteristics of flux linkages are presented in the paper. The flux linkages, electromagnetic torque and thrust are described in the form of Fourier series, where the relations among their Fourier coefficients are given in the form of equations.

The proposed approach to the modelling of synchronous machines, together with the experimental methods applied for determining magnetically nonlinear characteristic of flux linkages, is confirmed through the comparison of measured and calculated results, given for the low speed kinematic control of a linear SRM. The results presented clearly show the impact of considered order of higher harmonics in the thrust calculation on current and speed trajectories.

Author Contributions: G.Š. performed derivations, build up experimental system for linear machines, performed experiments and contributed in paper preparation. B.Š. designed rotary machines, provided prototypes of rotary machines and results of FEA. T.M. performed experimental work on rotary machines and contributed in the analysis of obtained results and in paper preparation.

Funding: This work was supported in part by the Slovenian Research Agency, project no. P2-0115 and J2-1742.

Conflicts of Interest: The authors declare no conflict of interest.

References

1. Kron, G. *Equivalent Circuits of Electric Machinery*; John Wiley & Sons: New York, NY, USA, 1951.
2. Kron, G. *Tensor for Circuits*; Dover: New York, NY, USA, 1959.
3. Kron, G. *Tensor Analysis of Networks*; MacDonald: London, UK, 1965.
4. Fitzgerlad, A.E.; Kingslley, C. *Electric Machinery*; McGraw-Hill Book Company: New York, NY, USA, 1961.
5. Krause, P.C.; Wasynczuk, O.; Sudhoff, S.D. *Analysis of Electric Machinery*; IEEE Press: New York, NY, USA, 2002.
6. Boldea, I.; Tutelea, L. *Electric Machines: Steady States, Transients, and Design with Matlab*; CRC Press: Boca Raton, FL, USA, 2010.
7. Vas, P.; Hallenius, K.; Brown, J. Cross-Saturation in Smooth-Air-Gap Electrical Machines. *IEEE Trans. Energy Convers.* **1986**, *1*, 103–112. [CrossRef]
8. Iglesias, I.; Garcia-Tabares, L.; Tamarit, J. A d-q model for the self-commutated synchronous machine considering magnetic saturation. *IEEE Trans. Energy Convers.* **1992**, *7*, 768–776. [CrossRef]
9. Levi, E.; Levi, V. Impact of dynamic cross-saturation on accuracy of saturated synchronous machine models. *IEEE Trans. Energy Convers.* **2000**, *15*, 224–230. [CrossRef]
10. Tahan, S.A.; Kamwa, I. A two-factor saturation model for synchronous machines with multiple rotor circuits. *IEEE Trans. Energy Convers.* **1995**, *10*, 609–616. [CrossRef]
11. Štumberger, B.; Štumberger, G.; Dolinar, D.; Hamler, A.; Trlep, M. Evaluation of saturation and cross-magnetization effects in interior permanent-magnet synchronous motor. *IEEE Trans. Ind. Appl.* **2003**, *39*, 1264–1271. [CrossRef]
12. Štumberger, G.; Štumberger, B.; Dolinar, D. Identification of Linear Synchronous Reluctance Motor Parameters. *IEEE Trans. Ind. Appl.* **2004**, *40*, 1317–1324. [CrossRef]
13. Dolinar, D.; Štumberger, G.; Milanovič, M. Tracking improvement of LSRM at low-speed operation. *Eur. Trans. Electr. Power* **2005**, *15*, 257–270. [CrossRef]
14. Melkebeek, J.A.; Willems, J.L. Reciprocity relations for the mutual inductances between orthogonal axis windings in saturated salinet-pole machines. *IEEE Trans. Ind. Appl.* **1990**, *26*, 107–114. [CrossRef]
15. Sauer, P.W. Constraints on saturation modelling in ac machines. *IEEE Trans. Energy Convers.* **1992**, *7*, 161–167. [CrossRef]
16. Vagati, A.; Pastorelli, M.; Scapino, F.; Franceschini, G. Impact of cross saturation in synchronous reluctance motors of the transverse-laminated type. *IEEE Trans. Ind. Appl.* **2000**, *36*, 1039–1046. [CrossRef]
17. Capecchi, E.; Guglielmi, P.; Pastorelli, M.; Vagati, A. Position-sensorless control of the transverse-laminated synchronous reluctance motor. *IEEE Trans. Ind. Appl.* **2001**, *37*, 1768–1776. [CrossRef]
18. Guglielmi, P.; Pastorelli, M.; Vagati, A. Impact of cross-saturation in sensorless control of transverse-laminated synchronous reluctance motors. *IEEE Trans. Ind. Electron.* **2006**, *53*, 429–439. [CrossRef]
19. Hadžiselimović, M. Magnetically Nonlinear Dynamic Model of a Permanent Magnet Synchronous Motor. Ph.D. Thesis, University of Maribor, Maribor, Slovenia, 2007.
20. Hadžiselimović, M.; Štumberger, G.; Štumberger, B.; Zagradišnik, I. Magnetically nonlinear dynamic model of synchronous motor with permanent magnets. *J. Magn. Magn. Mater.* **2007**, *316*, e257–e260. [CrossRef]
21. Hadžiselimović, M.; Štumberger, B.; Virtič, P.; Marčič, T.; Štumberger, G. Determining parameters of a two-axis permanent magnet synchronous motor dynamic model by finite element method. *Przegląd Elektrotechniczny* **2008**, *84*, 77–80.

22. Su, W.T.; Liaw, C.M. Adaptive positioning control for a LPMSM drive based on adapted inverse model and robust disturbance observer. *IEEE Trans. Power Electron.* **2006**, *21*, 505–517.

23. Zhang, Y.; Zhu, J. Direct torque control of permanent magnet synchronous motor with reduced torque ripple and commutation frequency. *IEEE Trans. Power Electron.* **2011**, *26*, 235–248. [CrossRef]

24. Cupertino, F.; Pellegrino, G.; Giangrande, P.; Salvatore, L. Sensorless Position Control of Permanent-Magnet Motors with Pulsating Current Injection and Compensation of Motor End Effects. *IEEE Trans. Ind. Appl.* **2011**, *47*, 1371–1379. [CrossRef]

25. Idkhajine, L.; Monmasson, E.; Maalouf, A. Fully FPGA-Based Sensorless Control for Synchronous AC Drive Using an Extended Kalman Filter. *IEEE Trans. Ind. Electron.* **2012**, *59*, 3908–3918. [CrossRef]

26. Shi, Y.; Sun, K.; Huang, L.; Li, Y. Online identification of permanent magnet flux based on extended Kalman filter for IPMSM drive with position sensorless control. *IEEE Trans. Ind. Electron.* **2012**, *59*, 4169–4178. [CrossRef]

27. Xu, Z.; Rahman, M.F. Comparison of a Sliding Observer and a Kalman Filter for Direct-Torque-Controlled IPM Synchronous Motor Drives. *IEEE Trans. Ind. Electron.* **2012**, *59*, 4179–4188. [CrossRef]

28. Xia, C.; Geng, Q.; Gu, X.; Shi, T.; Song, Z. Input–Output Feedback Linearization and Speed Control of a Surface Permanent-Magnet Synchronous Wind Generator with the Boost-Chopper Converter. *IEEE Trans. Ind. Electron.* **2012**, *59*, 3489–3500.

29. Wei, M.; Liu, T. A high-performance sensorless position control system of a synchronous reluctance motor using dual current-slope estimating technique. *IEEE Trans. Ind. Electron.* **2012**, *59*, 3411–3426.

30. Pellegrino, G.; Armando, E.; Guglielmi, P. Direct-flux vector control of IPM motor drives in the maximum torque per voltage speed range. *IEEE Trans. Ind. Electron.* **2012**, *59*, 3780–3788. [CrossRef]

31. Kwon, Y.; Kim, S.; Sul, S. Voltage feedback current control scheme for improved transient performance of permanent magnet synchronous machine drives. *IEEE Trans. Ind. Electron.* **2012**, *59*, 3373–3382. [CrossRef]

32. Lin, P.; Ali, Y. Voltage control technique for the extension of DC-link voltage utilization of finite-speed SPMSM drives. *IEEE Trans. Ind. Electron.* **2012**, *59*, 3392–3402. [CrossRef]

33. Rahman, K.; Hiti, S. Identification of Machine Parameters of a Synchronous Motor. *IEEE Trans. Ind. Appl.* **2005**, *41*, 557–565. [CrossRef]

34. Marčič, T.; Štumberger, G.; Štumberger, B.; Hadžiselimović, M.; Virtič, P. Determining Parameters of a Line-Start Interior Permanent Magnet Synchronous Motor Model by the Differential Evolution. *IEEE Trans. Magn.* **2008**, *44*, 4385–4388. [CrossRef]

35. Peretti, L.; Zanuso, G.; Sandulescu, P. Self-commissioning of flux linkage curves of synchronous reluctance machines in quasi-standstill condition. *IET Electr. Power Appl.* **2015**, *9*, 642–651. [CrossRef]

36. Hinkkanen, M.; Pescetto, P.; Molsa, E.; Saarakkala, S.E.; Pellegrino, G.; Bojoi, R. Sensorless Self-Commissioning of Synchronous Reluctance Motors at Standstill. In Proceedings of the XXII International Conference on Electrical Machines (ICEM), Lausanne, Switzerland, 4–7 September 2016.

37. Zhang, J.; Wen, X.; Wang, Y.; Li, W. Modeling and Analysis of Nonlinear Interior Permanent Magnet Synchronous Motors Considering Saturation and Cross-magnetization Effects. In Proceedings of the IEEE Transportation Electrification Conference, Busan, Korea, 1–4 June 2016.

38. Accetta, A.; Cirrincione, M.; Pucci, M.; Sferlazza, A. A Space-Vector State Dynamic Model of the Synchronous Reluctance Motor Including Self and Cross-Saturation Effects and its Parameters Estimation. In Proceedings of the IEEE Energy Conversion Congress and Exposition (ECCE), Portland, OR, USA, 23–27 September 2018.

39. Ibrahim, M.N.; Sergeant, P.; Rashad, E.M. Relevance of Including Saturation and Position Dependence in the Inductances for Accurate Dynamic Modeling and Control of SynRMs. *IEEE Trans. Ind. Appl.* **2017**, *53*, 151–160. [CrossRef]

40. Barcaro, M.; Morandin, M.; Pradella, T.; Bianchi, N.; Furlan, I. Iron Saturation Impact on High-Frequency Sensorless Control of Synchronous Permanent-Magnet Motor. *IEEE Trans. Ind. Appl.* **2017**, *53*, 5470–5478. [CrossRef]

41. Boll, R. *Weichmagnetische Werkstoffe: Einführung in den Magnetismus*; Vakuumschmelze GmbH: Hanau, Germany, 1990.

42. Daldaban, F.; Ustkoyuncu, N. A new double sided linear switched reluctance motor with low cost. *Energy Convers. Manag.* **2006**, *47*, 2983–2990. [CrossRef]

43. Daldaban, F.; Ustkoyuncu, N. New disc type switched reluctance motor for high torque density. *Energy Convers. Manag.* **2007**, *48*, 2424–2431. [CrossRef]

44. Daldaban, F.; Ustkoyuncu, N. A novel linear switched reluctance motor for railway transportation systems. *Energy Convers. Manag.* **2010**, *51*, 465–469. [CrossRef]
45. Parasiliti, F.; Villani, M.; Lucidi, S.; Rinaldi, F. Finite-element-based multiobjective design optimization procedure of interior permanent magnet synchronous motors for wide constant-power region operation. *IEEE Trans. Ind. Electron.* **2012**, *59*, 2503–3514. [CrossRef]
46. Barcaro, M.; Bianchi, N.; Magnussen, F. Permanent-magnet optimization in permanent-magnet-assisted synchronous reluctance motor for a wide constant-power speed range. *IEEE Trans. Ind. Electron.* **2012**, *59*, 2495–2502. [CrossRef]
47. Di Stefano, R.; Marignetti, F. Electromagnetic analysis of axial-flux permanent magnet synchronous machines with fractional windings with experimental validation. *IEEE Trans. Ind. Electron.* **2012**, *59*, 2573–2582. [CrossRef]
48. Hamler, A.; Trlep, M.; Hribernik, B. Optimal secondary segment shapes of linear reluctance motors using stochastic searching. *IEEE Trans. Magn.* **1998**, *34*, 3519–3521. [CrossRef]

Article

Neural Network-Based Model Reference Adaptive System for Torque Ripple Reduction in Sensorless Poly Phase Induction Motor Drive

S. Usha [1], C. Subramani [1,*] and Sanjeevikumar Padmanaban [2]

[1] Department of Electrical and Electronics Engineering, SRM Institute of Science and Technology, Kattankulathur 603203, India; ushakarthick@gmail.com

[2] Center for Bioenergy and Green Engineering, Department of Energy Technology, Aalborg University, 6700 Esbjerg, Denmark; san@et.aau.dk

* Correspondence: csmsrm@gmail.com; Tel.: +91-960-001-3988

Received: 22 February 2019; Accepted: 5 March 2019; Published: 9 March 2019

Abstract: This paper proposes the modified, extended Kalman filter, neural network-based model reference adaptive system and the modified observer technique to estimate the speed of a five-phase induction motor for sensorless drive. The proposed method is generated to achieve reduced speed deviation and reduced torque ripple efficiently. In inclusion, the result of speed performance and torque ripple under parameter variations were analysed and compared with the conventional direct synthesis method. The speed estimation of a five-phase motor in the four methods is analysed using MATLAB Simulink platform, and the optimum method is recognized using time domain analysis. It is observed that speed error is minimized by 60% and torque ripple is reduced by 75% in the proposed method. The hardware setup is carried out for the optimized method identified.

Keywords: induction motor; speed estimation; model reference adaptive system; kalman filter; luenberger observer

---

1. Introduction

An induction motor is the most commonly used motor in industries because it is reliable, robust and has low cost. Conventionally, the traction drive was operated with a direct current motor, but the maintenance cost is high. To overcome the above problem, a three-phase induction motor is used. The three- and poly-phase induction motor are modelled and analysed; the amplitude torque ripple is less in a polyphase machine. Even though the poly-phase machine has high efficiency, less torque pulsations, and higher fault tolerance due to the lack of power supply in earlier days, poly-phase has not been used. However, nowadays, due to the advancement of power electronics devices, control of the poly-phase induction motor also possible.

Since reliability is one of the important parameters to which the induction motor owes its operation, the reliability of the induction motor is improved under different operating conditions. To achieve this, the parameters of the induction motor need to be taken care of. To eliminate the design mistakes in the archetype construction and when testing the motor drive system, the dynamic simulation plays a major role in validating the design process of the system. A synchronously revolving rotor flux-oriented frame is taken as a reference, and the induction motor is modelled in it [1].

Accurate knowledge of a few of the parameters of the induction motor is important for the purpose of sensor-less vector control and the control schemes of the induction motor. If the original parameter values in the motor fail to match with the values used within the controller, it leads to the degradation of the presentation of the drive. The benefit of dynamic modelling is that it helps in understanding the behaviour of the motor in the transient and steady state in a better way [2].

The dynamic modelling comprises of all the mechanical equations including the torque and speed vs. time. The differential voltages, flux linkages, and currents between the moving rotor as well as the stationary stator can also be modelled by dynamic modelling. There are numerous schemes that show a narrow stable region for low speeds in the regenerative mode for high torques [3]. This paper focuses on analysing the parameters like flux and speed evaluations of an induction motor without a sensor using reduced order observers [4,5].

In the traction industries and electric vehicles, the induction motor is the most commonly used machine as it offers advantages like good performance, less primary cost, and low maintenance cost. Speed identification is needed for the induction motor drives [6]. However, installation of the speed sensor in the induction motor leads to disadvantages like less reliability, extra cost, large size, etc. Estimation of speed for sensorless induction motor drive can be done by various techniques. These techniques are designed by keeping the factors like accuracy and sensitivity adverse to the induction motor parameter variations in consideration [7–9].

The fault tolerance capability of Quinary phase machines is of a superior standard compared to those of the three-phase machines. The three-phase machine becomes a single-phase machine when one of the phases is short-circuited, which still allows the machine to workout but for initializations, external means are needed and must be de-rated. In the case of the open circuit of quinary phase machines, it still holds its self-start competence and runs with minimized de-rating [10].

The projected crop up through a dynamic modelling of a quinary phase induction motor on a Direct-Quadrature (d-q) axis and the speed guesstimate of the motor using two sensor-less guesstimate techniques and an evaluation is being shaped between those procedures centred on the tenets of the speed being judged. Exploration in the arena of a multi-phase machine zone has gifted significant scopes in the previous era. With the figure of conservative electrical machines unceasingly mounting, the curiosity in multiphase machines is also intensifying due to innate types such as power disbanding, better fault resilience, or lower torque ripple than three-phase machines. Quinary phase machines are more defensive than the counterparts of three-phase towards the time-harmonic element in the waveform with excitations which produce high ripples in the torque of the elementary frequency with excitations at even multiples. This paper proposes a dynamic modelling of a quinary phase induction motor and the speed guesstimation of the motor using two sensor-less method guesstimations, and a comparison is being formed with conventional methods based on the values of speed that are estimated [11].

The stationary part stimulation in a quinary phase machine creates a field with less harmonic content, so that the improved efficiency is more appreciative than in a traditional three-phase machine. In the conventional induction motor, one of the common methods for obtaining rotor speed is by using speed encoders which sense the speed signal and give the rotor speed. Besides the necessity for the extension of shaft and mounting arrangement, a speed encoder adds rate and reliability snags [12]. It is feasible to find the speed with the ease of a digital signal processor from machine terminal current and voltage.

In conventional methods, speed and rotor flux estimators are intended for sensor-less control of motion control structures with induction motors. More exactly, the estimators entail the conjunction of an adaptive speed guesstimation scheme and either a robust or standard descriptor-type Kalman filter. It is exposed that the descriptor structure of the Kalman filter permits for a direct transformation of parameter variations into coefficient disparities of the structure model, which leads to vulgarizations in the describing reservations and a stochastic guesstimation of inaccessible state variables, and the mysterious input of an electric vehicle driveline fortified with an innovative seamless clutch-less two speed transmission is being studied. However, the guesstimation is normally intricate and steadily dependent on the components of the machine [13]. Hence, sensor-less speed guesstimation techniques are being proposed to tackle these snags.

Although in the case of a flux estimator, the flux of motor cannot be measured immediately, the notion of comprehending a closed-loop structure is tranquil associable if the discrepancy flanked by a signal replacing the stable-state data of the source flux and the wave of the evaluated flux vector is back to the input signal that is a feedback signal. This reference data is normally attainable in rotor-flux-based regulations. In this proposed method, the filter equation is altered by including a sliding hyper plane in the modified Luenberger and extended Kalman filter method to improve the performance of the system. The conventional voltage model-based reference adaptive system replaced by a neural network-based system to overcome the integration problem in the conventional method is also robust to parameter variations.

In this paper, Section 2 explains the modelling of an induction motor and speed estimation methods. The simulation results of speed guesstimation of a quinary phase induction motor using a Direct Synthesis method, Model Reference Adaptive System, Luenberger observer, and Extended Kalman Filter are discussed in Section 3. Speed deviation and Torque ripple reduction of induction motor drive in parallel using an extended Kalman filter are discussed in Section 4. Section 5 delineates the hardware implementation's induction motor drive.

2. Modelling of Induction Motor and Speed Estimation Methods

The steady-state prototype and equivalent enlarged circuit are helpful for studying steady state interpretations of the machine. This involves all the transients being skipped during the variations in stator frequency and load. Such changes that emerge in implementation include changeable speed drives [14,15]. The output of converter fed changeable speed drives in their impotence to provide large transient power. Hence, it requires estimating the changing of converter support changeable speed drives to determine the fairness of the switches of the converter and for a particular motor and their interactions to find the expedition of torque and current in the motor and converter [16–18]. The first theory assumes that the Magneto motive force generated by different phases of the rotor and stator are expanded in a sinusoidal method with an air gap, when those windings are traversed by a stable current. A suitable distribution of the windings in the area allows extending this aim. The air gap of a machine is also assumed to be invariably thick: the notching results and originating space harmonic are disregarded. These hypotheses will permit regulating the modelling of the low frequency of alternative quantities [19,20].

2.1. Five Phase Induction Motor Model

The phase voltages of the quinary phase induction machine are V_a, V_b, V_c, V_e, and V_f. The phase angle between each phase is 72 degrees. The voltages are given by the Equations (1)–(5) respectively.

$$V_a = V_m \text{Sin} (\omega t) \tag{1}$$

$$V_b = V_m \text{Sin} \left(\omega t - \frac{2\pi}{5}\right) \tag{2}$$

$$V_c = V_m \text{Sin} \left(\omega t - \frac{4\pi}{5}\right) \tag{3}$$

$$V_e = V_m \text{Sin} \left(\omega t - \frac{6\pi}{5}\right) \tag{4}$$

$$V_f = V_m \text{Sin} \left(\omega t - \frac{8\pi}{5}\right) \tag{5}$$

The quinary phase voltages are converted into d-q axis using the transition matrix. The transition matrix is given by Equation (6).

$$
\begin{bmatrix} V_d \\ V_q \end{bmatrix} = \sqrt{\frac{2}{5}} \begin{bmatrix} \cos(\omega t)\cos(\omega t - \frac{2\Pi}{5})\cos(\omega t - \frac{4\Pi}{5})\cos(\omega t - \frac{6\Pi}{5})\cos(\omega t - \frac{8\Pi}{5}) \\ \sin(\omega t)\sin(\omega t - \frac{2\Pi}{5})\sin(\omega t - \frac{4\Pi}{5})\sin(\omega t - \frac{6\Pi}{5})\sin(\omega t - \frac{8\Pi}{5}) \end{bmatrix} \cdot \begin{bmatrix} Va \\ Vb \\ Vc \\ Vd \\ Ve \end{bmatrix} \tag{6}
$$

The d and q axis stator voltages are given by Equations (7) and (8), and the d and q axis rotor voltages are given by Equations (9) and (10) respectively.

$$
V_{ds} = R_s i_{ds} + \frac{d}{dt}\psi_{ds} - \omega_e\psi_{qs} \tag{7}
$$

$$
V_{qs} = R_s i_{qs} + \frac{d}{dt}\psi_{qs} + \omega_e\psi_{ds} \tag{8}
$$

$$
V_{dr} = R_r i_{dr} + \frac{d}{dt}\psi_{dr} - (\omega_e - \omega_r)\psi_{qr} \tag{9}
$$

$$
V_{qr} = R_r i_{qr} + \frac{d}{dt}\psi_{qr} + (\omega_e - \omega_r)\psi_{dr} \tag{10}
$$

The flux linkages in the d and q axis are expressed in terms of the direct axis and quadrature axis currents in the following equations given by (11)–(16).

$$
\psi_{ds} = L_{1s}i_{ds} + L_m,(i_{ds} + i_{dr}) \tag{11}
$$

$$
\psi_{dr} = L_{1r}i_{dr} + L_m,(i_{ds} + i_{dr}) \tag{12}
$$

$$
\psi_{qs} = L_{1s}i_{ds} + L_m,(i_{qs} + i_{qr}) \tag{13}
$$

$$
\psi_{qr} = L_{1r}i_{dr} + L_m,(i_{qs} + i_{qr}) \tag{14}
$$

$$
\psi_{dm} = L_m,(i_{ds} + i_{qr}) \tag{15}
$$

$$
\psi_{dm} = L_m,(i_{ds} + i_{qr}) \tag{16}
$$

The direct axis and quadrature axis currents in terms of flux and inductance are given by the following Equations (17)–(20).

$$
i_{ds} = \frac{\psi_{ds}(L_{1r} + L_m) - L_m\psi_{dr}}{(L_{1s}L_{1r} + L_{1s}L_m + L_{1r}L_m)} \tag{17}
$$

$$
i_{qs} = \frac{\psi_{qs}(L_{1r} + L_m) - L_m\psi_{qr}}{(L_{1s}L_{1r} + L_{1s}L_m + L_{1r}L_m)} \tag{18}
$$

$$
i_{dr} = \frac{\psi_{ds}(L_{1r} + L_m) - L_m\psi_{ds}}{(L_{1s}L_{1r} + L_{1s}L_m + L_{1r}L_m)} \tag{19}
$$

$$
i_{dr} = \frac{\psi_{ds}(L_{1r} + L_m) - L_m\psi_{ds}}{(L_{1s}L_{1r} + L_{1s}L_m + L_{1r}L_m)} \tag{20}
$$

The electrical torque developed in the rotor of a quinary induction motor is given by Equation (21).

$$
T_e = PL_m(i_{qs}i_{dr} - i_{ds}i_{qr}) \tag{21}
$$

The reference speed of the rotor is given by Equation (22).

$$
\omega_r = \int \frac{P}{2J}(T_e - T_1)dt \tag{22}
$$

2.2. Modelling of Conventional Direct Synthesis

To estimate the speed in the direct synthesis method for obtaining the rotor fluxes in the direct and quadrature axis, the voltage and the current model of rotating reference are used. Different regulation methods such as scalar regulation, field orient control approach, and regulation without a sensor are used. This method agonizes from parameter reactivity and partial presentations at a very low speed of operation. The combination of the direct synthesis method is extremely machine parameter-delicate and will move to give a less accurate guesstimation. The flux guesstimation can be done by using both the current regulation method and voltage regulation model.

The direct axis and quadrature axis rotor fluxes are given by the following Equations (23) and (24).

$$\psi^s{}_{dr} = \int \left(\frac{L_m}{T_r} i^s{}_{ds} - \omega_r \psi_{qr} - \frac{1}{T_r} \psi^s{}_{dr} \right) \tag{23}$$

$$\psi^s{}_{qr} = \int \left(\frac{L_m}{T_r} i^s{}_{qs} - \omega_r \psi_{dr} - \frac{1}{T_r} \psi^s{}_{dr} \right) \tag{24}$$

The direct axis and quadrature axis rotor fluxes are also given by the following Equations (25) and (26) respectively.

$$\psi^s{}_{dr} = \frac{L_r}{L_m} \left(\psi^s{}_{ds} - \sigma L_s i^s{}_{ds} \right) \tag{25}$$

$$\psi^s{}_{qr} = \frac{L_r}{L_m} \left(\psi^s{}_{qs} - \sigma L_s i^s{}_{qs} \right) \tag{26}$$

The Speed Equation for the Direct Synthesis method has been derived from the following Equations (27)–(32).

$$V^s{}_{ds} = i^s{}_{ds} R_s + L_{1s} \frac{d}{dt} i^s{}_{ds} + \frac{d}{dt} \psi^s{}_{dm} \tag{27}$$

$$V^s{}_{ds} = i^s{}_{ds} R_s + L_{1s} \frac{d}{dt} i^s{}_{ds} + \frac{d}{dt} \psi^s{}_{dm} \tag{28}$$

$$V^s{}_{ds} = \frac{L_m}{L_r} \frac{d}{dt} \left(\psi^s{}_{dr} \right) + \left(R_s + \sigma L_s s \right) i^s{}_{ds} \tag{29}$$

$$\sigma = 1 - \frac{L^2{}_m}{L_r L_s} \tag{30}$$

$$\frac{d}{dt} \left(\psi^s{}_{dr} \right) = \frac{L_r}{L_m} \left(V^s{}_{ds} \right) - \frac{L_r}{L_m} \left(R_s + \sigma L_s S \right) i^s{}_{ds} \tag{31}$$

$$\frac{d}{dt} \left(\psi^s{}_{qr} \right) = \frac{L_r}{L_m} V^s{}_{qs} - \frac{L_r}{L_m} \left(R_s + \sigma L_s S \right) i^s{}_{qs} \tag{32}$$

$$\omega_r = \frac{1}{\psi^2{}_r} \left[\left(\psi^s{}_{dr} \psi^s{}_{qr} - \psi^s{}_{qr} \psi^s{}_{dr} \right) - \frac{L_m}{T_r} \left(\psi^s{}_{dr} \psi^s{}_{qs} - \psi^s{}_{qr} \psi^s{}_{ds} \right) \right] \tag{33}$$

The stator and rotor flux are estimated. Stator voltage in the direct axis is derived in terms of flux and inductances. The rotor speed is estimated from the estimated flux in the direct and quadrature axis.

2.3. Modified Luenberger Observer Technique Model

A structure that observes the state of structure from the inference that there is no way of directly observing the states easily is called an observer. It is used to guess the unmeasurable state of a structure. Any structure that gives an evaluation of the intramural state of a stated real structure, from the survey of the given input and response of the real structure is termed as a state observer. An eloquent structure state is mandatory to solve many regulation concept technical hitches; for example, maintaining a structure using response. The regular state of the structure cannot be gritty by direct cognition in most practical cases. Instead, oblique consequences of the core state are pragmatic by way of the structure outputs. An estimator with a closed-loop system is grounded on the axiom that by giving back the discrepancy amongst the identified output of the perceived structure and the guessed output, and perpetually amending the prototype by the signal, the error has to be miniaturized. Although, in the case of a flux estimator, the flux of the motor cannot be measured immediately, the notion of comprehending a closed-loop structure is tranquil associable if the discrepancy flanked by a signal replacing the stable-state data of the source flux and the wave of the evaluated flux vector is back to the input signal that is a feedback signal. This reference data is normally attainable in rotor-flux-based regulations. In this proposed method, a filter equation is altered by including a sliding hyper plane to improve the performance of the system.

The Luenberger observer uses the electrical model in a d_s-q_s frame, where the state variables are currents of stator $i^s{}_{ds}$ and $i^s{}_{qs}$ and fluxes of the rotor are $\psi^s{}_{dr}$ and $\psi^s{}_{qr}$. The rotor voltage is given by Equations (34) and (35).

$$i^s{}_{dr}R_r + \frac{d}{dt}(\psi^s{}_{dr}) + \omega_r \psi^s{}_{qr} = 0 \tag{34}$$

$$i^s{}_{qr}R_r + \frac{d}{dt}(\psi^s{}_{qr}) + \omega_r \psi^s{}_{dr} = 0 \tag{35}$$

From the voltage model of flux guesstimation, the rotor axis flux is given as

$$\psi^s{}_{qr} = L_m i^s{}_{qs} + L_r i^s{}_{qr} \tag{36}$$

$$\psi^s{}_{dr} = L_m i^s{}_{ds} + L_r i^s{}_{dr} \tag{37}$$

After eliminating $i^s{}_{dr}$ and $i^s{}_{qr}$ from Equations (34) and (35) with the help of Equations (36) and (37), the result is

$$\frac{d}{dt}(\psi^s{}_{ds}) = -\frac{R_r}{L_r}\psi^s{}_{dr} - \omega_r \psi^s{}_{qr} + \frac{L_m R_r}{L_r}i^s{}_{ds} \tag{38}$$

$$\frac{d}{dt}(\psi^s{}_{dr}) = \frac{L_r}{L_m}V^s{}_{ds} - \frac{L_r}{L_m}(R_s + \sigma L_s S)i^s{}_{ds} \tag{39}$$

$$\frac{d}{dt}(\psi^s{}_{qr}) = -\frac{R_r}{L_r}\psi^s{}_{qr} - \omega_r \psi^s{}_{dr} + \frac{L_m R_r}{L_r}i^s{}_{qs} \tag{40}$$

$$\frac{d}{dt}(\psi^s{}_{qr}) = \frac{L_r}{L_m}V^s{}_{qs} - \frac{L_r}{L_m}(R_s + \sigma L_s S)i^s{}_{qs} \tag{41}$$

$$\frac{d}{dt}(i^s{}_{ds}) = \frac{(L^2{}_m R_r + L^2{}_r R_s)}{\sigma L_s L^2{}_r}i^2{}_{ds} + \frac{L_m R_r}{\sigma L_s L^2{}_r}\psi^s{}_{dr} + \frac{L_m \omega_r}{\sigma L_s L_r}\psi^s{}_{qr} + \frac{1}{\sigma L_s}V^s{}_{ds} \tag{42}$$

$$\frac{d}{dt}(i^s{}_{qs}) = \frac{(L^2{}_m R_r + L^2{}_r R_s)}{\sigma L_s L^2{}_r}i^2{}_{qs} + \frac{L_m R_r}{\sigma L_s L^2{}_r}\psi^s{}_{qr} - \frac{L_m \omega_r}{\sigma L_s L_r}\psi^s{}_{dr} + \frac{1}{\sigma L_s}V^s{}_{qs} \tag{43}$$

The change in d-q axis stator current is calculated from the estimated flux and voltages. Therefore, the desired state equations are given by

$$\frac{d}{dt}(X) = AX + BV_s \tag{44}$$

where,

$$X = [\; i_{ds} \quad i_{qs} \quad \Psi_{dr} \quad \Psi_{qr} \;]^T \tag{45}$$

$$V = [\; V_{ds} \quad V_{qs} \quad 0 \quad 0 \;]^T \tag{46}$$

$$A = \begin{bmatrix} -\dfrac{L_m^2 R_r + L_r^2 R_s}{\sigma L_s L_r^2} & 0 & \dfrac{L_m R_r}{\sigma L_s L_r} & \dfrac{L_m \omega_r}{\sigma L_s L_r^2} \\[2ex] 0 & -\dfrac{L_m^2 R_r + L_r^2 R_s}{\sigma L_s L_r^2} & -\dfrac{L_m \omega_r}{\sigma L_s L_r^2} & \dfrac{L_m R_r}{\sigma L_s L_r} \\[2ex] \dfrac{L_m R_r}{L_r} & 0 & -\dfrac{R_r}{L_r} & -\omega_r \\[2ex] 0 & \dfrac{L_m R_r}{L_r} & \omega_r & \dfrac{R_r}{L_r} \end{bmatrix} \tag{47}$$

$$B = \begin{bmatrix} \dfrac{1}{\sigma L_s} & 0 \\[2ex] 0 & \dfrac{1}{\sigma L_s} \\[2ex] 0 & 0 \\[1ex] 0 & 0 \end{bmatrix} \tag{48}$$

$$C = \begin{bmatrix} 1 & 0 & 0 & 0 \\ 0 & 1 & 0 & 0 \end{bmatrix} \tag{49}$$

The dynamic equation is modified by altering the hyper plane, and the filter equation is given by

$$\frac{d\hat{x}}{dt} = \left[\hat{A}\right]\hat{x} + [B]u + k_{sw}sat(\hat{i}_s - i_s - \hat{d}) \tag{50}$$

$$S = \hat{i}_s - i_s - \hat{d} \tag{51}$$

$$\hat{d} = k\widehat{T_{dis}} \ and \ \hat{y} = [C]\hat{x} \tag{52}$$

$$\widehat{T_{dis}} = T^*{}_e - j\frac{d\hat{w}}{dt} - B_v \ \hat{\omega} \tag{53}$$

The speed adaptation is given by Equations (54) and (55)

$$\omega_r = K_p \left(e_{ids} \ \psi^s{}_{qr} - e_{iqs} \ \psi^s{}_{dr}\right) + K_1 \int \left(e_{ids}\psi^s{}_{qr} - e_{iqs} \ \psi^s{}_{dr}\right)dt \tag{54}$$

$$\frac{dv}{dt} = e^T\left[\left(A + GC\right)^T + (A + GC)\right]e - \frac{2\Delta w_r \left(e_{ids}\widehat{\varphi^s{}_{qr}} - e_{iqs} \ \hat{\varphi}^s_{qr}\right)}{c} + \frac{2\Delta w_r}{\lambda} \frac{d\hat{\omega}_r}{dt} \tag{55}$$

The rotor speed is estimated from the estimated flux using a hyper plane-based state equation.

2.4. Neural Network Based Modified Model Reference Adaptive System Model

The model reference adaptive system (MRAS) estimators are the peak conservative method because of their modest structure and they have less associated calculation obligations than the other approaches. The elementary diagram of the MRAS contains an adjustable model, reference model, and an adaption mechanism. After the speed tuning, the error signal is given to the adaptation mechanism that is a neural network-based controller block. Here the output of both models is processed up to the errors linking the two models which depart to zero. The prototype accepts the voltage and current of the stator and determines the flux vector of the rotor. An adaptation mechanism with a neural network-based controller block is used to adapt the speed, so that the speed error is zero. The reference model equations are given by Equations (56) and (57).

$$\hat{\psi}^s_{dr} = \int \frac{-1}{T_r} \ \psi^s{}_{dr} - \omega_r \ \psi^s{}_{qr} + \frac{L_m}{T_r} \ i^s{}_{ds} \tag{56}$$

$$\hat{\psi}^s_{qr} = \int \omega_r \ \psi^s{}_{dr} - \frac{1}{T_r} \ \psi^s{}_{qr} + \frac{L_m}{T_r} \ i^s{}_{qs} \tag{57}$$

The neural network-based Adaptive model equations are given by the following equations.

$$O_k = \sum_{j=1}^{j} w_{kj} \ \varnothing_j(x) \tag{58}$$

$$\omega_{kj}(t+1) = \omega_{kj}(t) + \eta e(t)\varnothing_j \tag{59}$$

$$c_j(t+1) = c_j(t) + \frac{\eta e(t)\varnothing_j \omega_j (x - c_j)}{\sigma^2} \tag{60}$$

$$\varnothing_j(x) = exp\left[\frac{-||x - c_j||^2}{\sigma^2_j}\right] \tag{61}$$

$$\sigma_j(t+1) = \sigma_j(t) + \eta e(t)\varnothing_j \omega_j ||x - c_j|| \frac{1}{\sigma^3} \tag{62}$$

$$\hat{\psi}^s_{dr} = \int \frac{-1}{T_r}\psi^s{}_{dr} - \omega_r\psi^s{}_{qr} + \frac{L_m}{T_r} i^s{}_{ds} \tag{63}$$

$$\hat{\psi}^s_{qr} = \int \omega_r\,\psi^s{}_{dr} - \frac{1}{T_r}\psi^s{}_{qr} + \frac{L_m}{T_r} i^s{}_{qs} \tag{64}$$

The error signal is given by Equation (65). It is being fed to the P-I regulator to obtain the speed signal (66).

$$\xi = \hat{\psi}^s_{dr}\,\hat{\psi}^s_{qr} - \hat{\psi}^s_{dr}\,\psi^s{}_{qr} \tag{65}$$

$$\omega_r = \xi\left(K_p + \frac{K_1}{s}\right) \tag{66}$$

The rotor speed is estimated from the estimated flux of the neural network-based model reference adaptive system.

2.5. Modified Extended Kalman Filter Method

R.E. Kalman proposed the method of an Extended Kalman filter (EKF) method in the year 1960. He put out his exalted paper describing an iterative answer to the digital-data linear trickling issues. An iterative estimator is a Kalman filter. This implies that only the evaluated state from the precedent time step and the measurement of current are required to tally the evaluation for the present state. In contradiction to bundle guesstimation methods, no background of observations and evaluation is compulsory. It is bizarre in being solely a time domain filter. The Kalman filter has two incisive phases: predict and update. The forecast phase manipulates the state evaluation from the preceding time step to outgrow an evaluation of the state at the current time step. In the upgraded phase, calculated data at the prevailing time step is utilized to clarify this prophecy to appear at a recent, (hopefully) more authentic state estimate, again for the current time step. In the collected works, a trendy technique for the data guesstimation of IM is the EKF. Typical ambiguity and nonlinearities innate to the induction motor are highly beneficial for the speculative identity of EKF. Using this technique, it is likely to develop the networked approximation of states while acting on the concurrent interconnection of data in a reasonably smaller time gap, also captivating the structure and measurement noises exactly into annals. This is the logic behind the EKF, which has developed an ample utilization spectrum in the sensor-less regulation of IMs, even if it is estimated intricacy. In this proposed method, the filter equation is altered by including a sliding hyper plane, and this will improve the performance of the system. The rotor voltage equations are given by (67).

$$i^s{}_{qr}\,R_r + \frac{d}{dt}\left(\psi^s{}_{qr}\right) - \omega_r\,\psi^s{}_{dr} = 0 \tag{67}$$

From the voltage model of flux guesstimation, the rotor axis flux is given by

$$\psi^s{}_{qr} = L_m i^s{}_{qs} + L_r\,i^s{}_{qr} \tag{68}$$

$$i^s{}_{dr}\,R_r + \frac{d}{dt}(\psi^s{}_{dr}) + \omega_r\psi^s{}_{qr} = 0 \tag{69}$$

$$\psi^s{}_{dr} = L_m i^s{}_{ds} + L_r\, i^s{}_{dr} \tag{70}$$

After eliminating $i^s{}_{dr}$ and $i^s{}_{qr}$

$$\frac{d}{dt}\left(\psi^s{}_{dr}\right) = -\frac{R_r}{L_r}\,\psi^s{}_{dr} - \omega_r\psi^s{}_{qr} + \frac{L_m R_r}{L_r}\,i^s{}_{ds} \tag{71}$$

$$\frac{d}{dt}\left(\psi^s{}_{qr}\right) = -\frac{R_r}{L_r}\,\psi^s{}_{qr} + \omega_r\psi^s{}_{dr} + \frac{L_m R_r}{L_r}\,i^s{}_{qs} \tag{72}$$

$$\frac{d}{dt}\left(\psi^s{}_{dr}\right) = \frac{L_r}{L_m}\,V^s{}_{ds} - \frac{L_r}{L_m}\,(R_s + \sigma L_s S)i^s{}_{ds} + N(k) \tag{73}$$

$$\frac{d}{dt}\left(\psi^s{}_{qr}\right) = \frac{L_r}{L_m}\,V^s{}_{qs} - \frac{L_r}{L_m}\,(R_s + \sigma L_s S)i^s{}_{qs} + N(k) \tag{74}$$

$$\frac{d}{dt}\left(i^s{}_{ds}\right) = -\frac{\left(L^2{}_m R_r + L^2{}_r R_s\right)}{\sigma L_s L^2{}_r}\,i^2{}_{ds} + \frac{L_m R_r}{\sigma L_s L^2{}_r}\,\psi^s{}_{dr} + \frac{L_m \omega_r}{\sigma L_s L_r}\,\psi^s{}_{qr} + \frac{1}{\sigma L_s}\,V^s{}_{ds} + N(k) \tag{75}$$

$$\frac{d}{dt}\left(i^s{}_{qs}\right) = -\frac{\left(L^2{}_m R_r + L^2{}_r R_s\right)}{\sigma L_s L^2{}_r}\,i^2{}_{qs} + \frac{L_m R_r}{\sigma L_s L^2{}_r}\,\psi^s{}_{qr} - \frac{L_m \omega_r}{\sigma L_s L_r}\,\psi^s{}_{dr} + \frac{1}{\sigma L_s}\,V^s{}_{qs} + N(k) \tag{76}$$

Therefore, the desired state equation is given by

$$\frac{d}{dt}\left(X\right) = AX + BV_s + N(k) \tag{77}$$

$$X = \begin{bmatrix} i_{ds} & i_{qs} & \Psi_{dr} & \Psi_{qr} \end{bmatrix}^T \tag{78}$$

$$V = \begin{bmatrix} V_{ds} & V_{qs} & 0 & 0 \end{bmatrix}^T \tag{79}$$

$$A = \begin{bmatrix} -\frac{L^2_m R_r + L^2_r R_s}{\sigma L_s L^2_r} & 0 & \frac{L_m R_r}{\sigma L_s L_r} & \frac{L_m \omega_r}{\sigma L_s L^2_r} \\ 0 & -\frac{L^2_m R_r + L^2_r R_s}{\sigma L_s L^2_r} & -\frac{L_m \omega_r}{\sigma L_s L^2_r} & \frac{L_m R_r}{\sigma L_s L_r} \\ \frac{L_m R_r}{L_r} & 0 & -\frac{R_r}{L_r} & -\omega_r \\ 0 & \frac{L_m R_r}{L_r} & \omega_r & \frac{R_r}{L_r} \end{bmatrix} \tag{80}$$

$$B = \begin{bmatrix} \frac{1}{\sigma L_s} & 0 \\ 0 & \frac{1}{\sigma L_s} \\ 0 & 0 \\ 0 & 0 \end{bmatrix} \tag{81}$$

$$C = \begin{bmatrix} 1 & 0 & 0 & 0 \\ 0 & 1 & 0 & 0 \end{bmatrix} \tag{82}$$

The dynamic equation is modified by altering the hyper plane, and the Filter equation is given by (83)

$$\frac{d\hat{x}}{dt} = [\hat{A}]\hat{x} + [B]u + k_{sw}sat\left(\hat{i}_s - i_s - \hat{d}\right) \tag{83}$$

$$S = \hat{i}_s - i_s - \hat{d} \tag{84}$$

$$\hat{d} = k\widehat{T_{dis}} \text{ and } \hat{y} = [C]\hat{x} \tag{85}$$

$$\widehat{T_{dis}} = T^*{}_e - j\frac{d\hat{w}}{dt} - B_v\,\hat{w} \tag{86}$$

The speed adaptation algorithm is given by

$$\omega_r = K_p\left(e_{ids}\,\psi^s{}_{qr} - e_{iqs}\,\psi^s{}_{dr}\right) + K_1\int\left(e_{ids}\psi^s{}_{qr} - e_{iqs}\,\psi^s{}_{dr}\right)dt \tag{87}$$

The parameters of the induction motor illustrated in Table 1 are estimated for the 2.2 kW of the induction motor by a conventional direct current Resistance test, no load, blocked rotor test, and retardation test.

Table 1. Machine parameters.

Parameters	Values
power	2.2 kW
Current	4.4 A
resistance of stator	6.6 ohms
Resistance of rotor	5.5 ohms
Magnetising inductance	0.454 H
Frequency	50 Hz
No of poles	4
Moment of inertia	0.011787
Inductance of rotor	0.475 H
Stator inductance	0.475 H
torque	$2e^{-6}$ Nm

3. Simulation Results and Discussions

In sensor-less method regulation of an induction motor drive process the vector regulation without a speed sensor. For closed loop operations, the speed encoder is mandatory in both decoupling and scalar regulation drive. In the vector regulation with indirect form, a signal of speed is mandatory for the entire operation. A speed encoder is objectionable in the drive as it adds rate and consistency snags beyond the requirement for shaft enlargement and the mounting arrangements. It is important to evaluate the wave of speed from the currents and voltages of a machine terminal with the Digital signal processor (DSP). However, the guesstimation is commonly intricate and profoundly reliant on the parameters of the machine. While the regulation of without-sensor drives are free at this time, the change in parameter, predominantly closer to zero speed, enacts a provocation in the exactness of speed guesstimation [21].

For large routine, changeable speed solicitations, the poly-phase asynchronous motors are used broadly because of having less rate, powerfulness, and less maintenance, and thus, they substitute the direct current motor drives. To achieve better torque and efficiency, the speed regulation of the induction motor is important. The dynamic modelling of a quinary phase induction motor is simulated in Matlab software. The phase voltages are converted into d-q axis voltages in three phases to a two phase lock. The speed and torque of the quinary phase induction motor have been estimated. The dynamically modelled induction motor in the quinary phase has 66% reduced torque pulsations and better linearity in the speed of the rotor, and also speed has been increased by 50% when compared to the induction motor in three phases as illustrated in Figures 1 and 2 and Table 2.

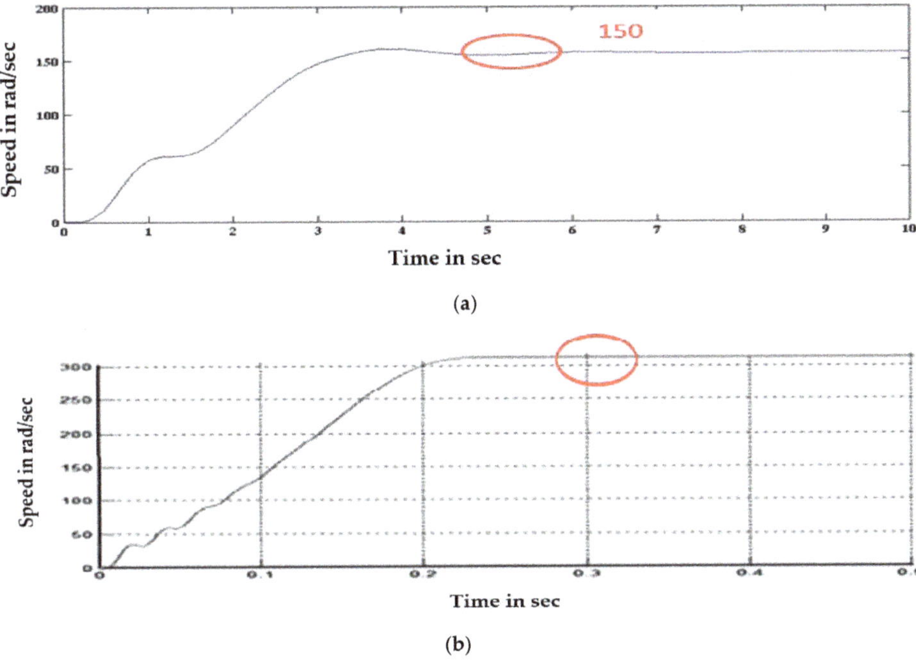

(a)

(b)

Figure 1. Speed response of dynamically modelled induction motor: (**a**) proposed five phase motor and (**b**) conventional three phase motor.

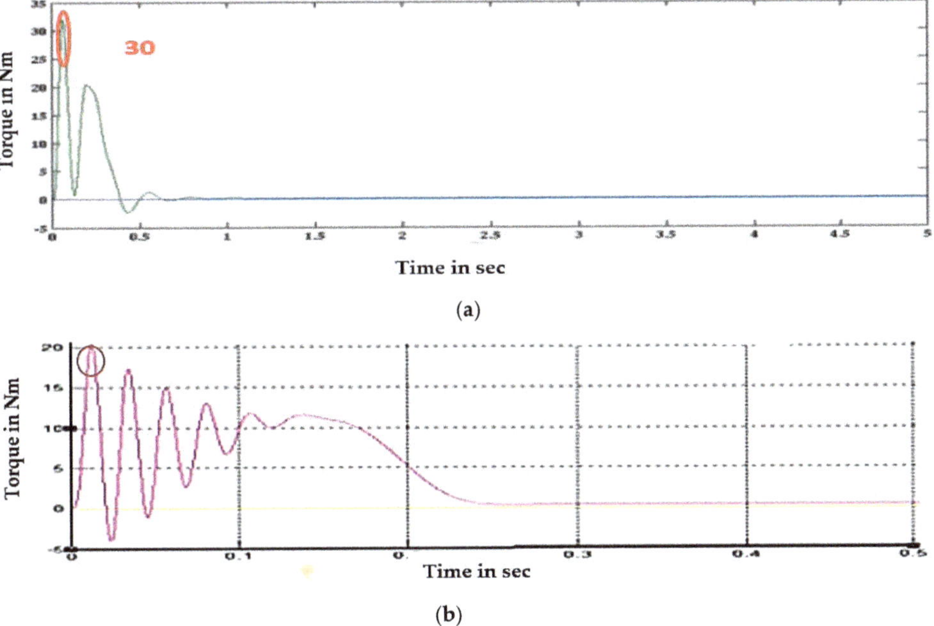

(a)

(b)

Figure 2. Torque response of dynamically modelled induction motor: (**a**) proposed five-phase motor and (**b**) conventional three-phase motor.

Table 2. Performance comparison of three- and poly-phase induction motor.

Three-Phase Induction Motor			Poly-Phase Induction Motor		
Speed in rad/s	Setting Time of Speed in s	Amplitude of Torque Ripple in Nm	Speed in Rad/s	Setting Time of Speed in s	Amplitude of Torque Ripple in Nm
150	2.5	30	300	0.2	20

3.1. Conventional Direct Synthesis Method

The speed guesstimation of the quinary phase induction motor using the direct synthesis method has been executed. The direct axis and quadrature axis fluxes are obtained using a current model block. The speed of the quinary phase motor has been approximated using the direct synthesis method. It can be seen that there is some difference between the estimated value and the reference value, so the accuracy of this method is low. Some speed pulsations of high magnitude are also present in the estimated speed. The estimated speed obtained using the direct synthesis method and reference speed has been demonstrated in Figure 3.

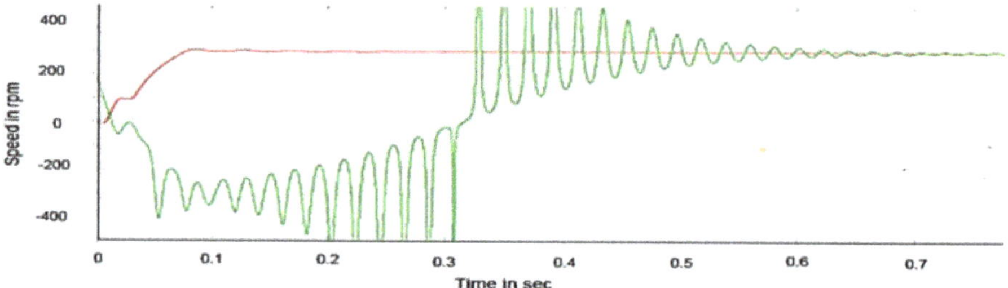

Figure 3. Estimated speed of a quinary phase induction motor obtained using Direct Synthesis Method.

3.2. Neural Network Based Model Reference Adaptive System

The speed guesstimation using MRAS is executed in MATLAB software. The d and q axis voltages and currents obtained from dynamic model block [22] are fed to MRAS regulation, and the estimated speed is feedback to the adjustable model as shown in Figure 4.

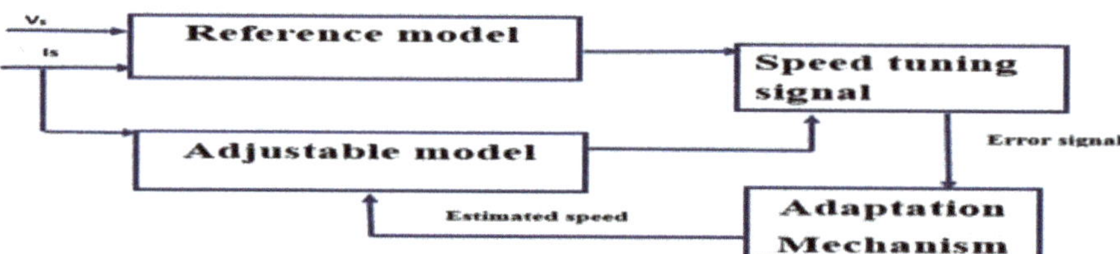

Figure 4. Model Reference Adaptive System.

The estimated speed of the quinary phase induction motor obtained using the MRAS method and the reference speed of the motor is demonstrated in Figure 5. The speed of the quinary phase motor has been approximated using the Model Reference Adaptive System method. It is observed that the estimated value and reference values are almost the same, thus offering good accuracy in guesstimation.

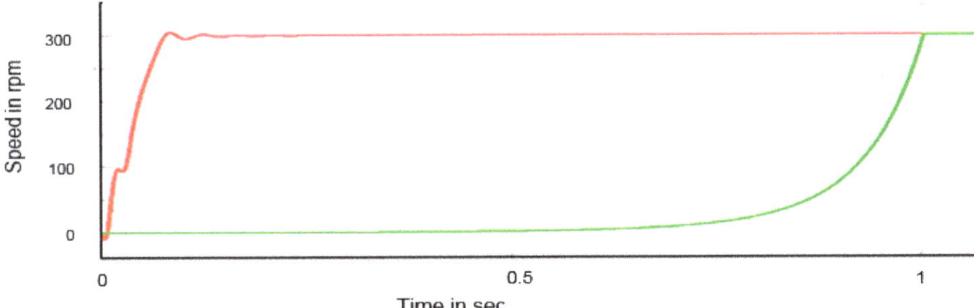

Figure 5. Estimated speed of the quinary phase induction motor obtained using Model Reference Adaptive Method.

3.3. Modified Luenberger Observer Technique

The speed guesstimation using the Luenberger observer method is executed in MATLAB software. This is illustrated in Figure 6. The speed response from the speed estimation using a Luenberger observer results in a small steady state error of 9 rpm, and the settling time took more than 2 s, which is high compared to all other methods.

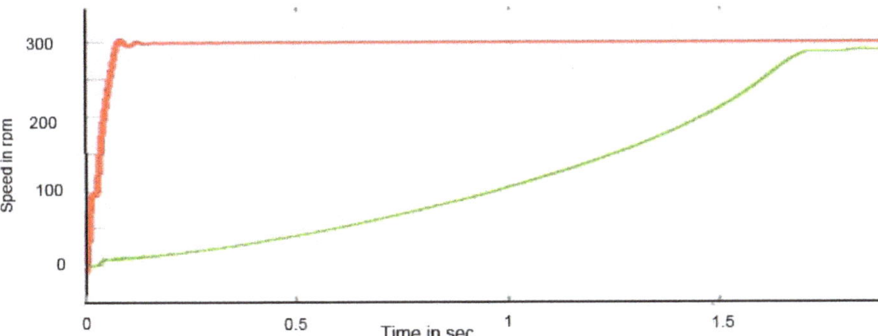

Figure 6. Estimated speed of the quinary phase induction motor obtained using a Luenberger observer method.

3.4. Modified Extended Kalman Filter

The speed guesstimation using the Kalman Filter method is being executed in MATLAB software. It can be observed from Figure 7 that the steady state error becomes zero and the settling time is also reduced drastically.

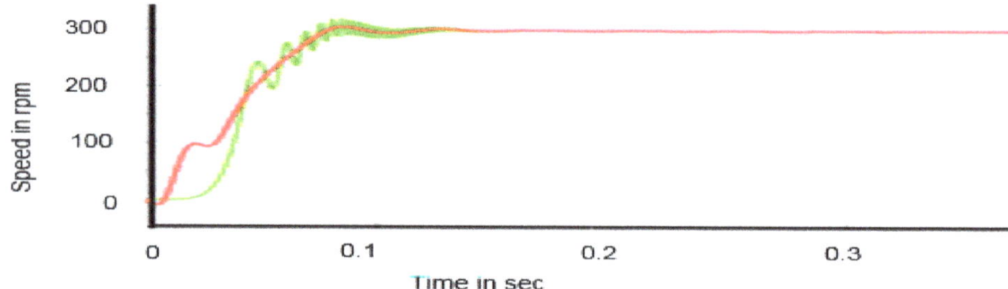

Figure 7. Estimated speed of the quinary phase induction motor obtained using Kalman Filter Method.

Speed pulsations of high magnitude are being observed in the estimated speed obtained by the direct synthesis method, which is not desirable. The rotor speed has been estimated using the direct synthesis method, model reference adaptive system, Luenberger observer method, and Kalman filter method. From the transient analysis, for the estimated speed obtained using these methods, the accuracy is better for the Kalman filter method because a steady-state error is zero and also the settling time is minimum compared to the other methods, so the speed reaches a stable state quickly. The Kalman filter enjoys very low settling time compared to all the above-mentioned guesstimation techniques that are depicted in Table 3.

The extended Kalman filter is the most popularly used observer in induction motor drive due to its nonlinearities and robustness of parameter variations. Conventionally, the covariance matrix and the parameters are tuned by a trial and error method. Due to the trial and error method, complexity and computational time is high. Based on the obtained model of a modified, extended Kalman filter, the computational time has been reduced.

Table 3. Comparison of the guesstimation techniques by time domain analysis.

	Rise Time (s)	Settling Time (s)	Delay Time (s)	Steady State Error	Peak Overshoot (%)
Model Reference Adaptive System	1	1	0.9	0	0
Luenberger Observer Method	1.5	2	1.25	9	7
Kalman Filter Method	0.03	0.15	0.4	0	−3.34

4. Speed Deviation and Torque Ripple Reduction of Induction Motor Drive in Parallel Using Extended Kalman Filter

It is observed in both the simulation and hardware that the speed control of the induction motor is a challenging problem in the absence of a power electronics component [23,24]. In the proposed method, the induction motor with estimated parameters for sensor-less drive is analysed and compared with a conventional space vector modulation-based induction motor drive with a speed sensor. As no sensors are used, it makes the system more rugged, stable, less costly, and reduce in size. The two bogies setup where considered in this proposed method for electric traction locomotive application.

The speed of the five phase induction motor is estimated by the direct synthesis method, modified model reference adaptive system, Luenberger observer, and extended Kalman filter. From the above-mentioned speed estimation techniques, the extended Kalman filter has better performance than the other three methods. The speed and torque performance of the proposed method were analysed under the change in stator resistance, rotor resistance, low speed, and low inertia and compared with the conventional method. Here the proposed method is considered for electric traction locomotive application.

The speed response of the proposed method under normal conditions and parameter variations is shown in Figures 8 and 9 respectively. In the rotor speed curve, the reference speed represented in yellow colour is matched to the estimated speed represented in red colour, and the actual speed is displayed in blue colour. The speed response of the conventional method under normal condition and parameter variations is as shown in Figures 10 and 11 respectively. The speed error is far less in the proposed method, and it is robust to parameter variation compared to the conventional method. The speed error has been reduced by 70% in the proposed method compared to the conventional method.

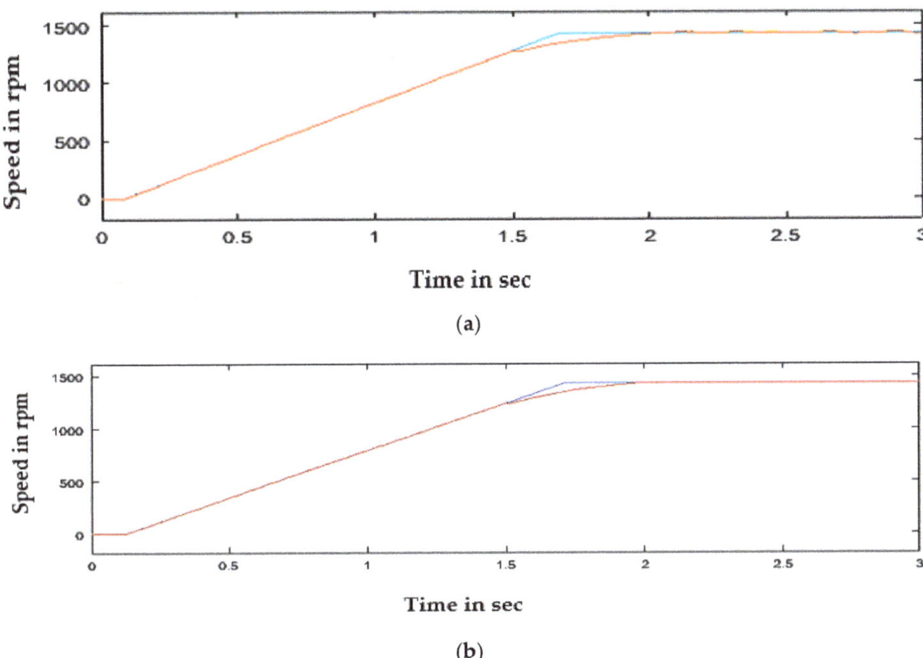

Figure 8. Speed response of proposed sensorless induction motor drive under normal running conditions using Extended Kalman filter: (a) induction motor 1 and (b) induction motor 2.

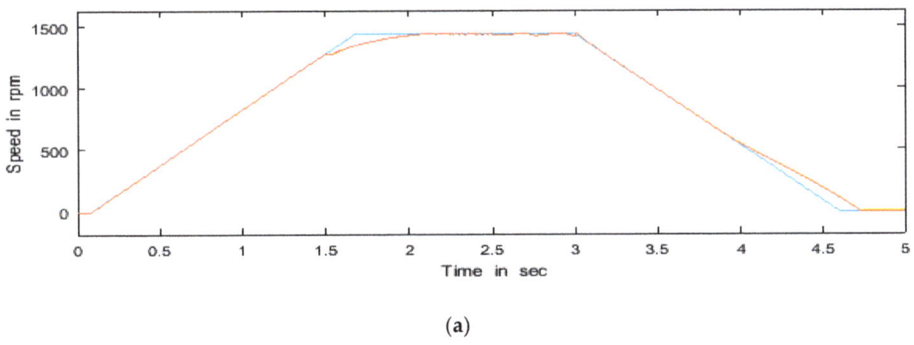

Figure 9. *Cont.*

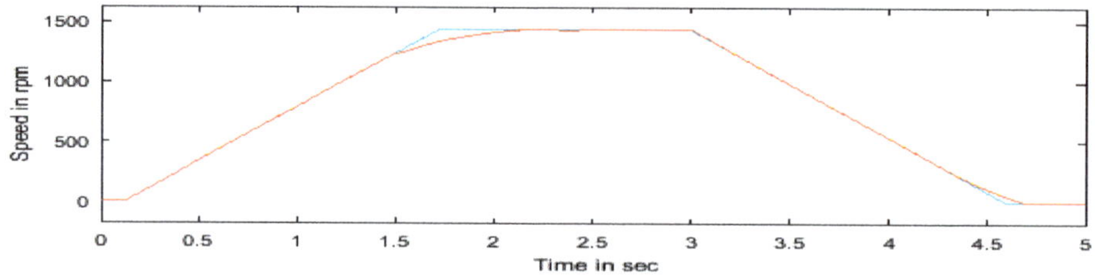

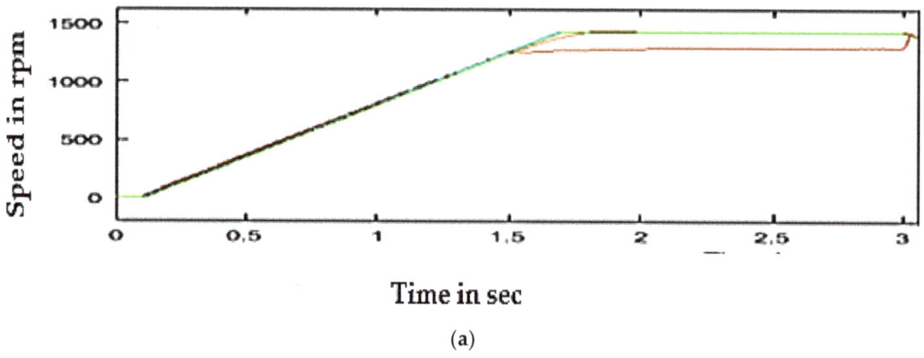

(b)

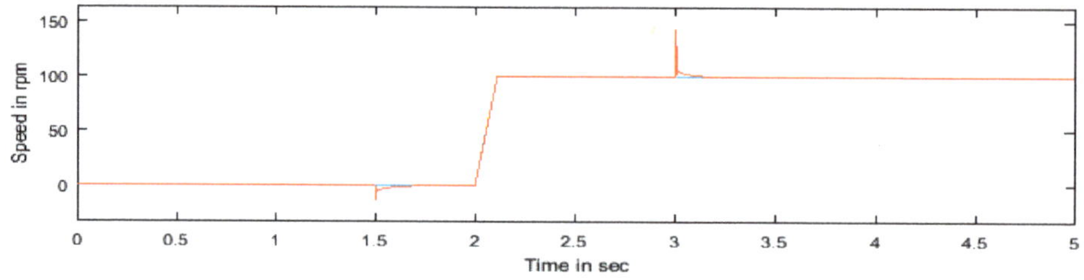

(c)

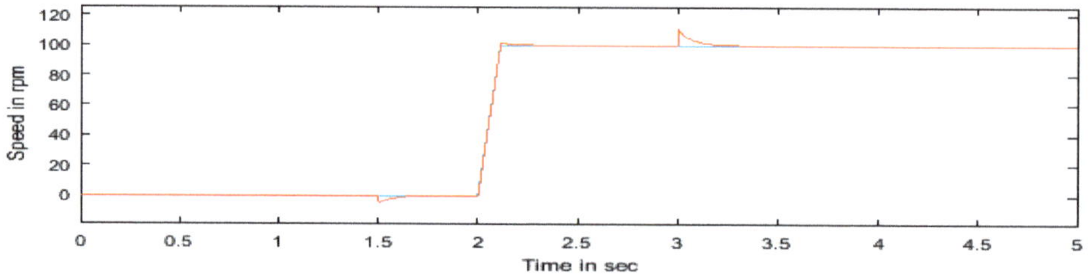

(d)

Figure 9. Speed response of proposed sensorless induction motor drive under parameter variations using Extended Kalman filter: (**a**) change in stator resistance, (**b**) change in rotor resistance, (**c**) low speed, and (**d**) low inertia.

(a)

Figure 10. *Cont.*

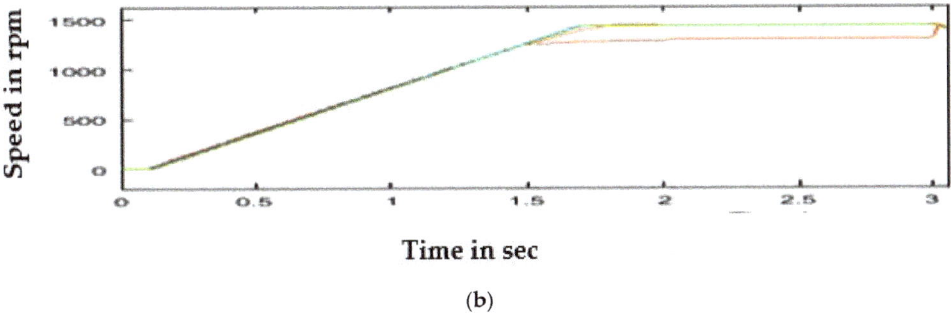

Speed in rpm (y-axis) — **Time in sec** (x-axis)

(b)

Figure 10. Speed response of inverter-fed induction under normal running conditions using a conventional space vector modulation-based induction motor drive with a speed sensor: (**a**) induction motor 1 and (**b**) induction motor 2.

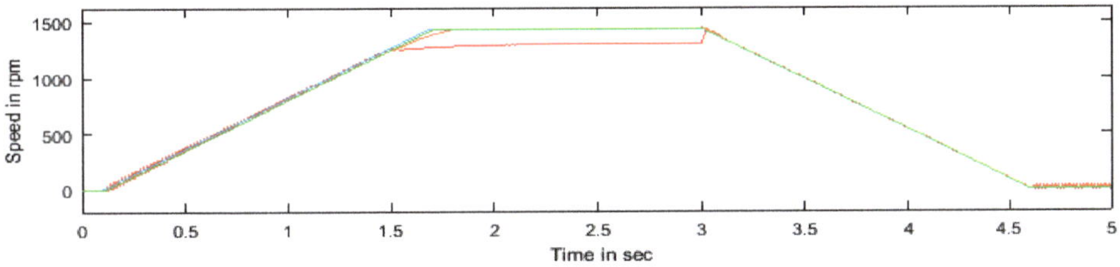

(a)

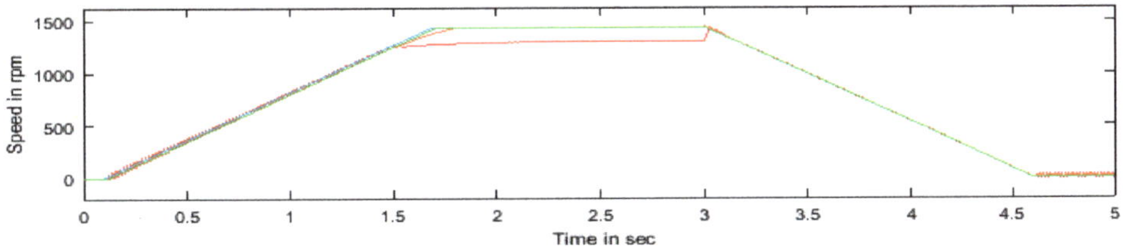

(b)

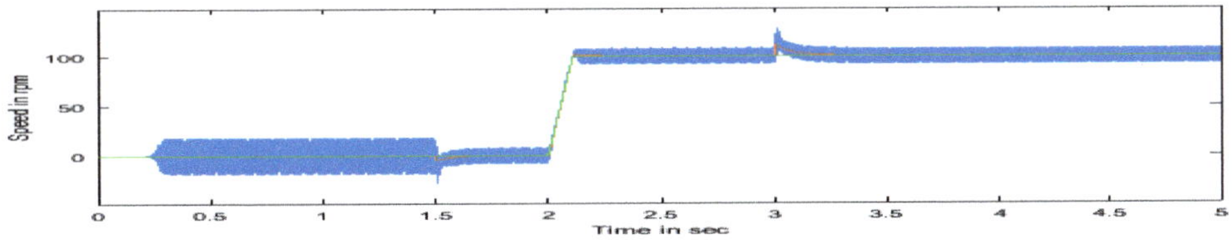

(c)

Figure 11. *Cont.*

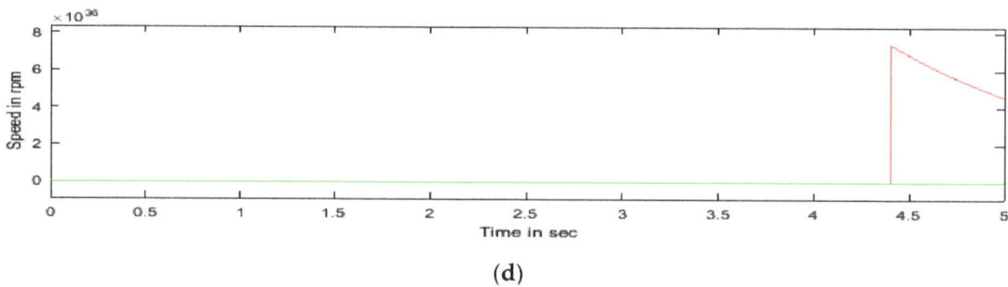

(d)

Figure 11. Speed response of inverter-fed induction motor under parameter variations using a conventional space vector modulation-based induction motor drive with a speed sensor: (**a**) change in stator resistance, (**b**) change in rotor resistance, (**c**) low speed, and (**d**) low inertia.

For the rotor speed, curve deviation is large compared to the proposed method. In the electromagnetic torque curve, the torque ripple is less in the proposed method compared to the conventional method as demonstrated in Figures 12–15 respectively. The torque ripple is reduced by 85% in the proposed method compared to the conventional method.

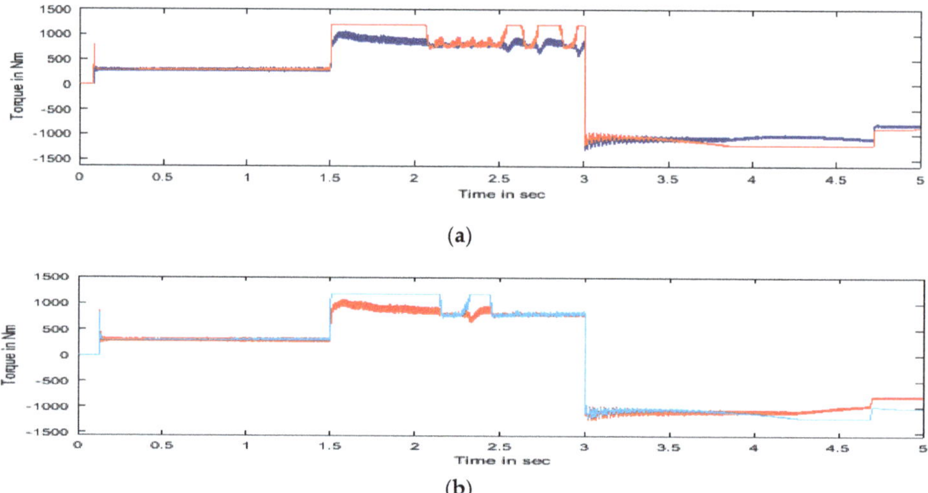

(a)

(b)

Figure 12. Torque response of induction motor drive under normal conditions for proposed method: (**a**) induction motor 1 and (**b**) induction motor 2.

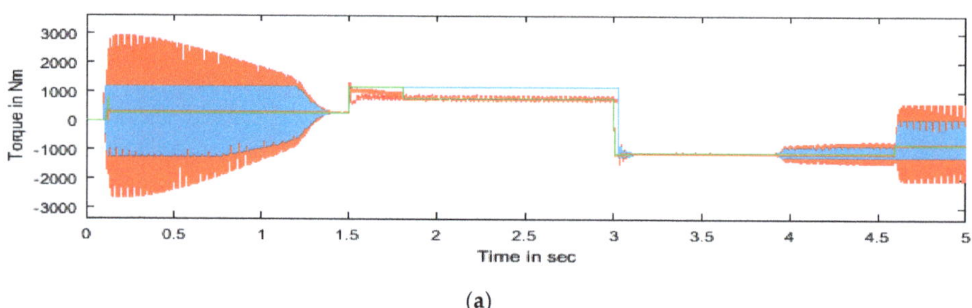

(a)

Figure 13. *Cont.*

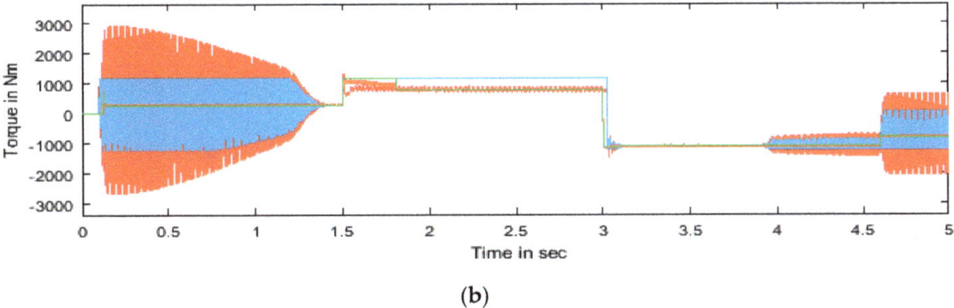

(**b**)

Figure 13. Torque response of induction motor drive under normal condition for conventional method: (**a**) induction motor 1 and (**b**) induction motor 2.

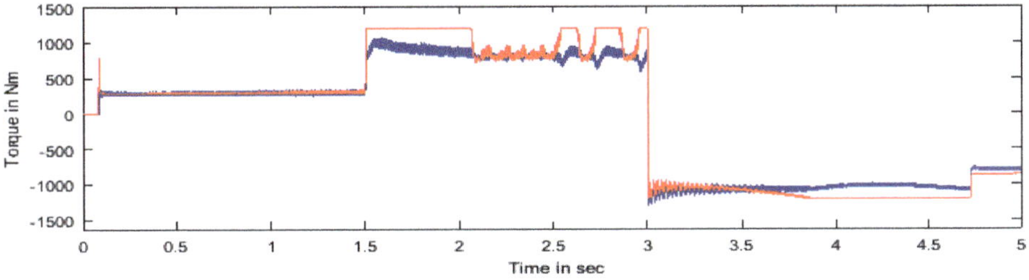

(**a**)

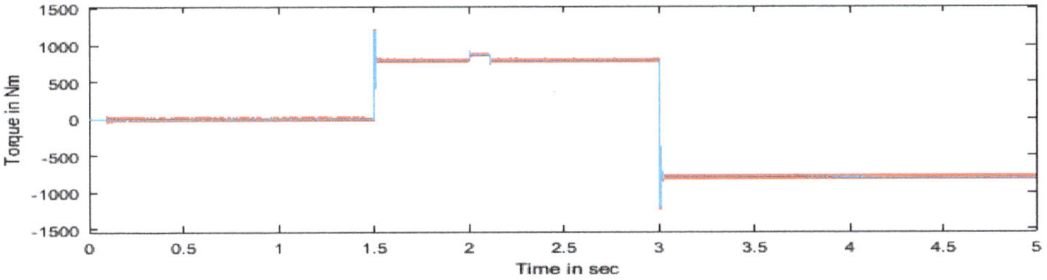

(**b**)

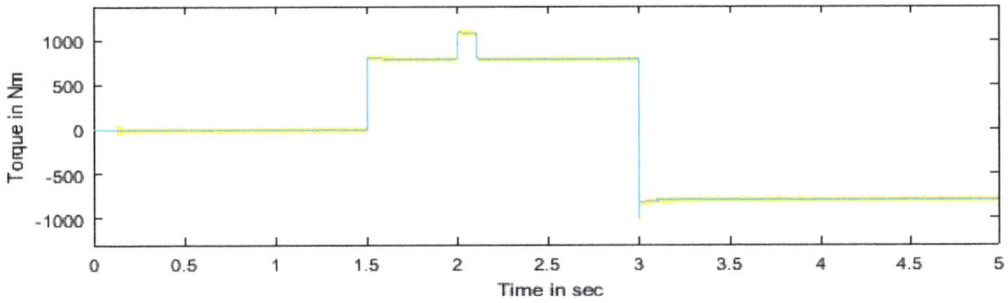

(**c**)

Figure 14. Torque response of induction motor drive under parameter variations for proposed method: (**a**) change in stator resistance, (**b**) low speed, and (**c**) low inertia.

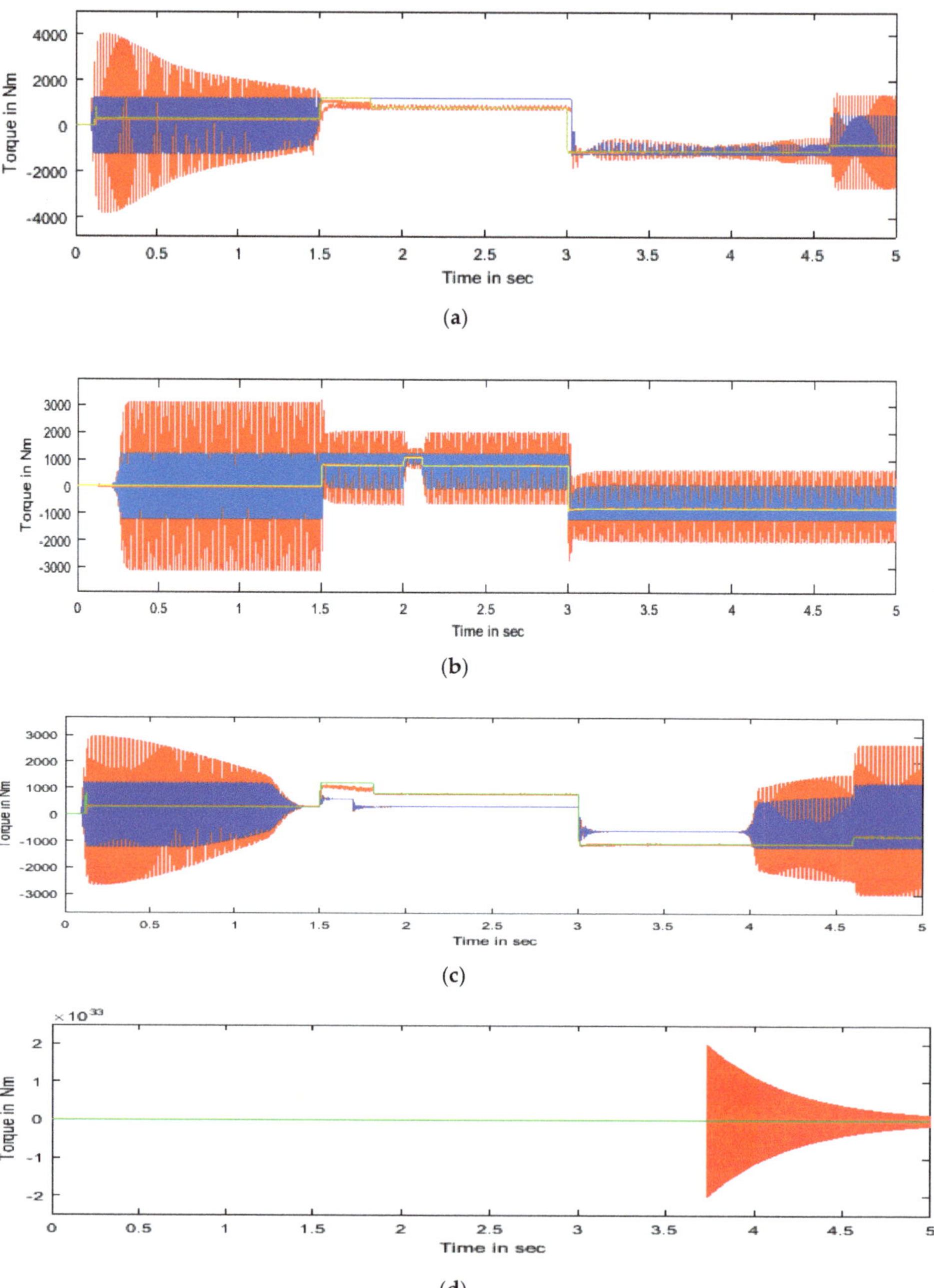

Figure 15. Torque response of induction motor drive under normal conditions for conventional method:
(**a**) change in stator resistance, (**b**) change in rotor resistance, (**c**) low speed, and (**d**) low inertia.

The robustness of the proposed extended Kalman filter-based sensorless induction motor drive is
compared with the conventional method. The robustness of the proposed method is proved by varying

the stator resistance, rotor resistances, inertia, and step change in speed; also, under the low-speed region, it is proved that even though parameters varied, the performances are not affected compared to the conventional method.

Inference of Result

In the proposed method, the torque ripple is reduced by 85% and speed deviation reduced by 70% more than the conventional one. The performance is robust for parameter variations like a change in stator resistance, rotor resistance, low frequency, low inertia, the step change in speed, and low speed. The five phase induction motor has better performance than three phase induction motor in terms of torque ripple and speed error.

Speed pulsations of high magnitude are observed in the estimated speed obtained by the conventional direct synthesis method presented in B.K. Bose, which is not desirable in terms of response time compared to the proposed methods presented in the article. The conventional space vector modulated induction motor drive with a speed sensor has a high amplitude of torque ripple, ultra-low resolution rotary encoder, and it is difficult to realize proper control not robust for faulty environmental situations. The above-mentioned difficulties have been corrected in the proposed method of a sensor-less drive presented in this article.

5. Hardware Results and Discussion

The performance acquired in the simulation is analysed experimentally in hardware setup also. The hardware setup for the sensor-based induction motor drive shown in Figure 16. The no-load test is performed on two motors connected in parallel. It is observed that the size and cost of the device are high. Thus, variable speed control becomes very difficult for this condition. The above-mentioned difficulties have been rectified in this proposed sensorless drive. The hardware setup for the sensorless drive is illustrated in Figure 17. Twenty-five percent of the size and cost of the system has been reduced in the proposed method.

The difference between set speed and actual speed becomes nearly zero, and the speed deviation between the two motors is less in the proposed method compared to the conventional method as depicted in Figures 18 and 19 and in Tables 4–7.

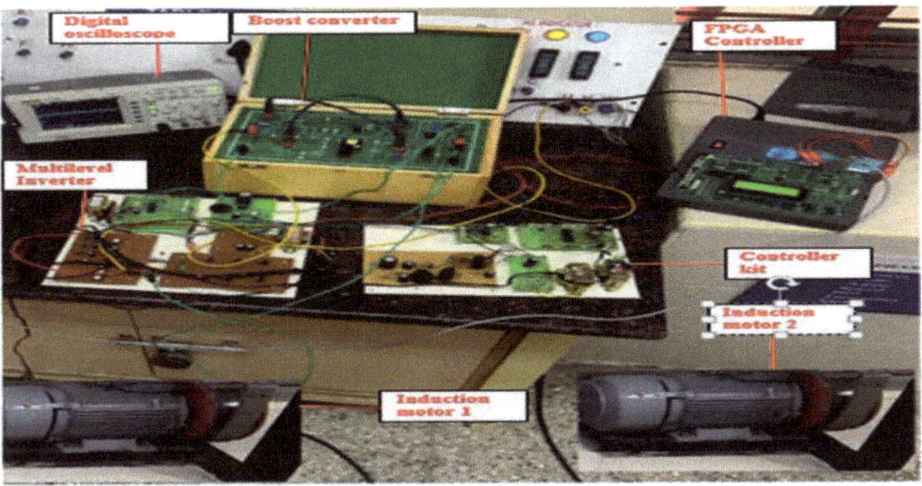

Figure 16. Hardware setup for conventional method.

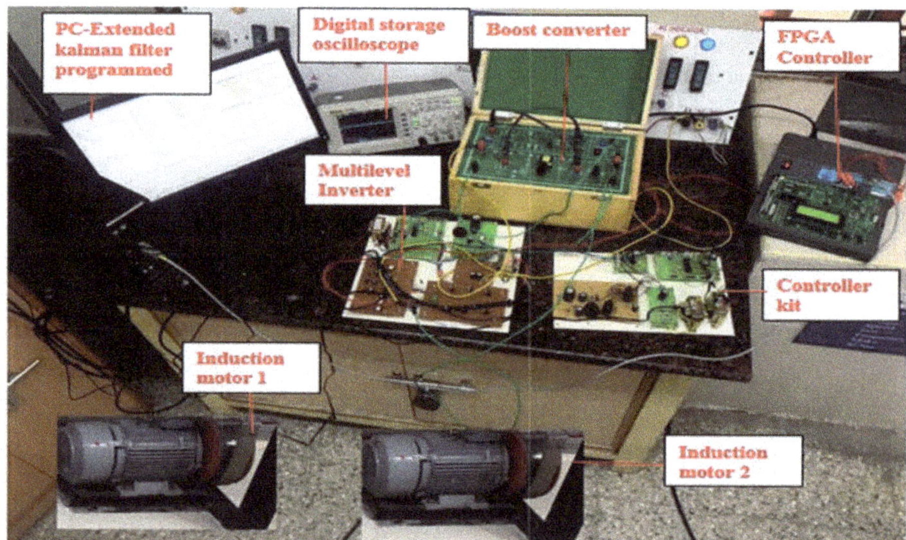

Figure 17. Hardware setup for proposed method.

Table 4. Speed response of conventional method under normal conditions.

Voltage (V)	Current (A)	Actual Speed (motor 1) (rpm)	Actual Speed (motor 2) (rpm)
	0.9	400	400
170	0.9	512	519
	0.9	615	610
	1.2	415	415
220	1.2	525	520
	1.3	627	627

Table 5. Speed response of conventional method under parameter variations.

Voltage (V)	Current (A)	Set Speed (rpm)	Actual Speed (motor 1) (rpm)	Actual Speed (motor 2) (rpm)
	0.9	425	395	393
170	0.9	540	514	518
	0.9	635	608	610
	1.2	425	400	398
220	1.2	540	526	530
	1.3	635	626	622

Table 6. Speed response of proposed method under normal conditions.

Voltage (V)	Current (A)	Set Speed (rpm)	Actual Speed (motor 1) (rpm)	Actual Speed (motor 2) (rpm)
	0.8	425	425	421
150	0.8	540	538	538
	0.8	635	630	630
	1.2	425	425	425
200	1.2	540	538	539
	1.3	635	635	635

Table 7. Speed response of proposed method under parameter variations.

Voltage (V)	Current (A)	Set Speed (rpm)	Actual Speed (motor 1) (rpm)	Actual Speed (motor 2) (rpm)
	0.9	425	424	424
170	0.9	540	540	540
	0.9	635	631	635
	1.2	425	420	420
220	1.2	540	535	535
	1.3	635	635	635

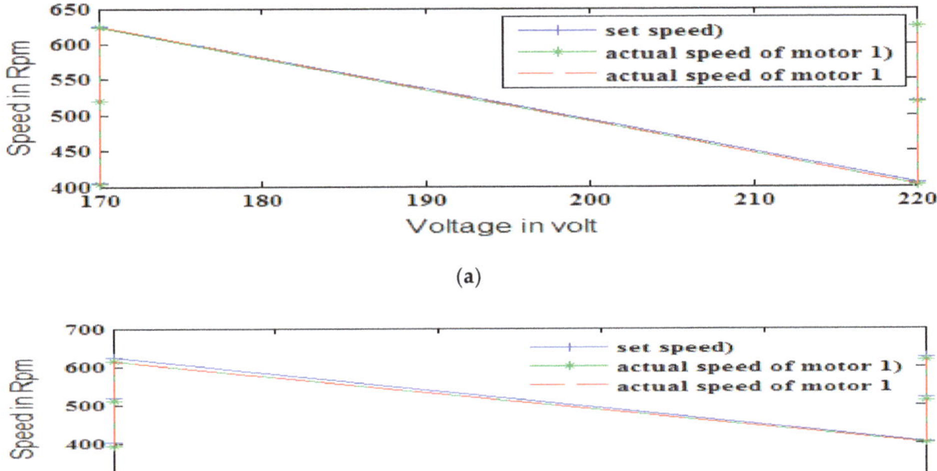

(**a**)

(**b**)

Figure 18. Speed response of proposed method: (**a**) normal conditions and (**b**) parameter variations.

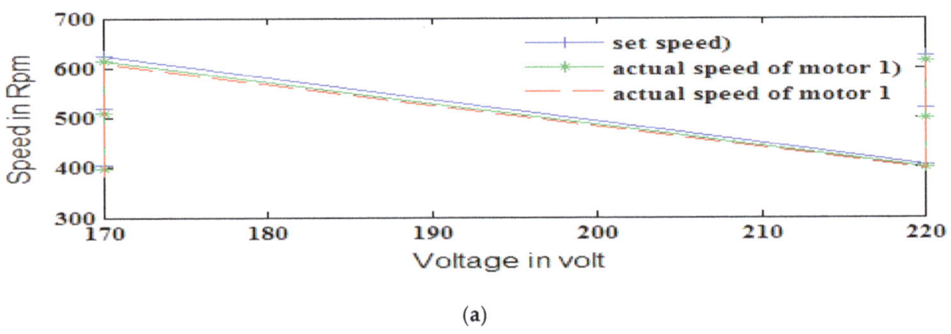

(**a**)

Figure 19. *Cont.*

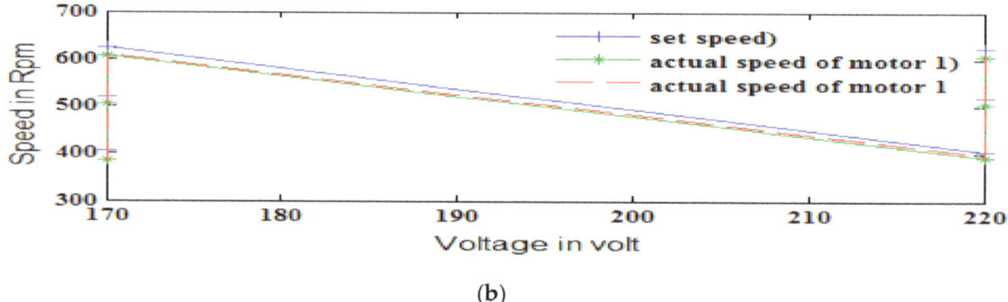

(**b**)

Figure 19. Speed response of conventional method: (**a**) normal conditions and (**b**) parameter variation.

6. Conclusions

The dynamically modelled quinary phase induction motor has reduced torque pulsations and better linearity in the rotor speed when compared to those of a three-phase induction motor. The rotor speed has been estimated using the conventional direct synthesis method, neural network-based model reference adaptive system, modified Luenberger observer, and extended Kalman filter methods. Comparing the estimated speed obtained using these methods, the accuracy is best for the Kalman filter method. Speed pulsations of high magnitude are observed in the estimated speed obtained by the direct synthesis method, which is not desirable. The Kalman filter employs very low settling time compared to all the above-mentioned guesstimation techniques. The minimization of speed error and torque ripple for the sensorless induction motor drive has been achieved. The hardware setup was implemented for the above-mentioned sensorless drive.

The current limiting factor of this proposed sensorless induction motor drive based on field-oriented control with proportional integral controller has ideal integration problems due to direct current offset, and, in addition, performance is not adequate at zero speed and low frequency. In future, the author will implement the modified model predictive control-based sensor-less induction motor drive for traction applications to improve the performance under zero speed and low-frequency conditions because of the speed characteristics of traction drive including the zero speed stat-up.

Author Contributions: Investigation, Software, Validation, Writing—original draft, Methodology, S.U.; Project administration, Supervision, Writing—review & editing, C.S.; Resources, S.P.

Funding: The research received no external funding.

Conflicts of Interest: The authors declare no conflict of interest.

References

1. Echavarria, R.; Horta, S.; Oliver, M. A three phase motor drive using IGBTs and constant V/F speed control with slip regulation. In Proceedings of the IEEE International Power Electronics Congress, San Luis Potosi, Mexico, 6–19 October 1995. [CrossRef]
2. Ho, T.J.; Chang, C.H. Robust speed tracking of induction motors: An arduino-implemented intelligent control approach. *Appl. Sci.* **2018**, *8*, 159. [CrossRef]
3. Suwankawin, S.; Somboon, S. A speed-sensorless IM drive with decoupling control and stability analysis of speed estimation. *IEEE Trans. Ind. Electron.* **2002**, *49*, 444–455. [CrossRef]
4. Hinkkanen, M. Reduced-Order Flux Observers with Stator-Resistance Adaptation for Speed-Sensorless Induction Motor Drives. *IEEE Energy Convers. Congr. Expo.* **2009**, *20*, 1–11. [CrossRef]
5. Holtz, J. Sensorless control of induction machines with or without signal injection. *IEEE Trans. Ind. Electron.* **2006**, *53*, 7–30. [CrossRef]
6. Edelbaher, G.; Jezernik, K.; Urlep, E. Low-speed sensorless control of induction machine. *IEEE Trans. Ind. Electron.* **2006**, *53*, 120–129. [CrossRef]

7. Castaldi, P.; Tilli, A. Parameter Estimation of Induction Motor at Standstill with Magnetic Flux Monitoring. *IEEE Trans. Control Syst. Technol.* **2005**, *13*, 386–400. [CrossRef]

8. Sumith, Y.J.; Siva Kumar, S.V. Estimation of parameter for induction motor drive. *IEEE Trans. Ind. Appl.* **2012**, *2*, 1668–1674. [CrossRef]

9. Stephan, J.; Bodson, M.; Chiasson, J. Real-time estimation of the parameters and fluxes of induction motors. *IEEE Trans. Ind. Appl.* **1994**, *30*, 746–759. [CrossRef]

10. Abdel-khalik, A.S.; Ahmed, S.; Massoud, A. A five-Phase Induction Machine Model Using Multiple DQ Planes Considering the Effect of Magnetic Saturation. In Proceedings of the IEEE Energy Conversion Congress and Exposition, Pittsburgh, PA, USA, 14–18 September 2014; pp. 287–293. [CrossRef]

11. Fnaiech, M.A.; Betin, F.; Capolino, G.A. MRAS applied to sensor-less regulation of a six phase induction machine. In Proceedings of the IEEE International Conference on Industrial Technology Version, Vina del Mar, Chile, 14–17 March 2010; pp. 1519–1524. [CrossRef]

12. de Lillo, L.; Empringham, L.; Wheeler, P.W.; Khwan-On, S.; Gerada, C.; Othman, M.N.; Huang, X.Y. Multiphase power converter drive for fault-tolerant machine development in aerospace applications. *IEEE Trans. Ind. Electron.* **2014**, *57*, 575–583. [CrossRef]

13. Levi, E. Multiphase Electric Machines for Variable-Speed Applications. *IEEE Trans. Ind. Electron.* **2013**, *55*, 1893–1909. [CrossRef]

14. Guglielmi, P.; Diana, M.; Piccoli, G.; Cirimele, V. Multi-n-phase electric drives for traction applications. In Proceedings of the IEEE International Electric Vehicle, Florence, Italy, 17–19 December 2014; pp. 1–6. [CrossRef]

15. Comanescu, M.; Xu, L. Sliding mode MRAS speed estimators for sensorless vector control of induction machine. *IEEE Trans. Ind. Electron.* **2006**, *53*, 146–153. [CrossRef]

16. Riveros, A.; Yepes, A.G.; Barrero, F.; Doval-Gandoy, J.; Bogado, B.; Lopez, O.; Jones, M.; Levi, E. Parameter Identification of five phase Induction Machines with Distributed Windings. *IEEE Trans. Energy Convers.* **2012**, *27*, 1067–1077. [CrossRef]

17. Levi, E. Advances in converter regulation and innovative exploitation of additional degrees of freedom for quinary phase machines. *IEEE Trans. Ind. Electron.* **2016**, *63*, 433–448. [CrossRef]

18. Kazuya, A.; Toru, I.; Matsuse, K.; Shigeru, I.; Nakajima, Y. Characteristics of speed sensorless vector controlled two induction motor drive fed by single inverter. In Proceedings of the 7th International Power Electronics and Motion Control Conference, Harbin, China, 2–5 June 2012. [CrossRef]

19. Subotic, I.; Bodo, N.; Levi, E. Single-phase on-board integrated battery chargers for EVs based on quinary phase machines. *IEEE Trans. Power Electron.* **2016**, *31*, 6511–6522. [CrossRef]

20. Slemon, G.R. Modelling of induction machines for electric drives. *IEEE Trans. Ind. Appl.* **1989**, *25*, 1126–1131. [CrossRef]

21. Ohyama, K.; Asher, G.M.; Sumner, M. Comparative analysis of experimental performance and stability of sensorless induction motor drives. *IEEE Trans. Ind. Electron.* **2006**, *53*, 178–186. [CrossRef]

22. Usha, S.; Subramani, C.; Geetha, A. Performance Analysis of H-bridge and T-Bridge Multilevel Inverters for Harmonics Reduction. *Inter. J. Power Electron. Driv. Syst.* **2018**, *9*, 231–239. [CrossRef]

23. Sengamalai, U.; Chinnamuthu, S. An experimental fault analysis and speed control of an induction motor using motor solver. *J. Electr. Eng. Technol.* **2017**, *12*, 761–768. [CrossRef]

24. Ang, J.; Yang, Y.; Blaabjerg, F.; Chen, J.; Diao, L.; Liu, Z. Parameter identification of inverter-fed induction motors: A Review. *Energies* **2018**, *11*, 2194. [CrossRef]

Article

Torque Distribution Algorithm for an Independently Driven Electric Vehicle Using a Fuzzy Control Method: Driving Stability and Efficiency

Jinhyun Park [1,2], In Gyu Jang [3] and Sung-Ho Hwang [1,*

[1] School of Mechanical Engineering, Sungkyunkwan University, Suwon 16419, Korea; dgb86@naver.com or jinhyun.park@doosan.com

[2] Vehicle Systems PowerTrain Integration Team, Doosan Infracore, 489, Injung-ro, Dong-gu, Incheon 22502, Korea

[3] Advanced technology, Development Mando Global R&D Center, Seongnam, Gyeonggi 13486, Korea; igjang@mando.com

* Correspondence: hsh0818@skku.edu; Tel.: +82-31-290-7464; Fax: +82-31-290-5889

Received: 30 October 2018; Accepted: 11 December 2018; Published: 13 December 2018

Abstract: In this paper, an integrated torque distribution strategy was developed to improve the stability and efficiency of the vehicle. To improve the stability of the low friction road surface, the vertical and lateral forces of the vehicle were estimated and the estimated forces were used to determine the driving torque limit. A turning stability index comprised of vehicle velocity and desired yaw rate was proposed to examine the driving stability of the vehicle while turning. The proposed index was used to subdivide turning situations and propose a torque distribution strategy, which can minimize deceleration of the vehicle while securing turning stability. The torque distribution strategy for increased driving stability and efficiency was used to create an integrated torque distribution (ITD) strategy. A vehicle stability index based on the slip rate and turning stability index was proposed to determine the overall driving stability of the vehicle, and the proposed index was used as a weight factor that determines the intervention of the control strategy for increased efficiency and driving stability. The simulation and actual vehicle test were carried out to verify the performance of the developed ITD. From these results, it can be verified that the proposed torque distribution strategy helps solve the poor handling performance problems of in-wheel electric vehicles.

Keywords: in-wheel electric vehicle; independent 4-wheel drive; torque distribution; fuzzy control; traction control; active yawrate control

---

1. Introduction

Eco-friendly vehicles have become a primary research issue due to problems like environmental pollution and energy resources. Hybrid electric vehicles have already gone beyond the commercialization stage and taken a large portion of the automobile market, and purely electric vehicles are expected to gradually appear in the automobile market after commercialization [1–5].

The biggest reasons for reluctance in purchasing electric vehicles are the expensive price and short driving distance on a single charge. However, these two factors are in a trade-off relationship with each other. Although battery capacity is the most important factor that determines driving distance on a single charge, increase of battery volume is the biggest factor that increases the price of vehicles. Therefore, important research topics for commercialization of electric vehicles would be to select a battery with appropriate capacity and maximize driving distance through efficient use of the selected battery [6–11].

Electric vehicles with in-wheel motors have advantages like fast response and easy embodiment of active driving safety systems like TCS (Traction Control System), ABS (Anti Brake-lock System)

and VDC (Vehicle Dynamic System) without adding additional actuators due to independent driving control of each wheel. Therefore, existing studies on the in-wheel system were mainly focused on improving linear driving performance, improving driving stability during turns, improving turn performance through torque vectoring, and controlling the slip on low-friction roads and asymmetric roads. However, such a vehicle dynamic control system has a problem that it influences the driving efficiency badly only considering the stability of the driving [1–11].

Gu proposed a method of optimizing efficiency of in-wheel electric vehicles through two-wheel/four-wheel conversion by considering the loss of motor and inverter [12]. However, there are many difficulties in maximizing the efficiency of a motor simply based on a two-wheel/four-wheel conversion. Lin proposed a driving strategy that uses DOE (Design of experiments) to derive the optimal driving torque ratio of a motor according to the velocity and accelerator pedaling [13], but failed to account for energy optimization while braking. Chen proposed a driving force distribution strategy to optimize the efficiency of a motor by predicting geographical conditions using GPS and GIS signal [14,15]. Such a control strategy has to be preceded by accurate positioning of vehicles, and it requires a high-precision GPS sensor. It is difficult to use a high-precision GPS sensor in mass-produced vehicles because of the high price. Xu proposed a regenerative braking control strategy to optimize the efficiency and braking performance of electric vehicles using a fuzzy control technique [16]. As such, several studies attempted to increase the efficiency of in-wheel electric vehicles, but most of these studies only focus on braking or driving and lack the consideration of reduced driving stability.

In this study, a torque distribution strategy considering driving stability and efficiency was proposed. In order to improve the stability of the low friction road surface, the vertical force and the lateral force of the vehicle were estimated and the limit driving torque was determined using the estimated force. A fuzzy-based cornering stability index was suggested, and a torque distribution strategy based on turning conditions was proposed. The vehicle stability index using the slip ratio and cornering stability index is proposed. An integrated torque distribution strategy was created using the proposed driving stability index and the torque distribution strategy for increased efficiency. A simulation and an actual vehicle test were conducted to verify the proposed algorithm

2. Vehicle Stability Control

The target vehicle is driven through four in-wheel motors. At this time, the driving/braking torque is determined by the driver's accelerator pedal or brake pedal. However, if the torque is controlled by merely reflecting the will of the driver, the vehicle may fall into an unstable state. Therefore, in this study, the stability of the vehicle is secured through slip control, yaw rate control.

2.1. Slip Control

Figure 1 shows Coulomb's friction circle [17]. The friction circle indicates that the vector sum of the longitudinal force and lateral force of the tire is less than or equal to the product of the normal force and the friction coefficient of the road surface [18–20].

$$\mu F_z \geq \sqrt{F_x^2 + F_y^2} \tag{1}$$

where μ is the coefficient of friction, F_z is the normal force of the tire, F_x is the longitudinal force of the tire, and F_y is the lateral force of the tire. In order for the vehicle to perform stable driving, the driving and braking forces should be kept not to exceed the friction circle, which is expressed as follows:

$$F_{x\_max} = \sqrt{(\mu F_z)^2 - F_y^2} \tag{2}$$

F_{x_max} denotes the maximum drive and braking force determined by the friction circle. In other words, the driving force limit can be obtained when the road friction coefficient, the normal force and the lateral force are known. However, the normal force cannot be measured from the sensor during the driving in real time, so Equation (3) is needed to get normal force values.

$$F_{zfl} = \frac{1}{2}\frac{L_r}{L}mg - \rho_f a_y m \frac{h_g}{t_f} - a_x m \frac{h_g}{L}$$
$$F_{zfr} = \frac{1}{2}\frac{L_r}{L}mg + \rho_f a_y m \frac{h_g}{t_f} - a_x m \frac{h_g}{L}$$
$$F_{zrl} = \frac{1}{2}\frac{L_f}{L}mg - \rho_r a_y m \frac{h_g}{t_r} + a_x m \frac{h_g}{L}$$
$$F_{zrr} = \frac{1}{2}\frac{L_f}{L}mg + \rho_r a_y m \frac{h_g}{t_r} + a_x m \frac{h_g}{L}$$

$$(3)$$

where F_{zfl}, F_{zfr}, F_{zrl} and F_{zrr} are the normal force of each tire. g is the gravitational acceleration. Front and rear roll stiffness distributions are defined as ρ_f and ρ_r. h_g means the height of the center of gravity, and the accelerations of longitudinal direction and lateral direction are a_x and a_y. t_f and t_r are the front and rear tread lengths of the vehicle. L is the wheelbase, L_f and L_r are the distance from the center of gravity to the front wheel and rear wheel.

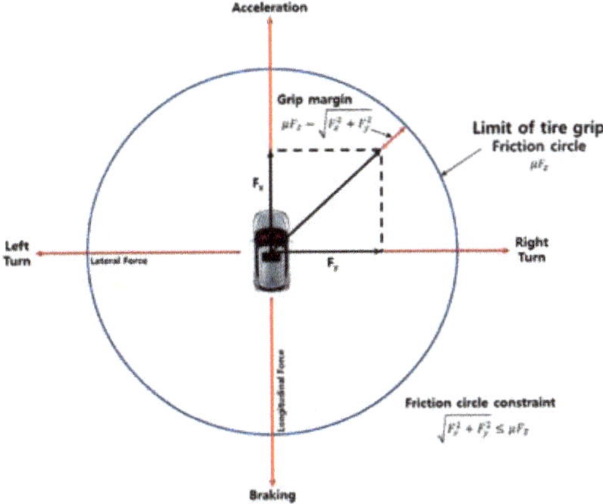

Figure 1. Concept of friction circle.

Lateral acceleration is generated in the vehicle when turning, and the lateral force is generated in the tire in order to cope with lateral acceleration. In this paper, the Dugoff tire model is used to formulate the nonlinear characteristics of the lateral forces in the tire [20,21].

The Dugoff tire model expresses the lateral force of a nonlinear tire as a function of the slip angle (α), the longitudinal slip rate (λ_x), the cornering stiffness (C_α) and the tire longitudinal stiffness (C_λ) that occur in each tire. Since the cornering stiffness and longitudinal stiffness of an actual tire are very different, the Dugoff tire model enables more precise tire behavior analysis compared with a linearly-expressed formula proportional to the cornering stiffness. In addition, in the actual Dugoff tire model, it is assumed that the vertical force of the tire is constant, but in this paper, the change is reflected including the previously estimated vertical force. However, the cornering stiffness is assumed to be a constant.

$$\alpha_f = \delta_f - \frac{V_y + L_f \gamma}{V_x}$$
$$\alpha_r = -\frac{V_y - L_r \gamma}{V_x}$$

$$(4)$$

$$\lambda_x = \frac{V_x - r\omega_w}{V_x} \ (during \ braking)$$
$$\lambda_x = \frac{r\omega_w - V_x}{r\omega_w} \ (during \ acceleration)$$

$$(5)$$

where δ_f is the front wheel steering angle, γ is the yawrate of the vehicle, r and ω_w are the effective radius of rotation and angular velocity of the tire respectively. Cornering stiffness and tire longitudinal stiffness are C_α and C_λ. The tire lateral force using the Dugoff tire model is expressed as follows:

$$F_y = C_\alpha \frac{\tan(\alpha)}{1 + \lambda_x} f(\kappa) \tag{6}$$

where $f(\kappa)$ and the variable κ are obtained as follows:

$$f(\kappa) = \begin{cases} (2 - \kappa)\kappa & (\kappa < 1) \\ 1 & (\kappa \geq 1) \end{cases}$$
$$\kappa = \frac{\mu F_z (1 + \lambda_x)}{2[(C_\lambda \lambda_x)^2 + (C_\alpha \tan(\alpha))^2]^{1/2}} \tag{7}$$

Finally, the lateral forces generated on each tire using the Dugoff tire model are defined as follows:

$$\begin{aligned} F_{yfl} &= C_{af} \frac{\tan(\alpha_{fl})}{1 + \lambda_{xfl}} f(\kappa_{fl}) \\ F_{yfr} &= C_{af} \frac{\tan(\alpha_{fr})}{1 + \lambda_{xfr}} f(\kappa_{fr}) \\ F_{yrl} &= C_{ar} \frac{\tan(\alpha_{rl})}{1 + \lambda_{xrl}} f(\kappa_{rl}) \\ F_{yrr} &= C_{ar} \frac{\tan(\alpha_{rr})}{1 + \lambda_{xrr}} f(\kappa_{rr}) \end{aligned} \tag{8}$$

A simulation was performed to verify the lateral forces estimated in this way. The simulation was performed using CarSim 8.0 and MATLAB/Simulink 2012a. Simulation parameters such as vehicle weight, wheelbase and tread lengths used for simulation were attached as an appendix, and parameters of vehicle velocity and acceleration were used assuming CarSim's data as sensor signals. Conditions of the simulation were set to a cruise driving situation after acceleration to 50 km/h with the steering angle fixed at 100°. Figure 2 shows the simulation results. FL, FR, RL and RR mean front left wheel, front right wheel, rear left wheel and rear right wheel respectively. The simulation on the left side is the result of assuming the normal force of the Dugoff tire model as the normal force on each wheel when the vehicle is at a stop. The simulation on the right side is the result of using normal force predicted by Equation (3). A moving average filter was applied to secure the reliability of the predicted normal force value. Despite the fact that the vehicle reached normal turning status with an estimator based on fixed normal force, the error of maximum lateral force was about 800 N, showing an accuracy of about 79%. On the contrary, when the vehicle reached normal turning state with an estimator based on the predicted normal force, the error of maximum lateral force was about 140 N, showing an accuracy of about 95.3%. Especially, outer wheels that can a have direct effect on the driving stability of the vehicle while turning were found to have high accuracy in an excessive turning situation.

Using the normal force and the lateral force obtained above, the limit driving torque can be obtained as follows:

$$\begin{aligned} T_{Slip\_limit\_fl} &= rF_{x\_max\_fl} = r\sqrt{(\mu F_{zfl})^2 - F_{yfl}^2} \\ T_{Slip\_limit\_fr} &= rF_{x\_max\_fr} = r\sqrt{(\mu F_{zfr})^2 - F_{yfr}^2} \\ T_{Slip\_limit\_rl} &= rF_{x\_max\_rl} = r\sqrt{(\mu F_{zrl})^2 - F_{yrl}^2} \\ T_{Slip\_limit\_rr} &= rF_{x\_max\_rr} = r\sqrt{(\mu F_{zrr})^2 - F_{yrr}^2} \end{aligned} \tag{9}$$

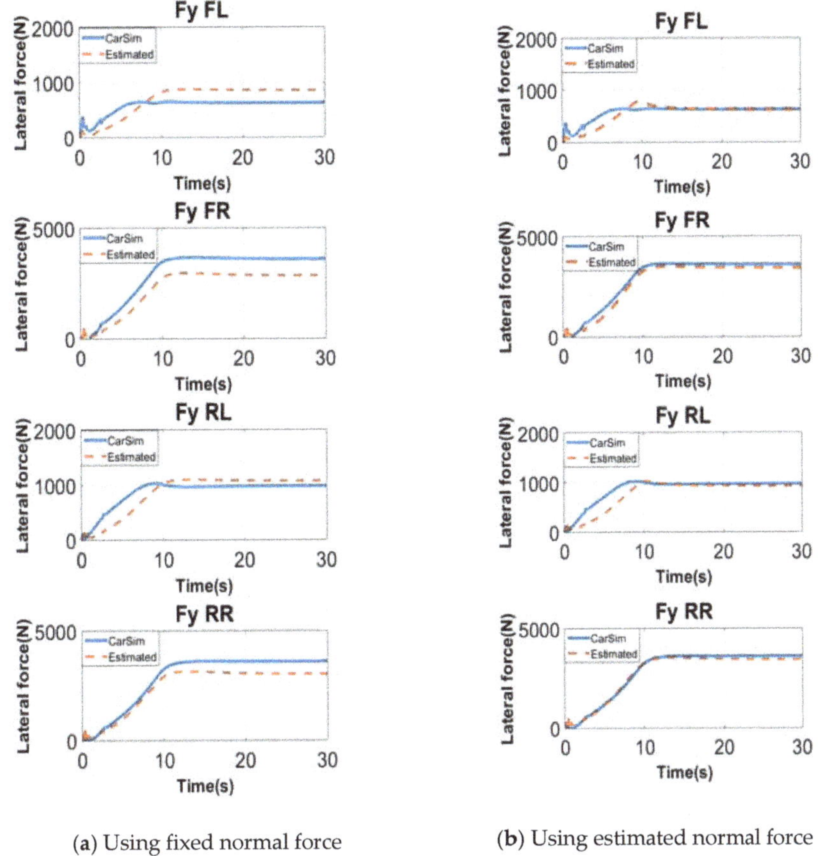

(a) Using fixed normal force   (b) Using estimated normal force

Figure 2. Simulation results of estimated lateral force.

2.2. Active Yawrate Control Considering Driving Efficiency

The bicycle model has been used for the design of direct yaw moment control (DYC) in many previous researches. The desired yaw rate can be easily and exactly calculated to guarantee yaw stability based on the bicycle model through the driver's steering intention. As shown in Figure 3, the bicycle model indicates vehicle lateral dynamics in an assumption that wheels are located at the vehicle center line.

The dynamic equations of the bicycle model in terms of force balance and moment balance are expressed as follows [22]:

$$mV_x(\dot{\beta}+\gamma)= C_f\left(\delta_f - \beta - \frac{L_f\gamma}{v}\right) + C_r\left(-\beta + \frac{L_r\gamma}{v}\right)$$
$$\quad_z\dot{\gamma}= L_fC_f\left(\delta_f - \beta - \frac{L_f\gamma}{V_x}\right) - L_rC_r\left(-\beta + \frac{L_r\gamma}{V_x}\right) + M_z \tag{10}$$

where β is the body side slip angle, γ is the yaw rate, m is the vehicle mass, I_z is the vehicle yaw moment of inertia, V_x is the vehicle longitudinal velocity, and M_z is the correction yaw moment. L_f and L_r are the CG-front and CG-rear axle distances, F_{yf} and F_{yr} are the lateral tire forces of the front and rear axle, respectively. C_f and C_r are the front and rear wheel cornering stiffness and δ_f is the steering angle. State-space expression of the bicycle model is given as follows:

$$
\begin{bmatrix} \dot{\beta} \\ \dot{\gamma} \end{bmatrix} =
\begin{bmatrix} \frac{-C_f - C_r}{mV_x} & \frac{-mV_x^2 - L_f C_f + L_r C_r}{mV_x^2} \\ \frac{-L_f C_f + L_r C_r}{I_z} & \frac{-L_f^2 C_f - L_r^2 C_r}{V_x I_z} \end{bmatrix}
\begin{bmatrix} \beta \\ \gamma \end{bmatrix} +
\begin{bmatrix} \frac{C_f}{mV_x} & 0 \\ \frac{L_f C_f}{I_z} & \frac{1}{I_z} \end{bmatrix}
\begin{bmatrix} \delta_f \\ M_z \end{bmatrix}
\tag{11}
$$

where L is the wheelbase. The desired yawrate can be expressed as a function of the steering angle δ_f and the vehicle longitudinal velocity V_x. It is represented as follows:

$$
\gamma_d = \frac{V_x}{L + \frac{mV_x^2(L_r C_r - L_f C_f)}{2C_f C_r L}} \delta_f
\tag{12}
$$

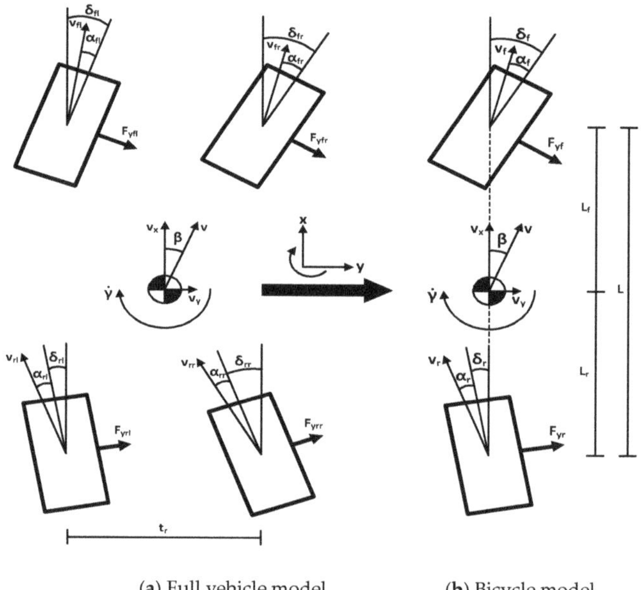

(a) Full vehicle model   (b) Bicycle model

Figure 3. Schematic diagram of full vehicle and bicycle lateral dynamics model.

The desired yaw moment to follow the desired yaw rate is defined as follows:

$$
M_z = I_z \dot{\gamma}_d + (C_f L_f - C_r L_r)\beta + \frac{C_f L_f^2 + C_r L_r^2}{V_x}\gamma - C_f L_f \delta - \eta I_z (\gamma - \gamma_d)
\tag{13}
$$

where η is a positive constant. In the lower controller, the control input in (13) consists of the individual tire forces generated by the in-wheel motor system. The correction yaw moment is obtained as follows:

$$
M_z = t_f \cos(\delta_f)(F_{xfr} - F_{xfl}) + L_f \sin(\delta_f)(F_{xfr} + F_{xfl}) + t_f \sin(\delta_f)(F_{yfl} - F_{yfr}) + t_r(F_{xrr} - F_{xrl})
\tag{14}
$$

Assuming that the steering angle is small, $\sin(\delta)$ is assumed to be zero. Equation (14) is rewritten as follows:

$$
M_z = t_f \cos(\delta_f)(F_{xfr} - F_{xfl}) + t_r(F_{xrr} - F_{xrl})
\tag{15}
$$

Since the VDC of an internal combustion engine vehicle or ordinary electric vehicle is difficult to control the driving force independently, the correction moment for over steering or under steering during a left turn is generated as shown in Figure 4. Such a correction moment can improve the driving stability of a vehicle, but it can decrease the velocity of the vehicle as well. Velocity decrease not only makes the driver feel odd while driving but also causes additional driving force, which leads to the low driving efficiency of the vehicle.

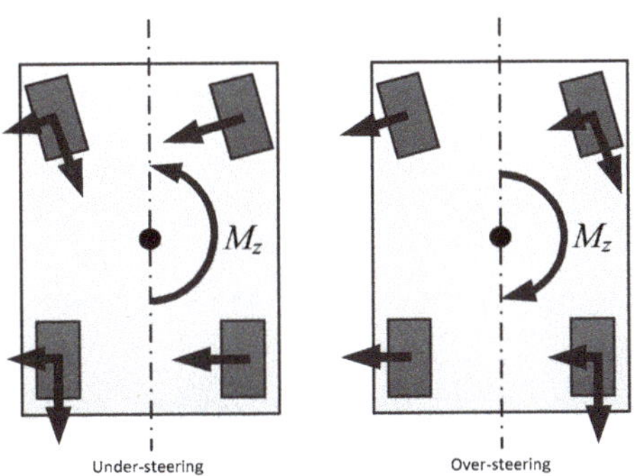

Figure 4. Correction yaw moment during left turn using brake.

To solve this problem, cornering situations were classified and a driving force distribution strategy was proposed as shown in Figure 5. Cornering situations are largely divided into three types, namely stable cornering, transient cornering and unstable cornering. Stable cornering is a driving situation in which the velocity of the vehicle is relatively slow and the desired yaw rate is low, so the driving torque of the motor is not expected to have a critical impact on driving stability. Thus, the correction yaw moment is generated through driving torque. Excessive cornering is a domain in which vehicle velocity and desired yaw rate have intermediate values. Since excessive driving torque can make the vehicle unstable, correction yaw moment is generated using both the driving force and braking force. Lastly, unstable cornering is a situation of high velocity or high desired yaw rate where the driving torque can greatly affect the driving stability. Correction yaw moment is generated only using braking force. To satisfy such rules, a controller needs to be comprised so that three yaw moment distribution strategies intervene according to the situation. However, there is a problem that excessive mode conversion occurs when the vehicle is operated around the mode conversion point. A Fuzzy controller was used to prevent excessive switching errors while fulfilling such a purpose. The fuzzy logic has been proposed to solve the problems of various logic which judged only the existing true and false, and it can output various values between 0 and 1. It is employed to handle the concept of partial truth, where the truth value may range between completely true and completely false [23]. Composition of the fuzzy controller is as presented in Table 1. The input membership function of the fuzzy controller includes the velocity of the vehicle and desired yaw rate. Output membership function outputs the cornering stability index (σ) with a value of 0~2. Cornering stability index indicates a stable cornering situation as it gets closer to 0 and an unstable cornering situation as it gets closer to 2. Final AYC (Active Yawrate Control) torque output is defined as Equation (16). The operating region of AYC is shown in Figure 6.

$$
\begin{aligned}
T_{fl\_AYC} &= \begin{array}{ll} T_{fl}(APS, BPS) \pm T_{fl\_stable}(1-\sigma) \pm T_{fl\_trans}(\sigma) & (\sigma \leq 1) \\ T_{fl}(APS, BPS) \pm T_{fl\_trans}(2-\sigma) \pm T_{fl\_unstable}(\sigma-1) & (1 < \sigma \leq 2) \end{array} \\
T_{fr\_AYC} &= \begin{array}{ll} T_{fr}(APS, BPS) \pm T_{fr\_stable}(1-\sigma) \pm T_{fr\_trans}(\sigma) & (\sigma \leq 1) \\ T_{fr}(APS, BPS) \pm T_{fr\_trans}(2-\sigma) \pm T_{fr\_unstable}(\sigma-1) & (1 < \sigma \leq 2) \end{array} \\
T_{rl\_AYC} &= \begin{array}{ll} T_{rl}(APS, BPS) \pm T_{rl\_stable}(1-\sigma) \pm T_{rl\_trans}(\sigma) & (\sigma \leq 1) \\ T_{rl}(APS, BPS) \pm T_{rl\_trans}(2-\sigma) \pm T_{rl\_unstable}(\sigma-1) & (1 < \sigma \leq 2) \end{array} \\
T_{rr\_AYC} &= \begin{array}{ll} T_{rr}(APS, BPS) \pm T_{rr\_stable}(1-\sigma) \pm T_{rr\_trans}(\sigma) & (\sigma \leq 1) \\ T_{rr}(APS, BPS) \pm T_{rr\_trans}(2-\sigma) \pm T_{rr\_unstable}(\sigma-1) & (1 < \sigma \leq 2) \end{array}
\end{aligned}
\tag{16}
$$

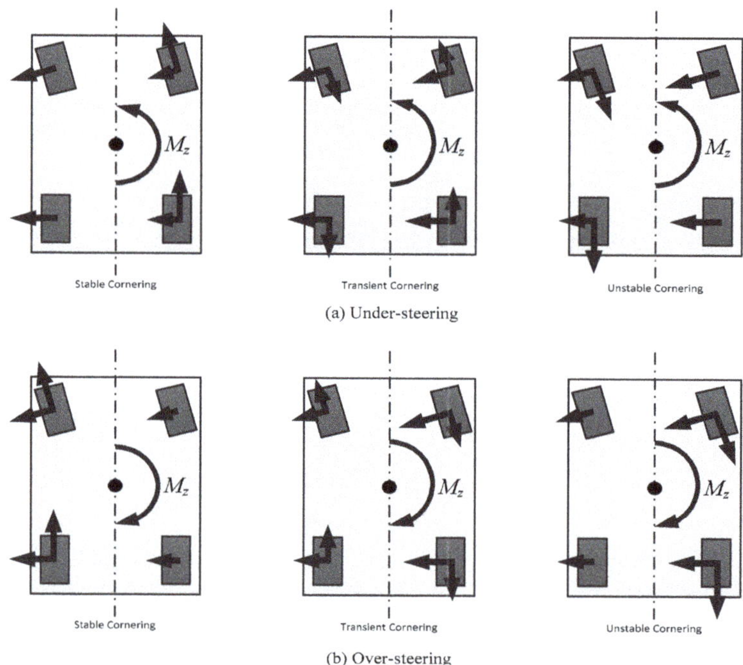

Figure 5. Correction yaw moment during left turn using in-wheel motor.

Table 1. Cornering stability Index.

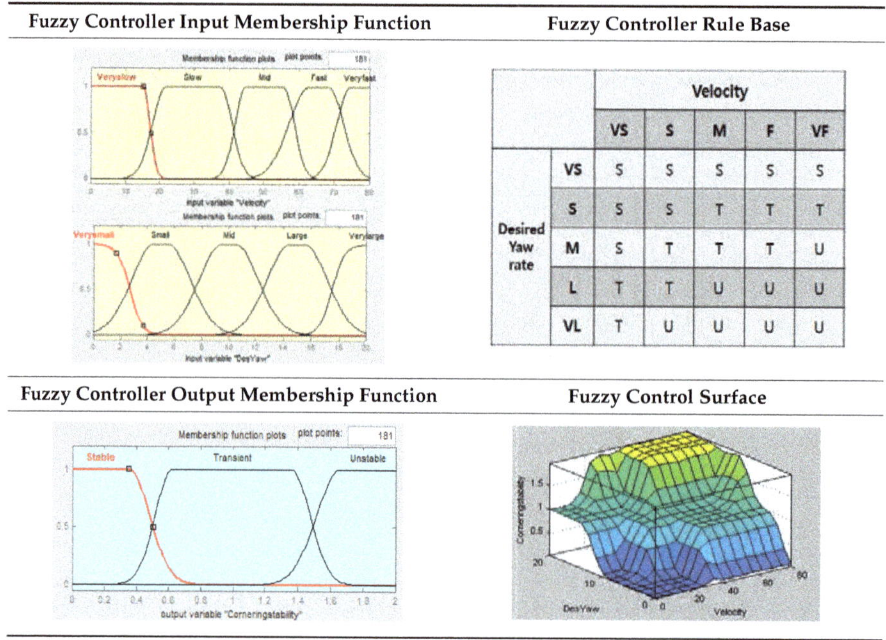

Fuzzy Controller Input Membership Function	**Fuzzy Controller Rule Base**

Fuzzy Controller Rule Base

		Velocity				
		VS	**S**	**M**	**F**	**VF**
Desired Yaw rate	**VS**	S	S	S	S	S
	S	S	S	T	T	T
	M	S	T	T	T	U
	L	T	T	U	U	U
	VL	T	U	U	U	U

Fuzzy Controller Output Membership Function	**Fuzzy Control Surface**

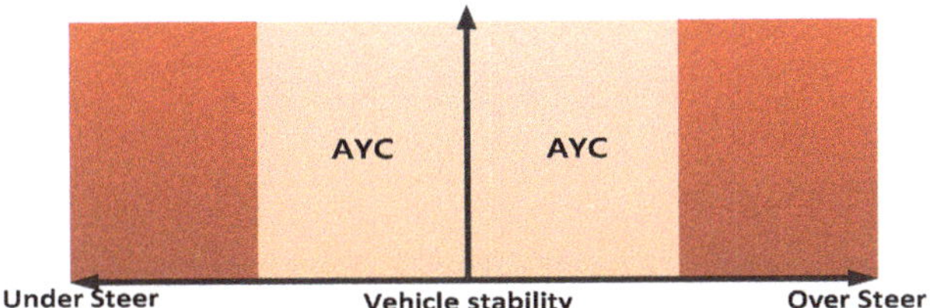

Figure 6. Operating region of acitve yaw rate control.

2.3. Integrated Driving Torque Distribution Strategy Considering Driving Efficiency and Stability

When a vehicle accelerates or climbs a hill while driving, it requires a large amount of driving force. However, the desire of the driver can be followed without using the maximum torque of the motor in case of cruise drive or downhill drive. Generating same driving torque on four wheels makes it difficult to drive the motor in the efficient region. As a solution to this problem, a driving torque distribution strategy that appropriately distributes the driving force and regenerative braking force of the front and rear wheels according to the situation to improve the efficiency of the vehicle was proposed as described in previous paper [24]. However, since the developed distribution strategy is a strategy that considers driving efficiency in a stable situation, there is a problem of reducing driving stability of the vehicle in a rapid cornering situation or low-friction road situation. The following driving stability index (ψ) was proposed to solve this problem.

$$\psi = \frac{1}{\tilde{\lambda}}(\lambda)(\lambda > 0.1) + \frac{1}{\tilde{\dot{\lambda}}}(\dot{\lambda})(\lambda > 0.1) + \frac{1}{\tilde{\sigma}}(\sigma) + \frac{1}{\tilde{\dot{\sigma}}}(\dot{\sigma}) \tag{17}$$

where λ is the slip rate, $\tilde{\lambda}$ is the slip rate limit, $\dot{\lambda}$ is the differential value of slip rate, $\tilde{\dot{\lambda}}$ is the differential value limit of slip rate, σ is the cornering stability index, $\tilde{\sigma}$ is the cornering stability index limit, $\dot{\sigma}$ is the differential value of cornering stability index, and $\tilde{\dot{\sigma}}$ is the differential value limit of the slip rate. Each term represents the ratio of the current value to the limit value that determines driving stability. For instance, if the slip rate limit is 0.2 and the current slip rate is 0.15, the stability index for slip rate of the vehicle is 0.75. The stability index is calculated for four variables, and the sum becomes the driving stability index of the vehicle. The inequality equation that constitutes the term that determines the stability of the slip rate was added to decide the stability of the slip rate only in the area where the slip rate becomes larger than 0.1 by considering the area with a slip rate smaller than 0.1 as a stable area. If one of four variables that constitute Equation (17) has a value larger than 1, the vehicle is considered to be in an unstable state and is controlled only using the torque for improving driving stability. However, if the value is between 0 and 1, the driving stability evaluation index is used as a weight factor to determine the intervention rate of the efficiency controller and driving stability controller. The final torque values determined are expressed by the equations as follows:

$$\begin{aligned}
T_{fl_final} &= T_{fl_STC}(\psi) \pm T_{fl_Eff}(1-\psi) \\
T_{fr_final} &= T_{fr_STC}(\psi) \pm T_{fr_Eff}(1-\psi) \\
T_{rl_final} &= T_{rl_STC}(\psi) \pm T_{rl_Eff}(1-\psi) \\
T_{rr_final} &= T_{rr_STC}(\psi) \pm T_{rr_Eff}(1-\psi)
\end{aligned} \tag{18}$$

where T_{STC} is torque determined to improve the driving stability of the vehicle, and T_{Eff} is the torque value determined to increase the efficiency of the vehicle. The vehicle stability index has values from 0 to 1, and the values close to 1 equate to unstable values. If the vehicle stability index is 1, only the T_{STC} is determined to be the output torque by $(1-\psi)$ term, and if the evaluation index is 0, only the T_{eff} is configured to be an output. Sometimes the final torque may exceed the T_{stc}, but the simulation found that the stability of the driving was less affected. It seems that when the final torque affects the driving stability of the vehicle, only the T_{stc} is output as the vehicle stability index becomes 1. The final torque, thus determined, controls the driving stability and efficiency of the vehicle through four in-wheel motors.

The block diagram of the integrated torque distribution strategy is shown in Figure 7. The torque distribution strategy for increased efficiency proposed in previous papers is located at the bottom of the block diagram. The slip controller and AYC proposed in this paper are located at the top of the block diagram. Driving efficiency and driving stability of the vehicle were secured in the end by determining the intervention rate of these two torque distribution strategies using the driving stability evaluation index.

The operating region of ITD is shown in Figure 8. In the area where the vehicle is determined to be stable, the distribution strategy for increased efficiency primarily intervenes. In the area where the instability of the vehicle increases, driving stability and driving efficiency are secured using the slip controller and AYC.

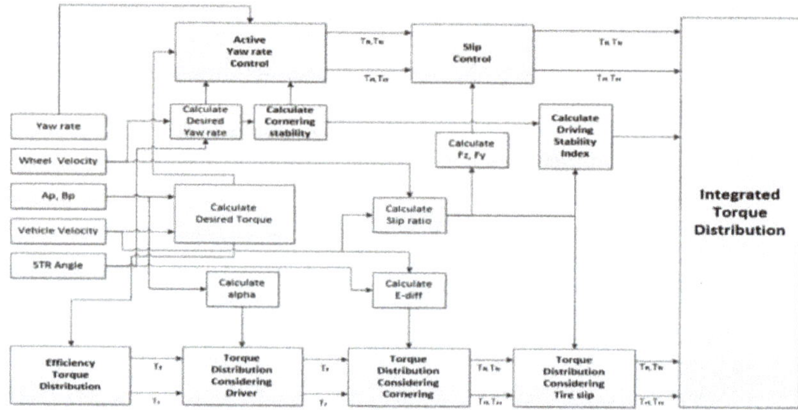

Figure 7. Integrated driving torque distribution strategy considering driving efficiency and stability.

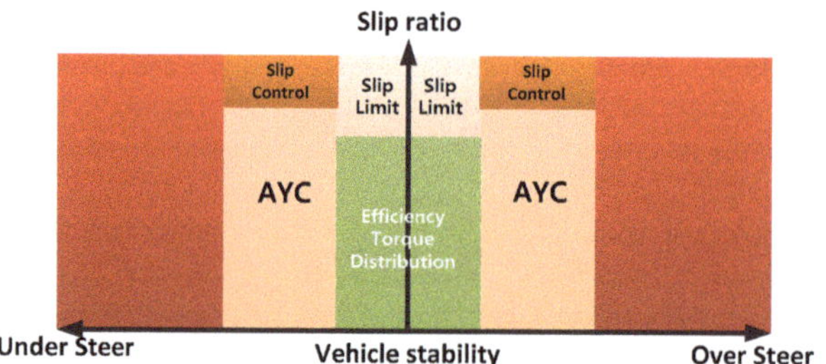

Figure 8. Operating region of integrated driving torque distribution strategy.

3. Simulation and Experimental Results

3.1. Double Lane Change Simulation

A double-lane change simulation was performed to identify the effect of the proposed torque distribution strategy on the driving stability of the vehicle. The ISO double lane change test consists of an entry and an exit lane and has a length of 12 m and a side lane with a length of 11 m. The width of the entry and side lane are dependent on the width of the vehicle; the width of the exit lane is constantly 3 m wide. The lateral offset between the entry and side lane is 1 m and the longitudinal offset is 13.5 m. For the same lateral offset, the side and exit lane have a slightly shorter longitudinal displacement of 12.5 m. The simulation was performed at a target speed of 50 km/h to verify the performance of the AYC. Driving trajectory is shown in Figure 9. The road friction condition was 0.5 of a wet road. According to the simulation results, vehicles without a torque distribution strategy showed a large over shoot of about 0.46 m, whereas vehicles applied with the torque distribution strategy showed a small over shoot of about 0.18~0.19 m. Vehicles applied with AYC have similar behavior as vehicles applied with integrated torque distribution (ITD). Since this is a low velocity situation with a high desired yaw rate, the final driving torque determined by the vehicle stability index (ψ) places a greater priority on the driving stability control than the efficiency control.

Detailed simulation results are shown in Figures 10 and 11. The x-axis of each graph represents the time, and the y-axis represents the data and units of each graph title. FL, FR, RL and RR mean front left wheel, front right wheel, rear left wheel and rear right wheel respectively. The trend line for comparison was based on vehicles without a torque distribution strategy. Based on the comparison of motor torque, vehicles applied with AYC and ITD showed relatively similar tendencies, but torque output differs between 0~1 seconds section and 5~6 seconds section. As this section is a section in which the vehicle drives straight, the torque distribution strategy intervenes to improve driving efficiency. The steering wheel input is largest in vehicles without a torque distribution strategy, and cases applied with AYC and ITD show similar values. Based on the yaw rate of 1~4 seconds section, vehicles without a torque distribution strategy have a low yaw rate despite having larger steering input compared to other vehicles. This is probably caused by correction yaw moment according to the application of the torque distribution strategy. However, vehicles without a torque distribution strategy have the highest yaw rate in 4~5 seconds section. This section is the 55~65 m region of driving trajectory shown in Figure 9. Excessive steering value results from increasing the deviation from the driving path. Vehicles applied with AYC and vehicles applied with ITD have similarly small values of side slip angle, which is used to determine the driving stability of a vehicle while cornering. This is probably because if the vehicle without ITD is not able to follow the driver's desired path, the driver enters a larger steering angle. On the other hand, if the vehicle to which the ITD is applied does not follow the driver's desired path, ITD will assist with the motor driving torque, so the driver will maintain the small steering input. Thus, if the driver maintains a small steering input, the probability that the vehicle will go into an unstable state is reduced.

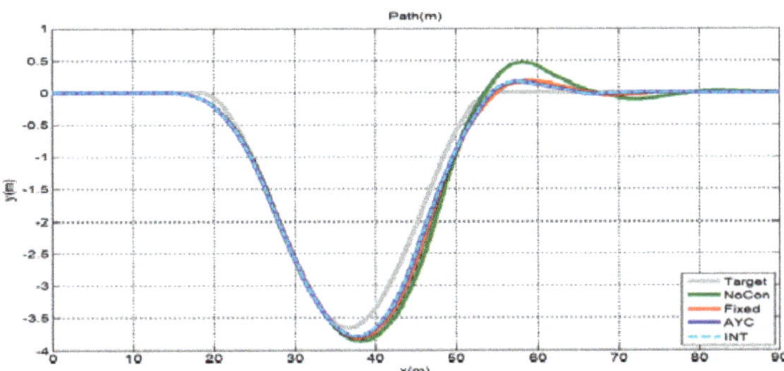

Figure 9. Target path and driving path of short double lane change (Target velocity: 50 km/h).

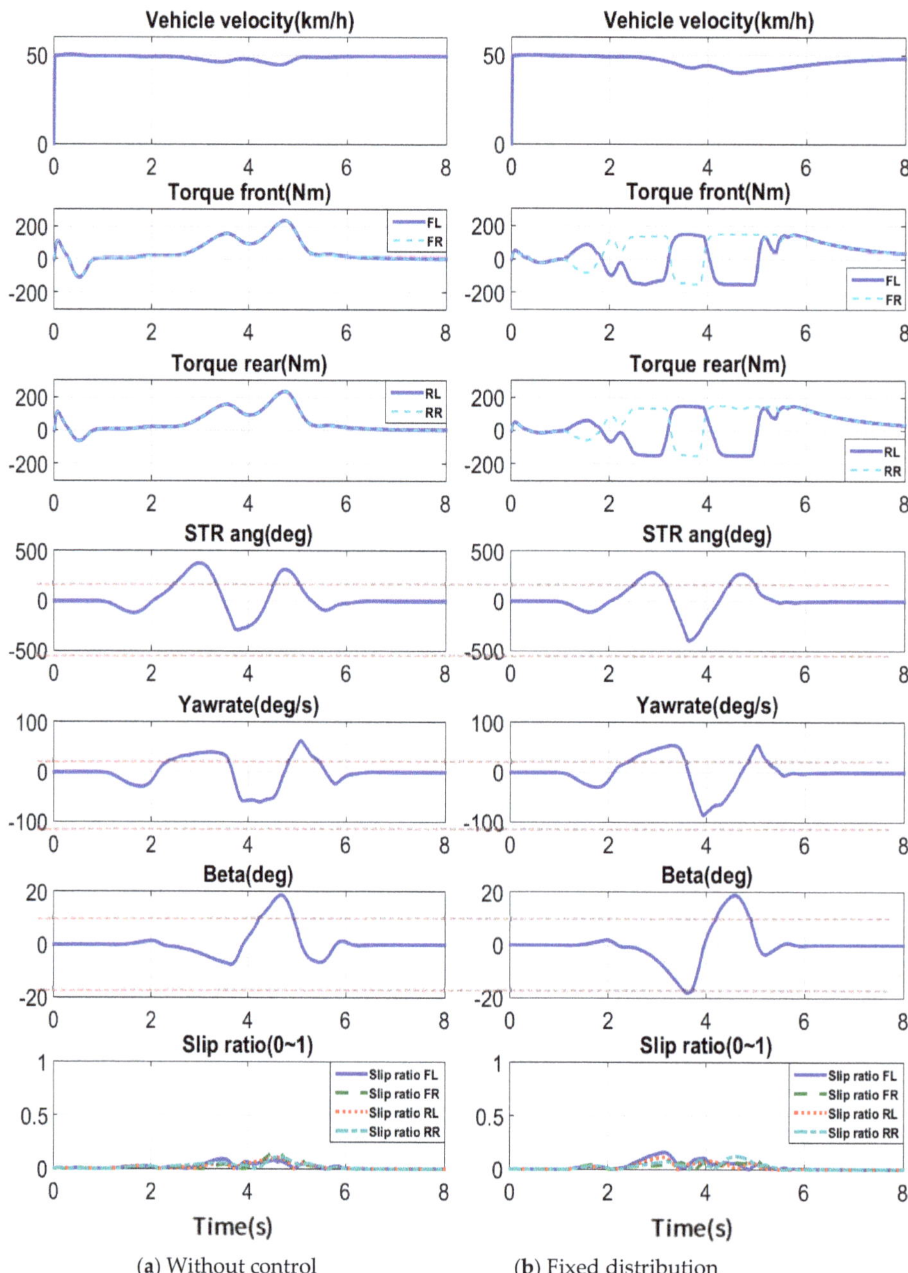

(**a**) Without control (**b**) Fixed distribution

Figure 10. Simulation results of short double-lane change (No control, Fixed distribution).

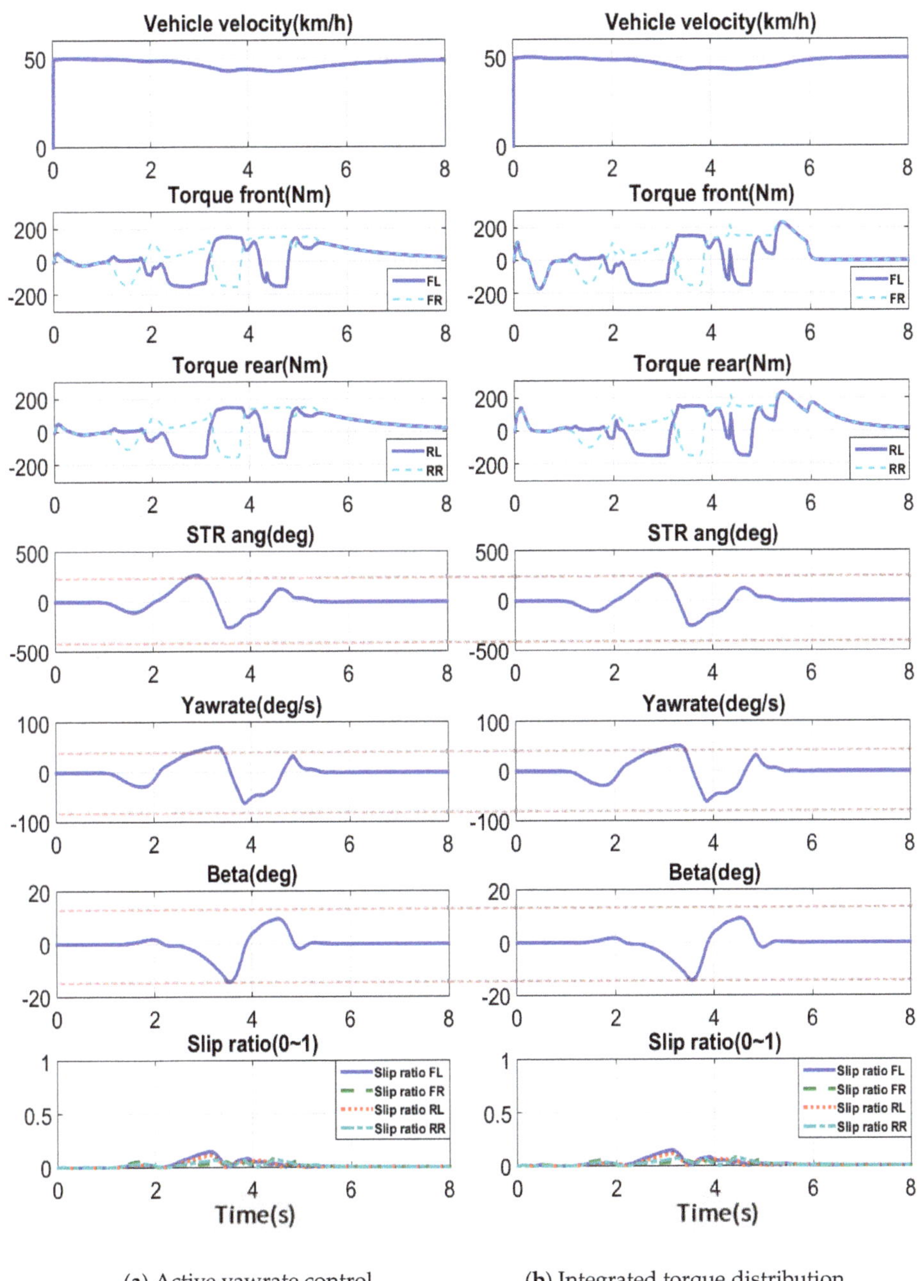

(**a**) Active yawrate control   (**b**) Integrated torque distribution

Figure 11. Simulation results of short double-lane change (AYC, ITD).

3.2. Complex Driving Road Simulation

A simulation was carried out to verify the overall performance of the proposed torque distribution strategy. Target driving path for the simulation and the result of driving according to the distribution strategy are shown in Figure 12. Road condition was configured as 0.5 of a wet road for driving stability assessment.

The simulation was performed on a case without a torque distribution strategy, a case with left and right fixing torque distribution, a case applied with active yaw rate control, and a case applied with integrated torque distribution. Comparing the results of driving trajectory, vehicles without a torque distribution strategy cannot follow the target driving path in the expanded region and roll over occurs. On the contrary, vehicles applied with the three torque distribution strategies show similar path following.

Figures 13 and 14 show detailed simulation results according to the application of each distribution strategy. The x-axis of each graph represents time, and the y-axis represents the data and units of each graph title. FL, FR, RL and RR mean front left wheel, front right wheel, rear left wheel and rear right wheel respectively. The results for points A, B and C were analyzed to compare the driving performance according to the distribution strategy. First, vehicles without the distribution strategy generate mostly identical torque in all motors despite a large amount of cornering at point A. Vehicles were found to deviate from the driving path due to a rapid increase of the yaw rate, side slip angle and slip ratio. The case applied with the left and right fixing torque distribution strategy and the case applied with AYC were found to maintain similar driving stability. However, the case applied with integrated torque distribution shows the reduced driving torque of the motor and maintains the slip ratio in a stable region compared to other distribution strategies because the slip controller intervenes at point A. However, it has a larger yaw rate and side slip angle compared to the other distribution strategies in section B-C. This is probably because the final driving torque determined by the vehicle stability index (ψ) places a greater priority on efficiency control. SOC (State of Charge) based on repeated driving of the driving path in Figure 12, intended to quantitatively compare driving efficiency according to the application of each distribution strategy, is shown in Figure 15. Driving distance based on SOC is shown in Table 2. The range of SOC is 0.8~0.7, and the driving distance from the use of 0.1 SOC was compared. The results for vehicles without a torque distribution strategy are not included because they could not be driven due to overturn. Looking at Table 2, the case applied with AYC showed total driving distance increased by about 3.87% compared to the case applied with the left and right fixing torque distribution strategy. Total driving distance was increased by about 10.93% in the case applied with integrated torque distribution. The proposed torque distribution strategy was found to increase driving efficiency while securing the driving stability of vehicles.

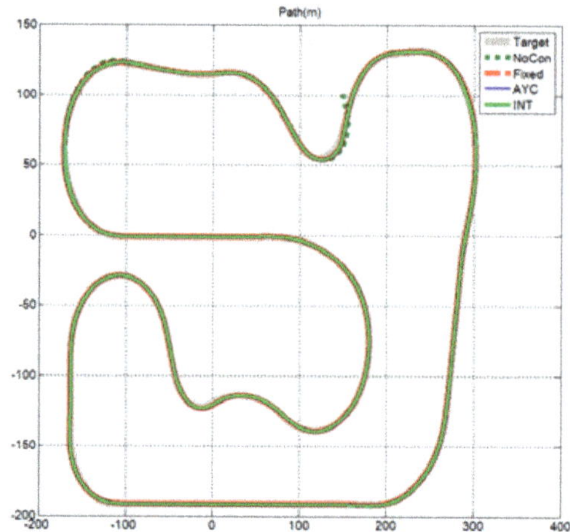

Figure 12. Target path and driving path for complex road.

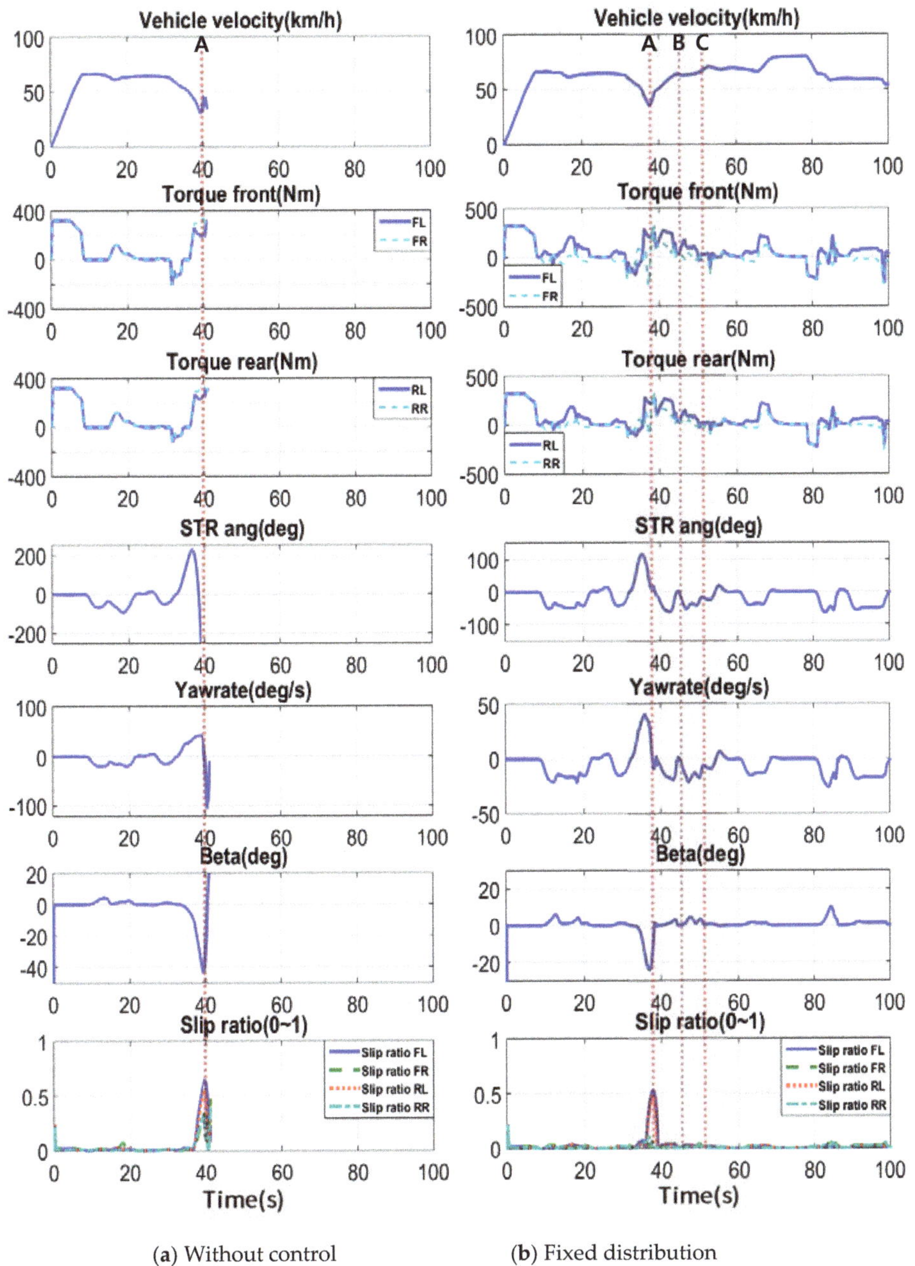

(**a**) Without control   (**b**) Fixed distribution

Figure 13. Simulation results for complex driving road (No control, Fixed distribution).

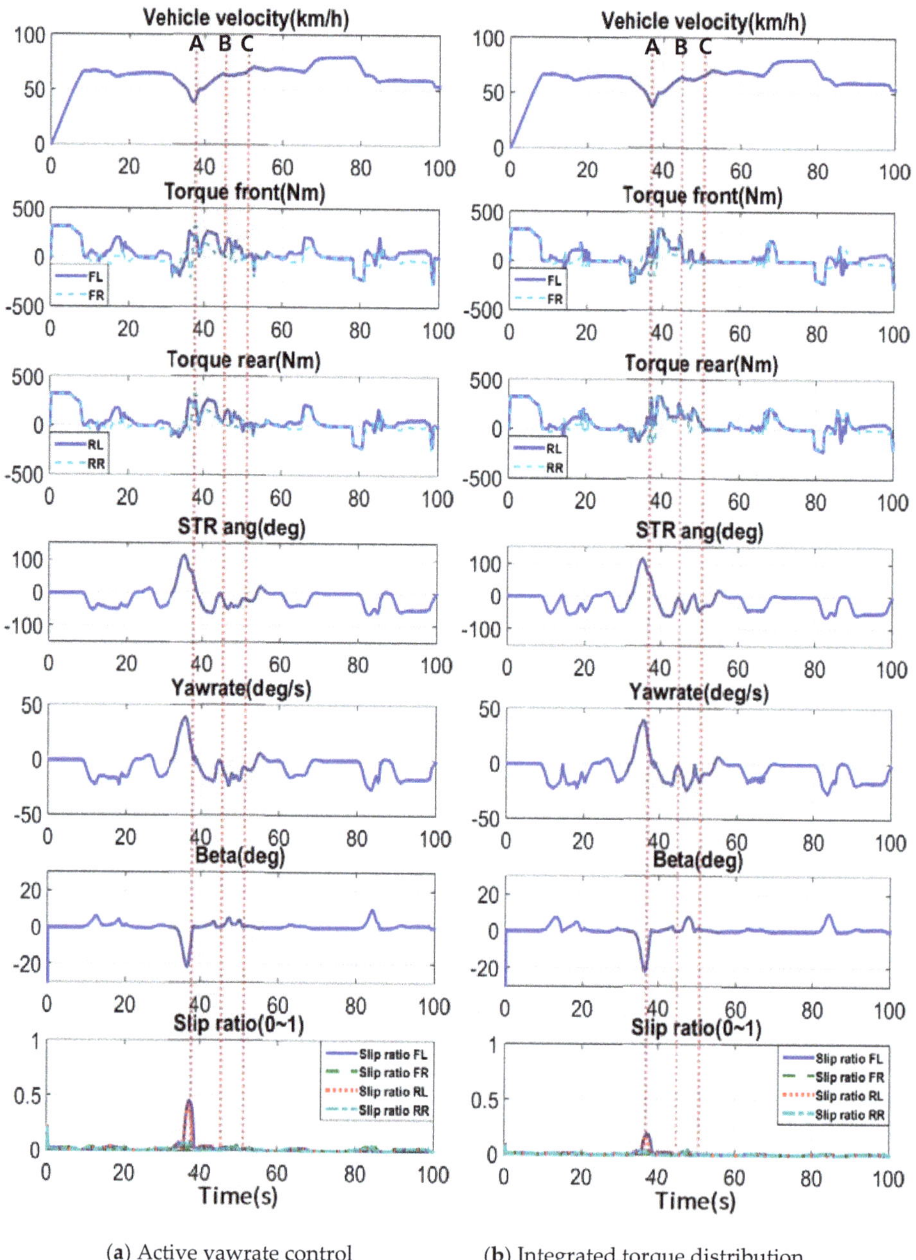

(**a**) Active yawrate control   (**b**) Integrated torque distribution

Figure 14. Simulation results for complex driving road (AYC, ITD).

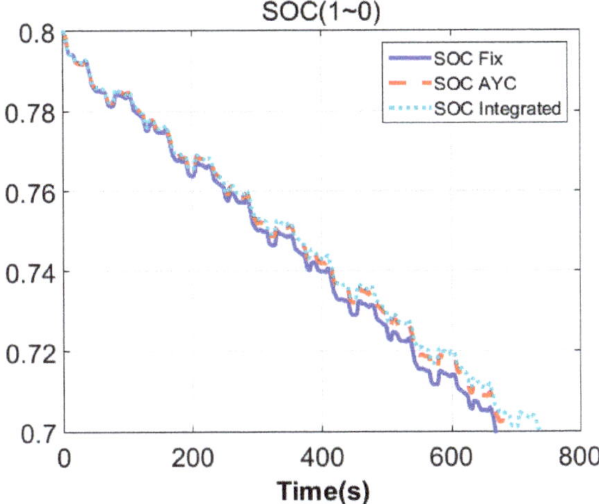

Figure 15. Simulation results of battery SOC for complex driving road.

Table 2. Simulation results of efficiency.

Performance Index	Fixed Distribution	Active Yawrate Control	Integrated Torque Distribution
Driving Distance	11.89 km	12.35 km	13.19 km
Rate of Increment	-	3.87%	10.93%

3.3. Vehicle Test

The actual vehicle experiment was conducted to determine the effect of the developed torque distribution strategy on the driving stability of the actual vehicle. Based on KIA automotive's gasoline ray model, the target vehicle was equipped with an in-wheel motor for independent driving. The experimental conditions were selected as follows.

- Test condition: Slalom course with pylon set at equal intervals of 30 m
- Driving method: Drive as close as possible without hitting the pylon
- Target Speed: 40 km/h
- Measurement data: vehicle velocity, wheel steering angle, motor drive torque, yaw rate, and lateral acceleration
- Measuring sensor: RT 3000

The velocity of the vehicle must be found accurately to apply the proposed control algorithm to the vehicle and to perform performance verification. In this paper, an experiment was conducted using the velocity signal of RT 3000 mentioned earlier. RT 3000 has an inertial navigation system that has position accuracy of about 40 cm and a velocity accuracy of 0.1 km/h. Data output rate is 100 Hz, which secures data in real time. Figure 16 shows the Slalom test course. Figure 17 shows the actual experimental scene and the sensor used during the experiment. The photograph on the top right side is the sensor attached to the vehicle.

Figure 18 shows the driving test results according to the application of ITD. The x-axis of each graph represents time, and the y-axis represents the data and units of each graph title. FL, FR, RL and RR mean front left wheel, front right wheel, rear left wheel and rear right wheel respectively. Comparing the velocity results, the two vehicles were driven while maintaining similar velocity. However, whereas the vehicle not applied with ITD maintained constant motor-driving torque, the vehicle applied with ITD showed left and right driving torques change to follow the desired yaw rate of the driver. Looking at the steering angle results, the vehicle applied with ITD showed a smaller steering angle compared to the vehicle not applied with ITD. This is probably because if the vehicle without ITD is not able to follow the driver's desired path, the driver enters a larger steering angle. On the other hand, if the vehicle to which the ITD is applied does not follow the driver's desired path, ITD will assist with the motor-driving torque, so the driver will maintain the small steering input.

To quantitatively compare the test results, the peak values of steering angle, yaw rate and lateral acceleration are presented in Table 3. The R.M.S value of the peak value of steering angle is 80.4 for the vehicle not applied with ITD and 55.8 for the vehicle applied with ITD, which is about 30.6% smaller. Also, since the vehicle applied with ITD was able to drive with a small radius of turning, the R.M.S value of the peak value of yaw rate was also decreased by about 6.5%. On the contrary, since the vehicle applied with ITD showed a sudden change of direction compared to the vehicle not applied with ITD, the R.M.S value of the peak value of lateral acceleration was increased by about 16.7%. The vehicle was confirmed to be driven in the path wanted by the driver with relatively small steering input through the application of ITD, contributing to improved driving stability of the vehicle.

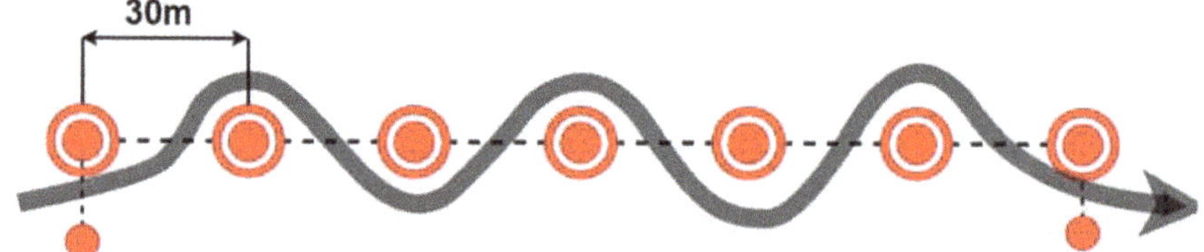

Figure 16. Slalom test course.

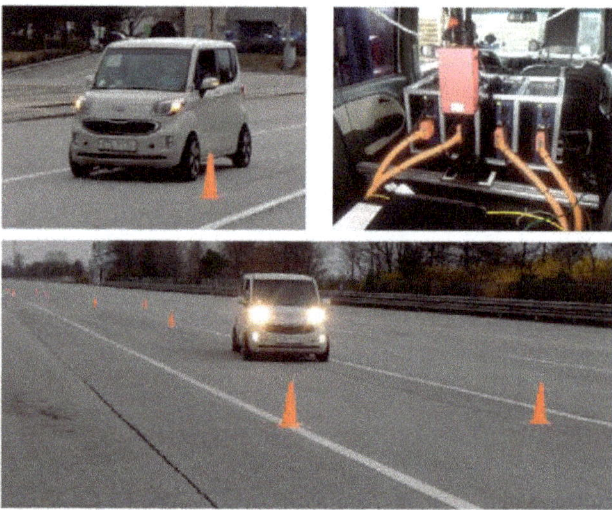

Figure 17. Slalom test and measuring sensor.

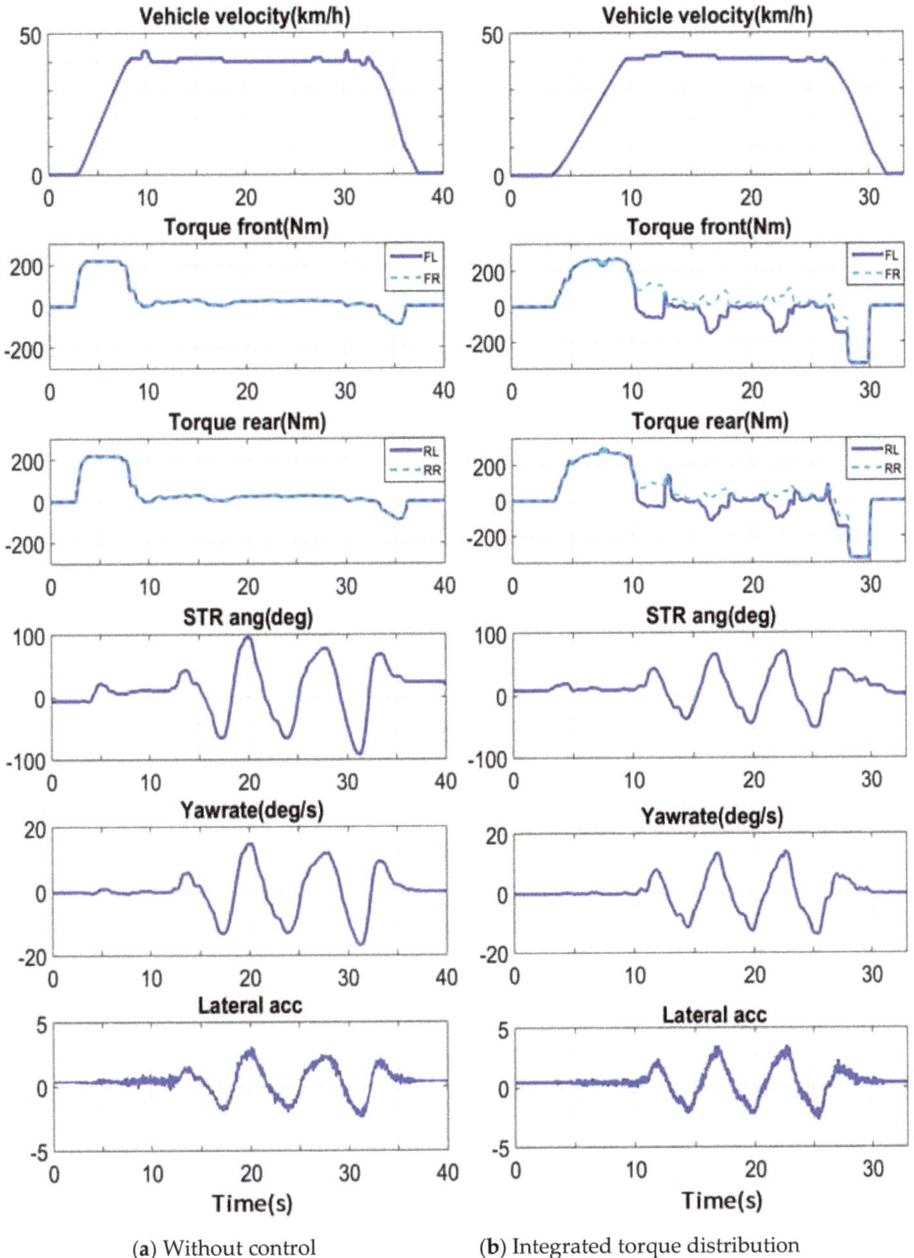

(**a**) Without control   (**b**) Integrated torque distribution

Figure 18. Slalom test results (Target velocity: 40 km/h).

Table 3. Peak values of slalom test results.

Signal Name	1st Peak Value	2nd Peak Value	3rd Peak Value	4th Peak Value	5th Peak Value	R.M.S Value
STR ang (deg) without ITD	−65.2	95.6	−66	77.5	−92.4	80.4
STR ang (deg) with ITD	−37.7	66.1	71.1	−45.2	−51.7	55.8
Yawrate (deg/s) without ITD	−13	15	−12.99	11.88	−17.04	14.11
Yawrate (deg/s) with ITD	−11.44	13.43	−12.73	14.16	13.97	13.2
Lateral acc without ITD	−1.95	3.13	−2.04	2.43	−2.37	2.4
Lateral acc with ITD	−2	3.49	−2.32	3.51	−2.31	2.8

4. Conclusions

An integrated torque distribution strategy was developed to improve the stability and efficiency of the vehicle. In order to improve the stability of the low friction road surface, the vertical force and the lateral force of the vehicle were estimated and the limit driving torque was determined using the estimated force. A turning stability index comprised of vehicle velocity and desired yaw rate of the driver was proposed to examine the driving stability of the vehicle while turning. The proposed index was used to subdivide the turning situations and to propose a torque distribution strategy which can minimize deceleration of the vehicle while securing turning stability. The torque distribution strategy for increased driving stability and the torque distribution strategy for increased efficiency proposed were used to create an integrated torque distribution strategy. A vehicle stability index based on the slip rate and turning stability index was proposed to determine the overall driving stability of the vehicle, and the proposed index was used as a weight factor that determines the intervention of the control strategy for increased efficiency and the control strategy for improved driving stability.

The simulation and actual vehicle test were carried out to verify the performance of the developed ITD. Based on the short double-lane change simulation on low-friction roads, the vehicle applied with ITD showed best path tracking and the smallest reduction of velocity. Based on the results of complex road simulation, the vehicles without torque distribution strategy were overturned around sharp corners, whereas the vehicles applied with the torque distribution strategy were driven stably. In addition, the vehicle applied with ITD showed the most desirable results for driving efficiency. An actual vehicle test was performed to evaluate the performance of ITD on an actual vehicle. The test conditions were selected for slalom driving at 40 km/h. As a result of the actual vehicle test, the vehicle applied with ITD was found to be driven in the path wanted by the driver with relatively small steering input compared to the vehicle not applied with ITD. From these results, it can be confirmed that the proposed driving torque distribution strategy improves the efficiency and driving stability of the independent driving electric vehicle. This is an improvement over the various studies mentioned in the introduction, focusing only on improving the driving efficiency of the vehicle.

Author Contributions: Conceptualization, J.P., S.-H.H. and I.G.J.; Methodology, J.P.; Software, J.P.; Validation, S.-H.H. and I.G.J.; Writing-Original Draft Preparation, J.P.; Writing-Review & Editing, S.-H.H.; Project Administration, S.-H.H.

Funding: Please add: This research was funded by the MSIT (Ministry of Science and ICT), Korea, under the ITRC (Information Technology Research Center) support program (IITP-2018-0-01426) supervised by the IITP (Institute for Information & communications Technology Promotion)

Acknowledgments: This research was supported by the MSIT (Ministry of Science and ICT), Korea, under the ITRC (Information Technology Research Center) support program (IITP-2018-0-01426) supervised by the IITP (Institute for Information & communications Technology Promotion)

Conflicts of Interest: The authors declare no conflict of interest.

Appendix A

Table A1. Parameters of the target in-wheel electric vehicle.

Component	Specification
Sprung mass	1231 kg
Unprung mass	161 kg
Vehicle height	1440 mm
Vehicle width	1600 mm
Wheel base	2567 mm
Tire radius	278 mm
Distance of CG to front wheel centerline	1312 mm
Roll inertia	506 kg·m^2
Pitch inertia	2012 kg·m^2
Yaw inertia	2065 kg·m^2
In-wheel motor	15 kW
In-wheel motor gear ratio	6
Battery	65 kW/50 Ah, Li-ion
SOC range	0.3~0.8

References

1. Hori, Y. Future Vehicle Driven by Electricity and Control-Research on Four-Wheel- Motored "UOT electric march II". *IEEE Trans. Ind. Electron.* **2004**, *51*, 954–962. [CrossRef]
2. Nasri, A.; Hazzab, A.; Bousserhane, I.; Hadjeri, S.; Sicard, P. Fuzzy-Sliding Mode Speed Control for Two Wheels Electric Vehicle Drive. *J. Electr. Eng. Technol.* **2009**, *4*, 499–509. [CrossRef]
3. Sekour, M.; Hartani, K.; Draou, A.; Allali, A. Sensorless Fuzzy Direct Torque Control for High Performance Electric Vehicle with Four In-Wheel Motors. *J. Electr. Eng. Technol.* **2013**, *8*, 530–543. [CrossRef]
4. Hartani, K.; Draou, A. A New Multimachine Robust Based Anti-skid Control System for High Performance Electric Vehicle. *J. Electr. Eng. Technol.* **2014**, *9*, 214–230. [CrossRef]
5. Cirovic, V.; Aleksendric, D. Adaptive neuro-fuzzy wheel slip control. *Expert Syst. Appl.* **2013**, *40*, 5197–5209. [CrossRef]
6. Kim, J.; Park, C.; Hwang, S.; Hori, Y.; Kim, H. Control Algorithm for an Independent Motor-Drive Vehicle. *IEEE Trans. Veh. Technol.* **2010**, *59*, 3213–3222. [CrossRef]
7. Peng, X.; Zhe, H.; Guifang, G.; Gang, X.; Binggang, C.; Zengliang, L. Driving and control of torque for direct-wheel-driven electric vehicle with motors in serial. *Expert Syst. Appl.* **2011**, *38*, 80–86. [CrossRef]
8. Wang, J.; Wang, Q.; Jin, L.; Song, C. Independent wheel torque control of 4WD electric vehicle for differential drive assisted steering. *Mechatronics* **2011**, *21*, 63–76. [CrossRef]
9. Lee, J.; Suh, S.; Whon, W.; Kim, C.; Han, C. System Modeling and Simulation for an In-wheel Drive Type 6 × 6 Vehicle. *Trans. KSAE* **2011**, *19*, 1–11.
10. Fujimoto, H.; Saito, T.; Tsumasaka, A.; Noguchi, T. Motion Control and Road Condition Estimation of Electric Vehicles with Two In-wheel Motors. In Proceedings of the 2004 IEEE International Conference on Control Applications, Taipei, Taiwan, 2–4 September 2004; pp. 1266–1271.
11. Park, J.; Choi, J.; Song, H.; Hwang, S. Study of Driving Stability Performance of 2-Wheeled Independently Driven Vehicle Using Electric Corner Module. *Trans. Korean Soc. Mech. Eng. A* **2013**, *37*, 937–943. [CrossRef]
12. Gu, J.; Ouyang, M.; Lu, D.; Li, J.; Lu, L. Energy Efficiency Optimization of Electric Vehicle Driven by in-Wheel Motors. *Int. J. Automot. Technol.* **2013**, *14*, 763–772. [CrossRef]
13. Lin, C.; Cheng, X. A Traction Control Strategy with an Efficiency Model in a Distributed Driving Electric Vehicle. *Sci. World J.* **2014**, 261085. [CrossRef]
14. Chen, Y.; Wang, J. Adaptive Energy-Efficient Control Allocation for Planar Motion Control of Over-Actuated Electric Ground Vehicles. *IEEE Trans. Control Syst. Technol.* **2014**, *22*, 1362–1373.
15. Chen, Y.; Wang, J. Design and Experimental Evaluations on Energy Efficient Control Allocation Methods for Overactuated Electric Vehicles: Longitudinal Motion Case. *IEEE-ASME Trans. Mechatron.* **2014**, *19*, 538–548. [CrossRef]

16. Xu, G.; Li, W.; Xu, K.; Song, Z. An Intelligent Regenerative Braking Strategy for Electric Vehicles. *Energies* **2011**, *4*, 1461–1477. [CrossRef]
17. Singh, K.; Taheri, S. Estimation of Tire–road Friction Coefficient and its Application in Chassis Control Systems. *Syst. Sci. Control Eng.* **2015**, *3*, 39–61. [CrossRef]
18. Bera, T.; Samantaray, A. Mechatronic Vehicle Braking Systems. *Mechatron. Innov. Appl.* **2012**, 3–36. [CrossRef]
19. Ko, S. A Study on Estimation of Core Parameters and Integrated Co-operative Control Algorithm for an In-wheel Electric Vehicle. Ph.D. Thesis, Sungkyunkwan University, Jongno-gu, Seoul, 2014.
20. Oudghiri, M.; Chadli, M.; Hajjaji, A. Robust Fuzzy Sliding Mode Control for Antilock Braking System. *Int. J. Sci. Tech. Autom. Control* **2007**, *1*, 13–28.
21. Bakker, E.; Pacejka, H.; Lidner, L. A New Tire Model with an Applicationin Vehicle Dynamics Studies. *SAE Trans. J. Passeng. Cars* **1989**, *98*, 101–113.
22. Thomas, D.G. *Fundamentals of Vehicle Dynamics*; SAE: Warrendale, PA, USA, 2009.
23. Zadeh, L. Fuzzy Sets. *Inf. Control* **1965**, *8*, 338–353. [CrossRef]
24. Park, J.; Jeong, H.; Jang, I.; Hwang, S. Torque Distribution Algorithm for an Independently Driven Electric Vehicle using a Fuzzy Control Method. *Energies* **2015**, *8*, 8537–8561. [CrossRef]

Article

Sensorless PMSM Drive Implementation by Introduction of Maximum Efficiency Characteristics in Reference Current Generation

Željko Plantić [1], Tine Marčič [2], Miloš Beković [3] and Gorazd Štumberger [3],*

[1] Energy Institute Hrvoje Požar, 10001 Zagreb, Croatia; zplantic@eihp.hr
[2] Energy Agency of the Republic of Slovenia, 2000 Maribor, Slovenia; tine.marcic@agen-rs.si
[3] Power engineering institute, Faculty of Electrical Engineering and Computer Science; University of Maribor, 2000 Maribor, Slovenia; milos.bekovic@um.si
* Correspondence: gorazd.stumberger@um.si; Tel.: +386-2220-7075

Received: 31 July 2019; Accepted: 5 September 2019; Published: 11 September 2019

Abstract: This paper presents the efficiency improvement in a speed closed-loop controlled permanent magnet synchronous machine (PMSM) sensorless drive. The drive efficiency can be improved by minimizing the inverter and the PMSM losses. These can be influenced by proper selection of DC-bus voltage and switching frequency of the inverter. The direct (d-) and quadrature (q-) axis current references generation methods, discussed in this paper, further improve the efficiency of the drive. Besides zero d-axis current reference control, the maximum torque per ampere (MTPA) characteristic is normally applied to generate the d- and q-axis current references in vector controlled PMSM drives. It assures control with maximum torque per unit of current but cannot assure maximum efficiency. In order to improve efficiency of the PMSM drive, this paper proposes the generation of d- and q-axis current references based on maximum efficiency (ME) characteristic. In the case study, the MTPA and ME characteristics are theoretically evaluated and determined experimentally by measurements on discussed PMSM drive. The obtained characteristics are applied for the d- and q-axis current references generation in the speed closed-loop vector controlled PMSM drive. The measured drive efficiency clearly shows that the use of ME characteristic instead of MTPA characteristic or zero d-axis current in the current references generation improves the efficiency of PMSM drive realizations with position sensor and without it—sensorless control.

Keywords: magnetic loss; maximum efficiency (ME) characteristic; maximum torque per ampere (MTPA) characteristic; modeling; permanent magnet synchronous machine (PMSM); sensorless control

1. Introduction

The energy savings during the entire exploitation time of an electric drive can be achieved only through a coordinated efficiency improvement of all components of the drive, including control algorithms. Nowadays, efficiency maximization is a normal design aspect for electric drives. Thus, more and more researchers are occupied with decreasing overall power losses of electric drives and simultaneously saving component and operational costs. In some applications, minimum loss is not only one of the design criteria, but is of paramount importance, e.g., in battery-powered road-vehicles [1,2] or even more importantly in autonomous aircrafts [3]. In such applications, even a small decrease of losses can have a significant impact on overall system performance in terms of ensuring adequate operating range. Due to its high efficiency, high power density and wide operating range of speed, the permanent magnet synchronous machine (PMSM) is used very often. By employment of a speed- or position-sensorless control, a more compact drive with less maintenance [4–6] can be obtained.

Besides designing a new machine, the efficiency of the drive can be improved by minimizing losses of the power inverter by proper selection of the DC-bus voltage and the switching frequency [7–9], and a proper control of the PMSM as well. Sensorless PMSM drive control algorithms vary in complexity and computational performance demanded by different target applications. The main difference between available sensorless control algorithms from the literature is that model-based methods [5,10–13] are more straightforward to implement but are not directly applicable at standstill, and therefore saliency-based methods [14–16] have been developed. Authors combine the aforementioned methods in various approaches [9,17–20], thus ensuring adequate sensorless performance from zero speed up to the field-weakening region.

In the case of vector controlled PMSM, a proper selection of the direct (d-) and quadrature (q-) axis current references can further improve drive efficiency. The simplest zero d-axis current reference control ($i_{dr} = 0$) is still employed in applications with surface mounted PMSMs [21], which exhibit similar values of d- and q-axis inductances. Although it is generally applicable, the maximum torque per ampere (MTPA) characteristic [10,15,19–27], which considers PMSM copper losses, is usually used for generation of current references in drives with interior PMSM, which exhibit different values of d- and q-axis inductances and thus produce a significant reluctance torque component in addition to the permanent magnet torque component. When MTPA is applied, the PMSM drive produces maximum torque per unit of stator current but cannot reach the maximum efficiency due to the neglected iron core losses. The maximum efficiency (ME) characteristic [21,23–28] considers copper losses as well as iron core losses of PMSM. When it is applied for generation of current references, the maximum efficiency of a PMSM drive can be achieved. The concept was first evaluated by Colby et al. [29] for a surface mounted PMSM. Later it was applied by Morimoto et al. [30] in control of an interior PMSM drive as well. More recently, Pohlenz et al. [23] and Cavallaro et al. [28] reported improved efficiencies by implementing ME in the control software. While applying ME, Hassan et al. [24] analyzed PMSM and inverter losses due to changed harmonic content. Ni et al. [21] developed an iron loss formula by introducing new coefficients. However, the needed detailed PMSM data might not always be available, as discussed later.

General forms of the MTPA characteristic and the ME characteristic are shown in Figure 1a, from where it is evident that ME characteristics require higher values of the negative d-axis current in comparison to the MTPA characteristics, and MTPA characteristics in turn require higher values of the negative d-axis current than the $i_{dr} = 0$ control. Figure 1b shows the stator current vector $\mathbf{I}_s$ and its components I_d and I_q, where $|\mathbf{I}_s|$ denotes the length of $\mathbf{I}_s$.

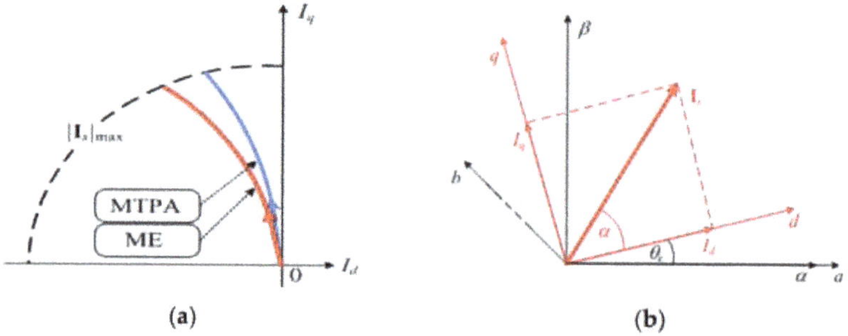

(a)   (b)

Figure 1. Maximum torque per ampere (MTPA) and maximum efficiency (ME) characteristics (**a**) and stator current vector in α-β and d-q reference frame (**b**).

In this paper, the origins of losses in a PMSM drive are introduced and evaluated in Section 2. The PMSM model is used to explain the differences between the MTPA and ME characteristics, and to obtain the equations which relate the ME characteristics to the stator currents I_d and I_q in Section 3. Although the ME characteristics are usually derived in relation to the magnetizing currents I_{md} and I_{mq}, the here presented Equation (20) provides the optimal current I_d for a given load torque by considering the ME characteristic-i.e., the copper and the iron losses of the PMSM. Furthermore, this paper focuses on the drive implementation from the control designer's point of view, where detailed information of the PMSM is not available, contrarily to machine designer's point of view, where all details of the PMSM are known [21]. Unfortunately, the aforementioned equations cannot be used directly in the control implementation because the needed iron core resistance parameter value R_c cannot be determined directly or to be modeled in form of a function. Therefore, this paper proposes an alternative, which is suitable for control design and applicable to all types of PMSM. There the ME characteristics are determined experimentally by measurements performed on the tested vector controlled PMSM drive. The existing control algorithm is extended by a current references generator, where the determined ME characteristics are applied to generate the d- and q-axis current references. The speed closed-loop vector control of PMSM, obtained in this way, is used to measure the efficiency of the entire drive. The control is realized with the position sensor and without it—sensorless control.

Moreover, in this work the ME characteristics are implemented in sensorless control for the first time and presented in Section 4. There the rotor speed and position are determined by nonlinear phase-locked loop (PLL) observer [9,31], whose inputs are calculated by the modified active flux method [32–35]. It is shown that the drive efficiency can be improved when ME characteristics are used to generate the d- and q-axis current references, which is true for the sensorless control as well as for the control realization with position sensor. In order to facilitate a direct comparison, the space vector modulation has been utilized without any advanced modulation techniques such as discussed in [36,37]. Section 5 then concludes the paper.

2. PMSM Drive Losses

A PMSM drive normally consists of the PMSM, power inverter, load, and the control system. The load is the element, which dictates the proper selection of the other PMSM drive elements. The energy conversion process inevitably includes losses, mostly in the power inverter and PMSM. These loss components substantially influence the drive efficiency. This paper focuses on the detailed research of PMSM losses, which are minimized in the speed closed-loop controlled PMSM drive by a proper generation of d- and q-axis current references.

The employed PMSM had surface mounted magnets and was rated for 400 V and 1.6 kW at 2250 rpm.

2.1. Power Losses in Power Inverter

The power losses in a power inverter consist of conduction or Ohm's losses and switching losses. In this work they are minimized by proper selection of the DC-bus voltage, where the impact of dead times is minimized [5,8,9] and by proper selection of the switching frequency [7,24,36,37] as well. The proper values were determined experimentally by measurements performed on the tested PMSM drive.

In order to evaluate the impact of switching frequency on drive efficiency, a series of tests was performed, where the PMSM load was varied, while the switching frequency was set to 4, 6, and 9 kHz, respectively. Figure 2 presents the PMSM drive efficiency characteristics in dependence of the current vector length $|i_s|$. Drive efficiencies vary because the inverter losses increase with the increase of switching frequency and simultaneously the PMSM losses decrease with the increase of switching frequency [38], and the drive efficiency represents their product. Figure 2 shows that the highest efficiency of this particular PMSM drive was obtained when the switching frequency of 6 kHz was employed. Thus, the aforementioned switching frequency value was used through all successive tests.

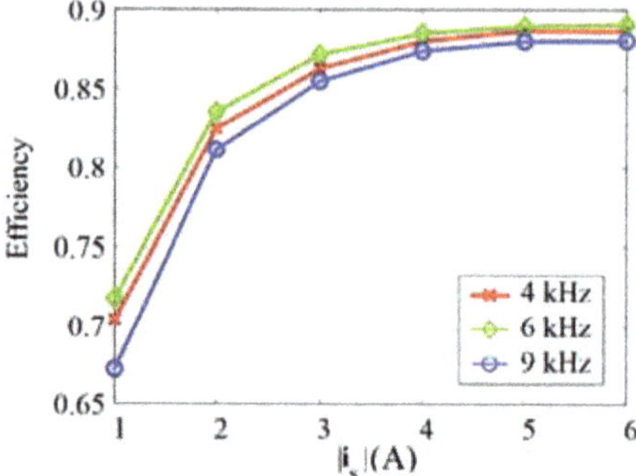

Figure 2. Measured efficiency characteristics of the permanent magnet synchronous machine (PMSM) drive at different switching frequencies.

In addition to the switching frequency, the inverter's DC-bus voltage influences the PMSM drive efficiency as well. When the desired inverter output voltage is much smaller than the DC-bus voltage, then the inverter output voltage pulses are unnecessarily narrow which increases switching losses. However, it is noted that the DC-bus voltage control is not possible if a diode rectifier is employed in the inverter. In order to evaluate the impact of DC-bus voltage on PMSM drive efficiency, a series of tests was performed, where the PMSM load was varied, while the rotor speed and DC-bus voltage were set to different values. The PMSM efficiency was not significantly affected by the change of DC-bus voltage. Thus, the significant variation of PMSM drive efficiency presented in Figure 3 was due to the inverter efficiency variation by the change of DC-bus voltage. The differences in efficiencies at low speeds were more expressed because the PMSM back-emf was smaller at low speeds. The implementation of variable DC-bus voltage could make sense in variable speed drives operating mostly at partial loads and in a broad range of speeds in applications where efficiency is crucial.

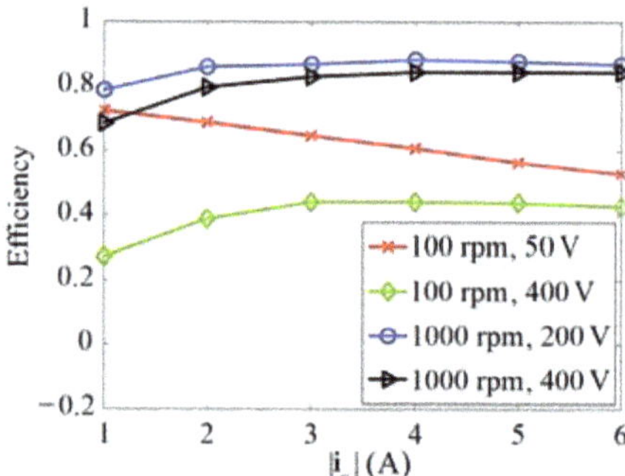

Figure 3. Measured efficiency characteristics of the PMSM drive at different rotor speed and different DC-bus voltage.

2.2. Power Losses in PMSM

The analysis of power losses in the PMSM is based on the two-axis steady-state PMSM model with lumped parameters, written in the d-q reference frame. The d-axis was aligned with the flux linkage vector due to the permanent magnets while the q-axis led for electrical $\pi/2$. The model is schematically shown in Figure 4.

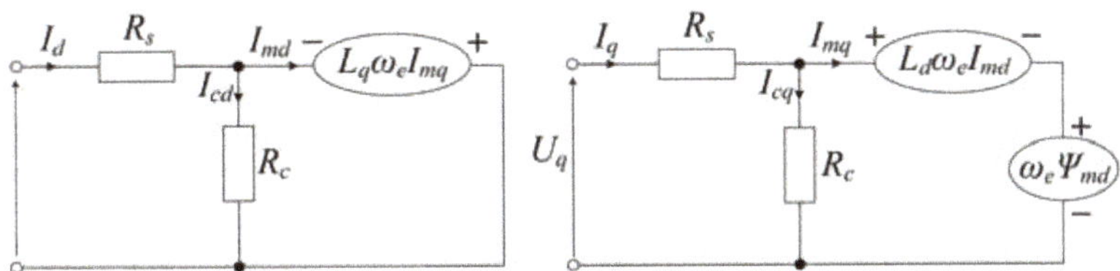

Figure 4. Applied steady-state model of PMSM in d-q reference frame.

The copper losses are considered by the stator resistance R_s while the iron core losses are accounted for by the iron core resistance R_c. In the steady-state model, written in the d-q reference frame, the indices d and q refer to the d- and q-axis, U_d and U_q are the stator voltages, I_d and I_q are stator currents, I_{md} and I_{mq} are the magnetizing currents, I_{cd} and I_{cq} are the iron core currents, L_d and L_q are the stator inductances, Ψ_{md} is the flux linkage due to the permanent magnets, T_e is the electromagnetic torque, ω_e is the electrical rotor speed and p_p is the number of pole pairs. Considering Figure 4, the voltage Equation (1) and the torque Equation (2) can be written:

$$U_d = R_s I_d - \omega_e L_q I_{mq}$$
$$U_q = R_s I_q + \omega_e L_d I_{md} + \omega_e \Psi_{md} \tag{1}$$

$$T_e = p_p I_{mq}\left(\Psi_{md} + \left(L_d - L_q\right)I_{md}\right) \tag{2}$$

where

$$I_d = I_{md} + I_{cd} = I_{md} - \frac{\omega_e L_q I_{mq}}{R_c}$$
$$I_q = I_{mq} + I_{cq} = I_{mq} + \frac{\omega_e}{R_c}\left(L_d I_{md} + \Psi_{md}\right) \tag{3}$$

The power losses in the $P_{MSM\,Ploss,PMSM}$ can be divided into controllable losses, which are the copper losses $P_{loss,Cu}$ and the iron core losses $P_{loss,Fe}$, and uncontrollable losses, which are the mechanical and additional losses [28]. Neglecting the uncontrollable losses, (4) describes the PMSM losses.

$$P_{loss,PMSM} = P_{loss,Cu} + P_{loss,Fe} \tag{4}$$

Considering Figure 4 and Equations (1) to (4), $P_{loss,Cu}$ and $P_{loss,Fe}$ can be expressed by (5) and (6).

$$P_{loss,Cu} = R_s\left(I_d^2 + I_q^2\right) = R_s\left(I_{md} - \frac{\omega_e L_q}{R_c}I_{mq}\right)^2 + R_s\left(I_{mq} + \frac{\omega_e\left(L_d I_{md} + \Psi_{md}\right)}{R_c}\right)^2 \tag{5}$$

$$P_{loss,Fe} = R_c\left(I_{cd}^2 + I_{cq}^2\right) = \frac{\omega_e^2}{R_c}\left[\left(L_q I_{mq}\right)^2 + \left(L_d I_{md} + \Psi_{md}\right)^2\right], \tag{6}$$

3. Improving Efficiency by Reference Current Generation

Assuming a given inverter and a given PMSM, the PMSM drive efficiency can be improved by minimizing inverter losses and PMSM losses. In addition to advanced modulation techniques [36,37], the inverter losses can be mostly influenced by the closed-loop control or a proper choice of DC-bus voltage and by an optimal choice of switching frequency [24]. The PMSM losses can be influenced by the voltage and current total harmonic distortion that depend on the switching frequency [7,24,38], and in the case of vector controlled PMSM by the generation of the d- and q-axis current references. The generation of current references is usually based $i_{dr} = 0$ or on the MTPA characteristic [10,15,19–23], which assures the maximum torque per unit of current but not the maximum efficiency of the entire drive [21,23]. The last can be achieved when the ME characteristic is applied for generation of the d- and q-axis current references.

3.1. Theoretical Background of the MTPA and ME Characteristics

Let us consider all parameters of the PMSM steady-state model, shown in Figure 4 and described by Equations (1) to (6), as constant. The starting point for derivation of the MTPA characteristic is the torque equation (Equation (2)) and the PMSM steady-state model shown in Figure 4, where the iron core resistance is neglected ($R_c \to \infty$). Consequently, according to Equation (3), I_{cd} and I_{cq} equal zero which leads to $I_d = I_{md}$ and $I_q = I_{mq}$. Considering the angle of the stator current vector α, shown in Figure 1b, the relation between the d- and q-axis currents, that provides the maximum torque per unit of current, can be determined from Equation (2) by Equation (7), which yields Equation (8) [22].

$$\max(T_e) \Rightarrow \frac{dT_e}{d\alpha} = 0 \tag{7}$$

$$I_{d_{1,2}} = \frac{-\Psi_{md}}{2(L_d - L_q)} \pm \sqrt{\frac{\Psi_{md}^2}{4(L_d - L_q)^2} + I_q^2} \tag{8}$$

Similarly, the ME characteristic describes the relation between the d- and q-axis current that provides the minimum of PMSM losses, which leads to the maximum efficiency of the PMSM drive. According to [24,28], the PMSM losses (Equation (4)) can be minimized for each combination of T_e, and ω_e. Considering Equation (2) in Equations (5) and (6), Equation (4) can be expressed as a function of I_{md}, T_e, and ω_e. In the case of steady-state operation, T_e, and ω_e are constants, which means that the PMSM power losses $P_{loss,PMSM}$ (Equation (4)) can be described as a function of I_{md}. However, because T_e (Equation (2)) depends on I_{md} and I_{mq}, the PMSM power losses can be minimized by Equation (9), which yields Equation (10) [24] describing the ME characteristic.

$$\min(P_{loss,PMSM}) \Rightarrow \left.\frac{dP_{loss,PMSM}(I_{md}, I_{mq}, \omega_e)}{dI_{md}}\right|_{T_e, \omega_e} = 0 \tag{9}$$

$$I_{md} = -\frac{\omega_e^2 L_d \Psi_{md}(R_s + R_c) + R_s R_c \omega_e (L_d - L_q) I_{mq}}{\omega_e^2 L_d^2 (R_s + R_c) + R_s R_c^2} \tag{10}$$

3.2. Evaluation of the Impact of Rc on the ME Characteristic

A sensitivity analysis was performed in order to evaluate the impact of the iron core resistance R_c on the ME characteristic. The parameter R_c was varied while the other PMSM parameters presented in Table 1 remained constant. The theoretical derivation of obtaining the necessary measurable current I_d for every corresponding combination of I_q and ω_e is presented hereinafter with the assumption of constant PMSM model parameters in the given operating points.

Table 1. Permanent magnet synchronous machine PMSM parameters.

Parameter	Value
p_p	5
R_s (Ω)	1.15
Ψ_{md} (Vs)	0.2415
L_d (H)	0.02654
L_q (H)	0.02865

The voltage equations derived from Figure 4 can be rewritten also in the form of Equation (11).

$$\begin{aligned} U_d &= R_s I_{md} - \frac{R_s + R_c}{R_c} \omega_e L_q I_{mq} \\ U_q &= R_s I_{mq} + \frac{R_s + R_c}{R_c} (\omega_e L_d I_{md} + \omega_e \Psi_{md}) \end{aligned} \tag{11}$$

Considering Equation (3), the currents I_{md} and I_{mq} can be expressed in terms of the measurable stator currents I_d and I_q (Equation (12)). Note that it is essential for control implementation that measurable quantities (such as I_d and I_q instead of I_{md} and I_{mq}) are included in the algorithm.

$$\begin{aligned}
I_{md}(I_d, I_q) &= \frac{R_c^2 I_d + R_c L_q \omega_e I_q - L_q \Psi_{md}\omega_e^2}{R_c^2 + L_d L_q \omega_e^2}\\[4pt]
I_{mq}(I_d, I_q) &= \frac{R_c^2 I_q - R_c L_d \omega_e I_d - R_c \Psi_{md}\omega_e}{R_c^2 + L_d L_q \omega_e^2}
\end{aligned}$$

$$(12)$$

Similarly, the currents I_{cd} and I_{cq} can be expressed in terms of I_d and I_q (13).

$$\begin{aligned}
I_{cd}(I_d, I_q) &= \frac{L_d L_q \omega_e^2 I_d - R_c L_q \omega_e I_q + L_q \Psi_{md}\omega_e^2}{R_c^2 + L_d L_q \omega_e^2}\\[4pt]
I_{cq}(I_d, I_q) &= \frac{R_c L_d \omega_e I_d + L_d L_q \omega_e^2 I_q + R_c \Psi_{md}\omega_e}{R_c^2 + L_d L_q \omega_e^2}
\end{aligned}$$

$$(13)$$

Then the PMSM losses can be expressed in terms of I_d and I_q as well (14).

$$\begin{aligned}
P_{loss,PMSM} &= A I_d^2 + B I_q^2 + C I_d + D I_q + E I_d I_q + F\\[4pt]
A &= \frac{L_d^2 L_q^2 \omega_e^4 (R_s + R_c) + R_c^2 \left(R_s R_c^2 + 2 R_s L_d L_q \omega_e^2 + R_c L_d^2 \omega_e^2\right)}{\left(R_c^2 + L_d L_q \omega_e^2\right)^2}\\[4pt]
B &= \frac{L_d^2 L_q^2 \omega_e^4 (R_s + R_c) + R_c^2 \left(R_s R_c^2 + 2 R_s L_d L_q \omega_e^2 + R_c L_q^2 \omega_e^2\right)}{\left(R_c^2 + L_d L_q \omega_e^2\right)^2}\\[4pt]
C &= \frac{2 R_c L_d \Psi_{md} \omega_e^2 \left(L_q^2 \omega_e^2 + R_c^2\right)}{\left(R_c^2 + L_d L_q \omega_e^2\right)^2}\\[4pt]
D &= \frac{2 R_c^2 L_q \Psi_{md} \omega_e^3 \left(L_d - L_q\right)}{\left(R_c^2 + L_d L_q \omega_e^2\right)^2}\\[4pt]
E &= \frac{2 R_c^2 L_d L_q \omega_e^3 \left(L_d - L_q\right)}{\left(R_c^2 + L_d L_q \omega_e^2\right)^2}\\[4pt]
F &= \frac{R_c \Psi_{md}^2 \omega_e^2 \left(L_q^2 \omega_e^2 + R_c^2\right)}{\left(R_c^2 + L_d L_q \omega_e^2\right)^2}
\end{aligned}$$

$$(14)$$

If we consider Equation (15) in Equation (14), then Equation (16) is obtained.

$$\begin{aligned}
I_d &= I_s \cos\alpha\\
I_q &= I_s \sin\alpha
\end{aligned}$$

$$(15)$$

$$P_{loss,PMSM} = A I_s^2 \cos^2\alpha + B I_s^2 \sin^2\alpha + C I_s \cos\alpha + D I_s \sin\alpha + E I_s^2 \sin\alpha \cos\alpha + F \tag{16}$$

The PMSM losses are minimized by Equation (17), yielding Equation (18).

$$\frac{dP_{loss,PMSM}}{d\alpha} = 0 \tag{17}$$

$$\begin{aligned}
&-2 A I_s^2 \sin\alpha \cos\alpha + 2 B I_s^2 \sin\alpha \cos\alpha - C I_s \sin\alpha +\\
&+ D I_s \cos\alpha + E I_s^2 \cos^2\alpha - E I_s^2 \sin^2\alpha = 0\\
&-2 A I_d I_q + 2 B I_d I_q - C I_q + D I_d + E I_d^2 - E I_q^2 = 0
\end{aligned}$$

$$(18)$$

By rearranging Equation (18) we obtain Equation (19) and consequently Equation (20) which provides the necessary current I_d for every corresponding combination of I_q and ω_e.

$$0 = E I_d^2 + \left(D + 2 I_q (B - A)\right) I_d - \left(E I_q^2 + C I_q\right) \tag{19}$$

$$I_{d1,2} = \frac{-D + 2 I_q (A - B)}{2E} \pm \frac{\sqrt{\left(D + 2 I_q (B - A)\right)^2 + 4E\left(E I_q^2 + C I_q\right)}}{2E} \tag{20}$$

Figure 5 presents the obtained MTPA characteristic by Equation (8), and the obtained ME characteristics by Equation (20) at different values of rotor speed and different values of the iron core resistance R_c. Although, this may not be evident from Figure 5 and similarly to the MTPA characteristics, the ME characteristics are nonlinear. It is however evident from Figure 5 that the iron core resistance R_c plays a crucial role in shaping the ME characteristic. Unfortunately, in practice its value changes depending on operating conditions and it cannot be determined directly or to be given in form of a function.

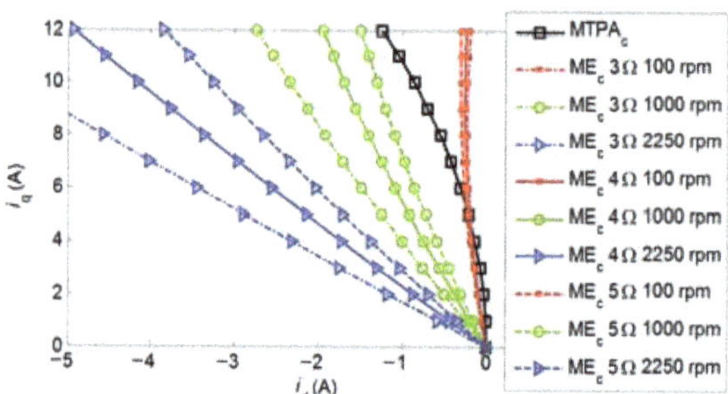

Figure 5. Calculated MTPA and ME characteristics at different rotor speeds and different constant values of the iron core resistance R_c.

Therefore, the authors have turned to an implicit but general approach for determination of the ME characteristic based on experiments, which is described hereinafter. The determined ME characteristics are then used to improve the efficiency of the PMSM drive in sensorless control as well as for the control realization with position sensor. The approach proposed here is novel, and can easily be adopted by control designers when of-the-shelf (or unknown) PMSMs are employed—contrarily to methods based, e.g., on finite element analyses which depend on detailed information about the employed geometry and materials in the PMSM.

3.3. Experimental Determination of ME Characteristics

The ME characteristics derived theoretically in the previous subsections are based on the steady-state PMSM model with constant parameters. With the aim to improve the PMSM drive efficiency and to evaluate the previously presented results, the characteristics were determined also by measurements performed on the PMSM drive, schematically shown in Figure 6. The presented approach includes all effects that are also present during real drive operation in the ME characteristics, these being effects of magnetic nonlinearities, saturation, cross-coupling, end-winding leakage, permanent magnets and slotting, converter nonlinearities and dead times in transistor switching. The PMSM was torque controlled and fed by the inverter. This means that during all tests the d- and q-axis currents were close-loop controlled, following their references set to different constant values. The vector control of PMSM was realized on the control system dSPACE PPC 1103. The rotor speed was closed-loop controlled by an active load IM. The input and output power of the inverter and PMSM, and their efficiencies, were measured by the power analyzers NORMA I (Norma D5255M) and NORMA II (Norma D6100), as shown Figure 6. The equal equipment was used throughout all successive tests, thus enabling a direct and comprehensive comparison of the PMSM drive efficiency operating in closed-loop control with or without the position sensor.

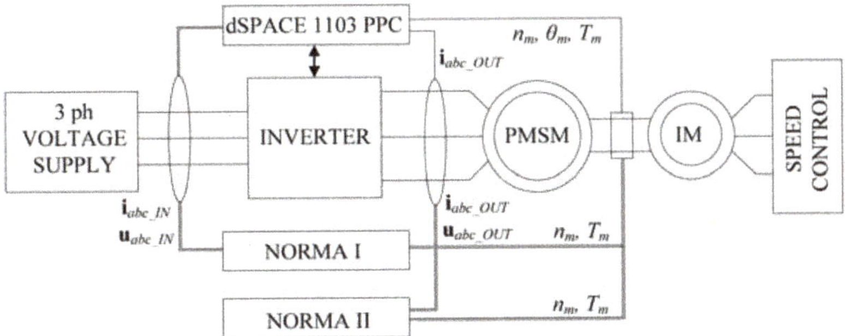

Figure 6. Experimental set-up—schematic presentation.

The ME characteristics were determined by changing the length and angle of the stator current vector at different values of the rotor speed. In individual steady-states, based on the measurements of voltages, currents, powers, torque and speed, the efficiencies are determined. A three-dimensional matrix of measurement results was generated, which includes length and angle of the stator current vector as well as the rotation speed. The experimentally obtained drive efficiency curves at 2250 rpm in dependence of the length and angle of the stator current vector are shown in Figure 7. The ME characteristic at 2250 rpm was then determined by connecting the points of maximum efficiencies of respective efficiency curves (black line). The determined ME characteristics at different rotor speeds are then shown in Figure 8, where they are presented in dependence of d- and q-axis currents. From Figure 8 it is evident that if we want to control the drive with maximum efficiency, the d-axis current component has to be increased by increase of the rotor speed, which is also known as flux-weakening. Unfortunately, due to hardware limitations, only currents of up to 6 A could be achieved during the experiments.

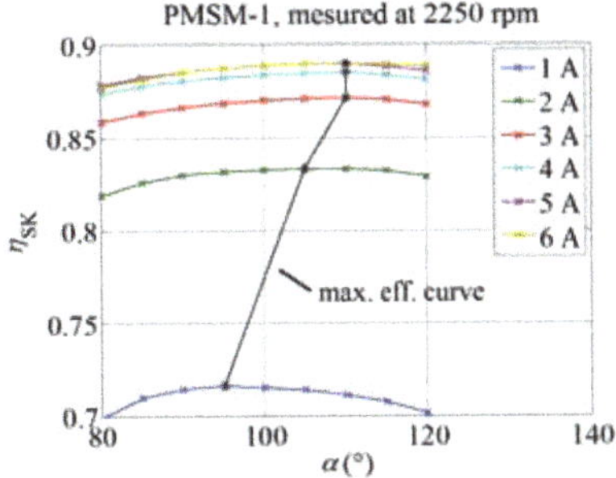

Figure 7. ME characteristic at 2250 rpm in dependence of the length and angle of the stator current vector.

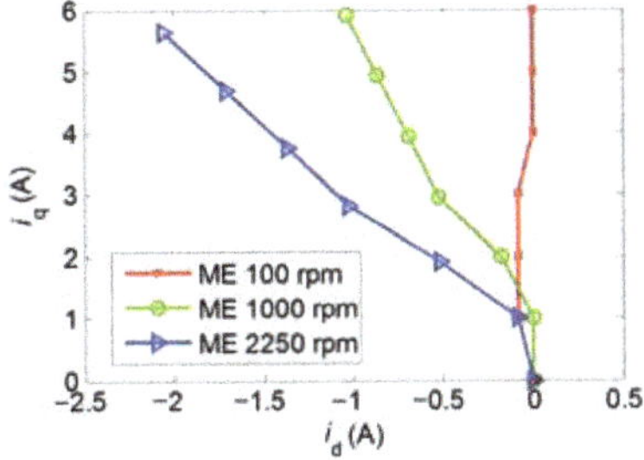

Figure 8. Experimentally determined ME characteristics.

4. Results and Discussion

This section shows how the employment of the previously experimentally determined ME characteristics influences the efficiency of the discussed speed closed-loop controlled PMSM drive, when it is used to generate the d- and q-axis current references. The rotor position and speed required for the control realization are obtained in two different ways: firstly, by using sensors and secondly, by the sensorless control. In the last case, the measured PMSM currents and DC-bus voltage are used together with the PMSM dynamic model to calculate the rotor position and speed. The active flux based sensorless control method is applied. The active flux Ψ_{AF} is defined as a hypothetical flux, which multiplied by the instantaneous q-axis current i_q produces the measured instantaneous electro-magnetic torque t_e [32–35]. Consequently, the effects which are neglected in the PMSM model are thus considered by employment of Ψ_{AF}.

$$t_e = p_p \left| \hat{\mathbf{\Psi}}_{AF} \right| i_q \tag{21}$$

The active flux concept uses the flux linkage vector estimates $\hat{\mathbf{\Psi}}_{\alpha\beta u}$ and $\hat{\mathbf{\Psi}}_{\alpha\beta i}$ obtained by the voltage (index u) and current models (index i) (Equation (22)), written in the α-β reference frame [32]. The difference between both estimates gives the compensation voltage u_k, which is added in the voltage model to eliminate the DC-offset and other errors in measured stator currents, variation of stator resistance, and integrator drifts in the calculation procedure.

$$
\begin{aligned}
\hat{\mathbf{\Psi}}_{\alpha\beta u} &= \int_0^t \left(\mathbf{u}_{\alpha\beta}(\tau) - R_s \mathbf{i}_{\alpha\beta}(\tau) + u_k(\tau) \right) d\tau + \hat{\mathbf{\Psi}}_{\alpha\beta u}(0) \\
\hat{\mathbf{\Psi}}_{\alpha\beta i} &= \mathbf{L}_{\alpha\beta} \begin{bmatrix} 1 & 0 \\ 0 & 1 \end{bmatrix} \mathbf{i}_{\alpha\beta} + \mathbf{\Psi}_{md} \begin{bmatrix} \cos\theta_e \\ \sin\theta_e \end{bmatrix}
\end{aligned}
\tag{22}
$$

Considering zero initial conditions, the required estimates of the electrical rotor speed $\hat{\omega}_e$ and position $\hat{\theta}_e$ are calculated by the nonlinear PLL-observer (Equation (23)) [31], where ε represents the error signal; k_1 and k_2 represent the observer gains.

$$
\begin{aligned}
\hat{\omega}_e &= \int_0^t k_1 \varepsilon(\tau) d\tau \\
\hat{\theta}_e &= \int_0^t \left(\hat{\omega}_e(\tau) + k_2 \varepsilon(\tau) \right) d\tau
\end{aligned}
\tag{23}
$$

Thus, in the case of sensorless control, the estimates $\hat{\omega}_e$ and $\hat{\theta}_e$ (Equation (23)) are used instead measured speed and position in the control algorithm, enabling closed loop-speed control without mechanical speed sensor.

Figure 9 shows the schematic presentation of the speed closed-loop control realizations with sensors (position and speed) and without them. These control realizations are applied in the experimental set-up, shown in Figure 6, which is used to measure drive efficiencies in cases when $i_{dr} = 0$, MTPA and ME characteristics, respectively. These characteristics (MTPA or ME) are used to determine the d- and q-axis current references i_{dr} and i_{qr}.

Figure 10 shows the measured PMSM drive efficiencies (η_{ALL}) at the rated speed of 2250 rpm and different loads in the case when $i_{dr} = 0$ and in the cases when i_{dr} and i_{qr} are determined with the MTPA and ME characteristics, respectively. The results are given for the control realization with position and speed sensors and for the sensorless control. They are completed by the presentation of differences between different measured efficiency curves.

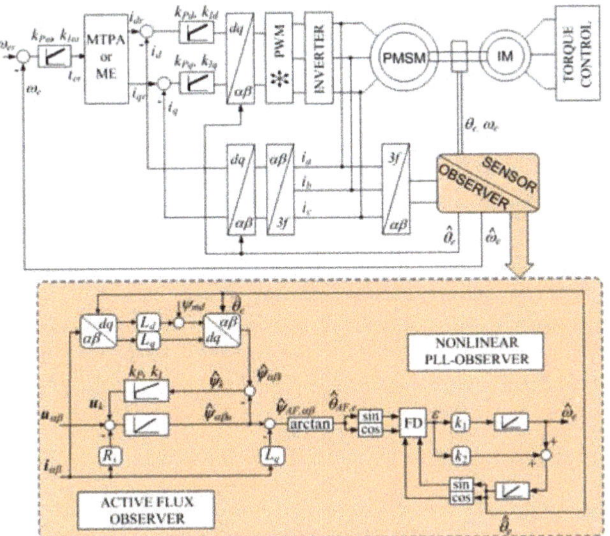

Figure 9. Schematic presentation of the applied speed closed-loop control.

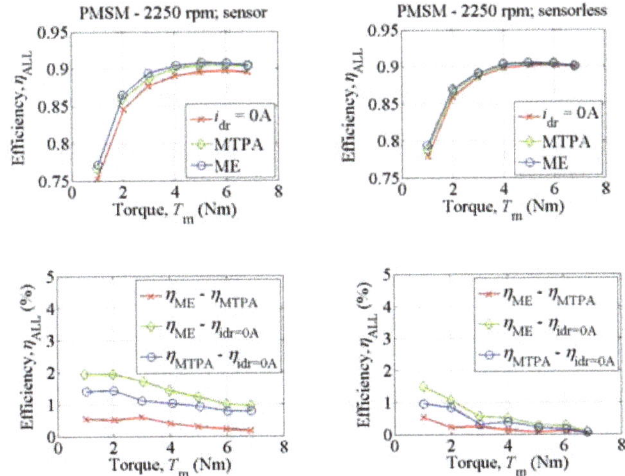

Figure 10. Measured efficiencies of the speed closed-loop controlled PMSM drive and their differences for the control realization with the mechanical sensor and for the sensorless control in the cases when i_{dr} and i_{qr} are determined with MTPA and ME characteristics and in the case when $i_{dr} = 0$.

The presented results clearly show that the lowest efficiencies are achieved when $i_{dr} = 0$. The drive efficiency is improved when the MTPA characteristic is used for the generation of i_{dr} and i_{qr}. When it is replaced with the ME characteristic, the drive efficiency is further improved. Similar findings have been reported in [21,23,28] for an IPMSM drive with position sensor. The efficiency improvements decrease with the increase of the load torque (T_m). This is true for the control realization with the position and speed sensors as well as for the sensorless control realization.

5. Conclusions

This paper presents a way of improving the PMSM drive efficiency, where the more usual $i_{dr} = 0$ and MTPA characteristic for generation of reference current in vector controlled PMSM drives are compared with the proposed ME characteristic. There, both copper and iron core losses of the PMSM are considered. The equations which relate the ME characteristics to the measurable stator currents I_d and I_q which are needed by control designers (instead to the magnetizing currents I_{md} and I_{mq}) are derived, and consequently the equation which provides the optimal current I_d for a given load torque by considering the ME characteristic is obtained for the first time. The problem of adequately determining the iron core resistance parameter value R_c is effectively mitigated by employing experimentally determined ME characteristics on the real PMSM drive system. Differences between the MTPA and ME characteristics are detailed; they are used in speed closed-loop vector control realizations with position and speed sensors and without them. The ME characteristic is implemented in sensorless control for the first time, where a nonlinear PLL observer is applied. Its inputs are calculated by the modified active flux method. The work clearly demonstrates that the use of ME characteristic instead of MTPA characteristic or $i_{dr} = 0$ in the current references generation improves the efficiency of PMSM drive realizations with position and speed sensors and without them—sensorless control.

This particular comparative case study showed that choosing an adequate switching frequency might improve the drive efficiency by 1.5–5.0%. At small load and speed values where the voltage value is small as well, an adequate DC-bus voltage control may improve the drive efficiency up to 30%. Ultimately, this particular PMSM drive efficiency was improved by 0.1–0.5% when the proposed ME was implemented in comparison to the MTPA, and by 1.0–2.0% in comparison to $i_{dr} = 0$, respectively. Therefore, it should be preferred that the PMSM drive control solutions proposed here are implemented in applications where efficiency is crucial. Moreover, the ME implementation can cut down the annual energy consumption of the studied drive by up to approximately 130 kWh, depending on the load and control case presented in Figure 11. This fact certainly cannot be neglected, since its implementation requires merely the modification of a few lines of code in the control realization.

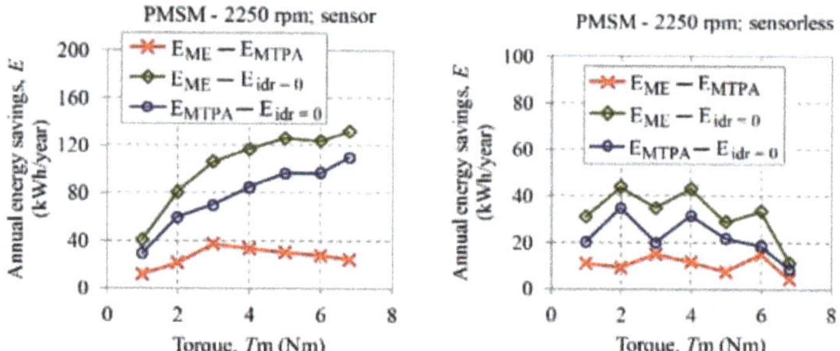

Figure 11. Annual energy savings achieved by different methods of reference current generation in the speed closed-loop controlled PMSM drive for the realization with the mechanical sensor and for the sensorless control.

Author Contributions: Ž.P. performed derivations, experimental work and analyses. T.M. and M.B. prepared the final form of the paper, including editing of the text, equations and figures. G.Š. prepared the experiments and coordinated the research work and paper preparation.

Funding: This work was supported by the Slovenian Research Agency [grant numbers J2-1742, P2-0115].

Conflicts of Interest: The authors declare no conflict of interest.

References

1. Yilmaz, M. Limitations/capabilities of electric machine technologies and modeling approaches for electric motor design and analysis in plug-in electric vehicle applications. *Renew. Sustain. Energy Rev.* **2015**, *52*, 80–99. [CrossRef]
2. Wang, E.; Guo, D.; Yang, F. System design and energetic characterization of a four-wheel-driven series-parallel hybrid electric powertrain for heavy-duty applications. *Energy Convers. Manag.* **2015**, *106*, 1264–1275. [CrossRef]
3. Gao, X.Z.; Houn, Z.X.; Guo, Z.; Chen, X.Q. Reviews of methods to extract and store energy for solar-powered aircraft. *Renew. Sustain. Energy Rev.* **2015**, *44*, 96–108.
4. Daili, Y.; Gaubert, J.P.; Rahmani, L. Implementation of a new maximum power point tracking control strategy for small wind energy conversion systems without mechanical sensors. *Energy Convers. Manag.* **2015**, *97*, 298–306. [CrossRef]
5. Finch, J.W.; Giaouris, D. Controlled AC electrical drives. *IEEE Trans. Ind. Electron.* **2008**, *55*, 481–491. [CrossRef]
6. Chaoui, H.; Khayamy, M.; Okoye, O. MTPA based operation point speed tracking for PMSM drives without explicit current regulation. *Electr. Power Syst. Res.* **2017**, *151*, 125–135. [CrossRef]
7. Ancuti, R.; Boldea, I.; Andreescu, G.D. Sensorless V/f control of high-speed surface permanent magnet synchronous motor drives with two novel stabilising loops for fast dynamics and robustness. *IET Electr. Power Appl.* **2010**, *4*, 149–157. [CrossRef]
8. Krüner, S.; Hackl, C.M. Experimental Identification of the Optimal Current Vectors for a Permanent-Magnet Synchronous Machine in Wave Energy Converters. *Energies* **2019**, *12*, 862. [CrossRef]
9. Wang, G.; Yang, R.; Xu, D. DSP-based control of sensorless IPMSM drives for wide-speed-range operation. *IEEE Trans. Ind. Electron.* **2013**, *60*, 720–727. [CrossRef]
10. Chi, W.C.; Cheng, M.Y. Implementation of a sliding-mode-based position sensorless drive for high-speed micro permanent-magnet synchronous motors. *ISA Trans.* **2014**, *53*, 444–453. [CrossRef]
11. Toman, M.; Dolinar, D.; Štumberger, G. Sensorless control of the permanent magnet synchronous motor. *Electrotech. Rev.* **2004**, *71*, 277–282.
12. Kung, Y.S.; Thanh, N.P.; Wang, M.S. Design and simulation of a sensorless permanent magnet synchronous motor drive with microprocessor-based PI controller and dedicated hardware EKF estimator. *Appl. Math. Model.* **2015**, *39*, 5816–5827. [CrossRef]
13. Lin, S.; Zhang, W. An adaptive sliding-mode observer with a tangent function-based PLL structure for position sensorless PMSM drives. *Electr. Power Energy Syst.* **2017**, *88*, 63–74. [CrossRef]
14. Rostami, A.; Asaei, B. A novel method for estimating the initial rotor position of PM motors without the position sensor. *Energy Convers. Manag.* **2009**, *50*, 1879–1883. [CrossRef]
15. Zhu, Z.Q.; Gong, L.M. Investigation of effectiveness of sensorless operation in carrier-signal-injection-based sensorless-control methods. *IEEE Trans. Ind. Electron.* **2011**, *58*, 3431–3439. [CrossRef]
16. Wu, X.; Huang, S.; Liu, P.; Wu, T.; He, Y.; Zhang, X.; Chen, K.; Wu, Q. A reliable initial rotor position estimation method for sensorless control of interior permanent magnet synchronous motors. *ISA Trans.* **2019**, in press. [CrossRef] [PubMed]
17. Song, Z.; Hou, Z.; Jiang, C.; Wei, X. Sensorless control of surface permanent magnet synchronous motor using a new method. *Energy Convers. Manag.* **2006**, *47*, 2451–2460. [CrossRef]
18. Omrane, I.; Etien, E.; Dib, W.; Bachelier, O. Modeling and simulation of soft sensor design for real-time speed and position estimation of PMSM. *ISA Trans.* **2015**, *57*, 329–339. [CrossRef] [PubMed]
19. Foo, G.; Rahman, M.F. Sensorless sliding-mode MTPA control of an IPM synchronous motor drive using a sliding-mode observer and HF signal injection. *IEEE Trans. Ind. Electron.* **2010**, *57*, 1270–1278. [CrossRef]
20. Chen, S.Y.; Luo, Y.; Pi, Y.G. PMSM sensorless control with separate control strategiesand smooth switch from low speed to high speed. *ISA Trans.* **2015**, *58*, 650–658. [CrossRef] [PubMed]
21. Ni, R.; Xu, D.; Wang, G.; Ding, L.; Zhang, G.; Qu, L. Maximum efficiency per ampere control of permanent-magnet synchronous machines. *IEEE Trans. Ind. Electron.* **2015**, *62*, 2135–2143. [CrossRef]
22. Pan, C.T.; Sue, S.M. A linear maximum torque per ampere control for IPMSM drives over full-speed range. *IEEE Trans. Energy Convers.* **2005**, *20*, 359–366. [CrossRef]

23. Pohlenz, D.; Böcker, J. Efficiency improvement of an IPMSM using maximum efficiency operating strategy. In Proceedings of the 14th International Power Electronics and Motion Control Conference (EPE/PEMC), Ohrid, Macedonia, 6–8 September 2010; pp. T5-15–T5-19.

24. Hassan, W.; Wang, B. Efficiency optimization of PMSM based drive system. In Proceedings of the 7th International Power Electronics and Motion Control Conference (IPEMC), Harbin, China, 2–5 June 2012; pp. 1027–1033.

25. Wang, M.S.; Hsieh, M.F.; Lin, H.Y. Operational Improvement of Interior Permanent Magnet Synchronous Motor Using Fuzzy Field-Weakening Control. *Electronics* **2018**, *7*, 452. [CrossRef]

26. Yu, L.; Zhang, Y.; Huang, W. Accurate and Efficient Torque Control of an Interior Permanent Magnet Synchronous Motor in Electric Vehicles Based on Hall-Effect Sensors. *Energies* **2017**, *10*, 410. [CrossRef]

27. Li, H.; Gao, J.; Huang, S.; Fan, P. A Novel Optimal Current Trajectory Control Strategy of IPMSM Considering the Cross Saturation Effects. *Energies* **2017**, *10*, 1460. [CrossRef]

28. Cavallaro, C.; Di Tommaso, A.O.; Miceli, R.; Raciti, A.; Galluzzo, G.R.; Trapanese, M. Efficiency enhancement of permanent-magnet synchronous motor drives by online loss minimization approaches. *IEEE Trans. Ind. Electron.* **2005**, *52*, 1153–1160. [CrossRef]

29. Colby, R.S.; Novotny, D.W. Efficient operation of surface-mounted PM synchronous motors. *IEEE Trans. Ind. Appl.* **1987**, *IA-23*, 1048–1054. [CrossRef]

30. Morimoto, S.; Tong, Y.; Takeda, Y.; Hirasa, T. Loss minimization control of permanent magnet synchronous motor drives. *IEEE Trans. Ind. Electron.* **1994**, *41*, 511–517. [CrossRef]

31. Harnefors, L.; Nee, H.P. A general algorithm for speed and position estimation of AC motors. *IEEE Trans. Ind. Electron.* **2000**, *47*, 77–83. [CrossRef]

32. Boldea, I.; Paicu, M.C.; Andreescu, G.D. Active flux concept for motion-sensorless unified AC drives. *IEEE Trans. Power Electron.* **2008**, *23*, 2612–2618. [CrossRef]

33. Boldea, I.; Paicu, M.C.; Andreescu, G.D.; Blaabjerg, F. Active flux" orientation vector sensorless control of IPMSM. In Proceedings of the 11th International Conference on Optimization of Electrical and Electronic Equipment (OPTIM), Brasov, Romania, 22–24 May 2008; pp. 161–168.

34. Boldea, I.; Paicu, M.C.; Andreescu, G.D.; Blaabjerg, F. "Active Flux" DTFC-SVM sensorless control of IPMSM. *IEEE Trans. Energy Convers.* **2009**, *24*, 314–322. [CrossRef]

35. Chen, J.; Chen, S.; Wu, X.; Tan, G.; Hao, J. A Super-Twisting Sliding-Mode Stator Flux Observer for Sensorless Direct Torque and Flux Control of IPMSM. *Energies* **2019**, *12*, 2564. [CrossRef]

36. Di Piazza, M.C.; Pucci, M. Techniques for efficiency improvement in PWM motor drives. *Electr. Power Syst. Res.* **2016**, *136*, 270–280.

37. Ghaeb, J.A.; Smadi, M.A.; Ababneh, M. Progressive decrement PWM algorithm for minimum mean square error inverter output voltage. *Energy Convers. Manag.* **2011**, *52*, 3309–3318. [CrossRef]

38. Marčič, T.; Štumberger, G.; Hadžiselimović, M.; Zagradišnik, I. The impact of modulation frequency on losses, current harmonic distortion and motor temperature in an induction motor drive. *Electrotech. Rev.* **2006**, *73*, 125–130.

Article

Induction Machine Control for a Wide Range of Drive Requirements

Bojan Grčar [1,*], Anton Hofer [2] and Gorazd Štumberger [1]

[1] Faculty of Electrical Engineering and Computer Science, University of Maribor, Koroška 46, 2000 Maribor, Slovenia; gorazd.stumberger@um.si

[2] Institute of Automation and Control, Graz University of Technology, Inffeldgasse 21/B, 8010 Graz, Austria; anton.hofer@tugraz.at

* Correspondence: bojan.grcar@um.si

Received: 24 October 2019; Accepted: 23 December 2019; Published: 31 December 2019

Abstract: In this paper, a method for induction machine (IM) torque/speed tracking control derived from the 3-D non-holonomic integrator including drift terms is proposed. The proposition builds on a previous result derived in the form of a single loop non-linear state controller providing implicit rotor flux linkage vector tracking. This concept was appropriate only for piecewise constant references and assured minimal norm of the stator current vector during steady-states. The extended proposition introduces a second control loop for the rotor flux linkage vector magnitude that can be either constant, programmed, or optimized to achieve either maximum torque per amp ratio or high dynamic response. It should be emphasized that the same structure of the controller can be used either for torque control or for speed control. Additionally, it turns out that the proposed controller can be easily adapted to meet different objectives posed on the drive system. The introduced control concept assures stability of the closed loop system and significantly improves tracking performance for bounded but arbitrary torque/speed references. Moreover, the singularity problem near zero rotor flux linkage vector length is easily avoided. The presented analyses include nonlinear effects due to magnetic saturation. The overall IM control scheme includes cascaded high-gain current controllers based on measured electrical and mechanical quantities together with a rotor flux linkage vector estimator. Simulation and experimental results illustrate the main characteristics of the proposed control.

Keywords: induction machines; nonlinear control; energy efficiency; torque/speed control

---

1. Introduction

Squirrel cage induction machines (IM) are widely used in various industrial applications, machining, transportation, and electrical vehicles. They are relatively simply constructed and therefore easy to produce but quite difficult to control, especially if high performance and energy efficiency are required. Field oriented control (FOC) and direct torque control (DTC) schemes with many modifications introduced to fulfill different control objectives are widely accepted by the industry. Following the progress in non-linear control theory, several propositions based on feedback linearization, passivity, and flatness [1–4] were introduced in the past. These concepts facilitate deeper understanding of nonlinear IM dynamics and improve feedback performance and robustness but are often quite difficult to design and to implement. As a consequence, a great number of modifications and extensions have been proposed for basic FOC and DTC to improve overall performance, efficiency, and robustness along with new estimation techniques [5–10].

It is a long term goal to develop IM control strategies which are able to meet increasing efficiency requirements, i.e., controllers which can produce prescribed torque trajectories while minimizing the power losses. For example, in [11] an optimal torque controller for an IM with a linear magnetic characteristic was proposed based on the calculus of variations. A quite different approach was presented in [12]. Introducing a general *steady state* model suitable for induction motors, permanent-magnet synchronous motors, reluctance synchronous motors, and DC motors, the problem of power loss minimization including core saturation is considered. In [13] a control strategy is proposed based on ideas similar to [11]. The problem of missing information about the future reference torque trajectory is removed assuming simple first order transients between constant reference torque phases. The online implementation requires the solution of a small static optimization problem. The problem of power loss minimization for step changes of the torque reference trajectory is also considered in [14]. In addition to [13] saturation effects of the main inductance are taken into account and a simplified suboptimal solution avoiding an online optimization is presented. Recently a quite different control scheme for a permanent magnet synchronous machine was given in [15]. Based on an linearized parameter dependent mathematical model for the motor which also includes magnetic saturation, a model predictive control concept is developed. The online optimization problem is solved by the projected fast gradient method.

In this paper, a new control concept based on 3-D non-holonomic (NH) integrator including drift terms is proposed. Using standard modeling assumptions and assuming "perfect" current control, a reduced IM (current) model can be written in the general NH integrator form. From the control standpoint NH systems, which originate in mechanics, are quite challenging since they fail to fulfill the necessary conditions for stabilization with a smooth state feedback [16]. For this class of systems, control schemes based on discontinuous feedback (hybrid, sliding mode, and time optimal), time-varying state feedback (trajectory planning, back-stepping, pattern generation) were proposed in [17–20]. In [21] the authors introduced a non-linear state controller based on NH system analysis applied to IM torque control. Considering the non-holonomic constraint requiring periodic orbits of the rotor flux linkage and stator current vectors, the proposed feedback assured global asymptotic stability and maximal torque/amp ratio in steady states for the nominal parameter case. An essential characteristic of the proposed control scheme is implicit rotor flux linkage tracking provided by the adjusted amplitude and frequency modulation of the stator current vector. Two main limitations were observed for the proposed control: only piecewise *constant* references were allowed and the problem of singularity (zero rotor flux) required an additional time optimal flux controller and a switching logic to provide soft transitions between the two regimes [22].

This contribution is based on the results given in [23] for the control of the general 3-D-nonholonomic integrator with drift. The main focus of the present paper is the adaption of the control concept presented in [23] for the usage in IM torque and speed control applications. For this purpose a nonlinear flux model is included and the problem of flux optimization in order to achieve the maximum torque per ampere feature is considered from the practical point of view. It is shown that a simple suboptimal strategy for the adjustent of the flux magnitude can be applied to a wide range of torque reference scenarios as well as to speed control tasks. The control input is consequently composed of two orthogonal components, the first providing the required rotor flux linkage vector and the second producing the desired torque. In this way the overall control structure becomes partially similar to FOC but without relying on the uncertain rotor flux reference frame. The proposed structure enables the rotor flux linkage vector to be either constant, programmed, or optimized for energy efficiency with respect to drive characteristics and requirements. Due to similarities with widespread FOC, the proposed control scheme can be easily implemented on a standard industrial hardware.

The presentation of the results is structured as follows. In Section 2, the basic IM model is introduced in the rotor reference frame. The resulting reduced current model along with a dynamic torque estimation constitute a general 3-D non-holonomic integrator that is used to introduce the basic nonlinear state controller. Section 3 shows how the basic control structure can be extended in order to provide improved tracking performance. A proof of the stability of the resulting closed loop system is given. In Section 4 the mathematical description of the IM is extended by a simple nonlinear flux model covering the phenomenon of magnetic saturation. Next, the problem of rotor flux magnitude optimization is addressed, such that torque tracking of the IM is performed efficiently. Different classes of torque reference signals are considered and appropriate solutions are given. Section 5 is dedicated to extensions of the controller structure and the task of speed control. In Section 6, implementation aspects and experimental results confirming the improved tracking performance for torque and speed control are discussed. Concluding remarks are given in Section 7.

2. Im Model and Basic Controller

Using a standard modeling approach the two-axis dynamic model of a 2-pole IM model written in an arbitrary d-q reference frame is given by

$$\begin{aligned}
\dot{\mathbf{i}}_{d,q} &= \left(-\tau_\sigma^{-1}\mathbf{I} - \mathbf{J}\omega_s\right)\mathbf{i}_{d,q} + \left(\frac{M}{L_r L_\sigma \tau_r}\mathbf{I} - \frac{\omega_m M}{L_r L_\sigma}\mathbf{J}\right)\boldsymbol{\psi}_{d,q} \\
&\quad + L_\sigma^{-1}\mathbf{u}_{d,q} \\
\dot{\boldsymbol{\psi}}_{d,q} &= \left(-\tau_r^{-1}\mathbf{I} - \mathbf{J}(\omega_s - \omega_m)\right)\boldsymbol{\psi}_{d,q} + \frac{M}{\tau_r}\mathbf{i}_{d,q} \\
T_{el} &= \frac{M}{L_r}\left(\psi_d i_q - \psi_q i_d\right),
\end{aligned} \tag{1}$$

where $\mathbf{i}_{d,q}$, $\mathbf{u}_{d,q}$, $\boldsymbol{\psi}_{d,q}$ are the stator current vector, the stator voltage vector, and the rotor flux linkage vector written in the d-q reference frame. The mutual inductance is denoted by M, L_r is the rotor inductance, R_r is the rotor resistance, ω_m is the rotor speed, ω_s is the speed of the d-q reference frame with respect to the stator, T_{el} is the electric torque while $\mathbf{I}$ is the 2×2 identity matrix and $\mathbf{J}$ is the 2×2 skew symmetric rotational matrix. The leakage inductance L_σ, the resistance R_σ, the rotor time constant τ_r and the leakage time constant τ_σ are defined as $L_\sigma = L_s - M^2/L_r$, $R_\sigma = R_s - (R_r M)^2/L_r^2$, $\tau_r = L_r/R_r$ and $\tau_\sigma = L_\sigma/R_\sigma$. Throughout the text references, estimated variables and errors values are denoted as $(.)^*, \widehat{(.)}, \widetilde{(.)}$.

Under the assumption of a perfect current control, i.e., $\mathbf{i}_{d,q} = \mathbf{i}_{d,q}^*$, the simplified IM (current) model written in the rotor reference frame γ-δ is obtained from the model written in (1) by considering $\omega_m = \omega_s$ and changing d to γ and q to δ:

$$\dot{\boldsymbol{\psi}}_{\gamma,\delta} = -\tau_r^{-1}\boldsymbol{\psi}_{\gamma,\delta} + \frac{M}{\tau_r}\mathbf{i}_{\gamma,\delta} \tag{2}$$

where $\mathbf{i}_{\gamma,\delta}$ are control inputs, while the machine torque represents the controlled variable. For torque control the output error is defined as $\tilde{T}_{el} := T_{el}^* - \hat{T}_{el}$, where the actual machine torque T_{el} in (1) is replaced by a filtered estimate

$$\tau_f \dot{\hat{T}}_{el} = -\hat{T}_{el} + \frac{M}{L_r}\left(\psi_\gamma i_\delta - \psi_\delta i_\gamma\right). \tag{3}$$

In this model $\tau_f \ll \tau_r$ is the filter time constant, that is introduced to capture the estimation dynamics and to remove higher frequency components generated by estimation process and measurement of stator currents. A standard singular perturbation argument involving the estimation time constant τ_f recovers the algebraic expression for the actual machine torque. Equations (2) and (3) constitute the general 3-D non-holonomic integrator that includes drift terms.

Using the output error $\tilde{T}_{el}$ and the abbreviation $s_3 := \text{sign}(T_{el}^*)$ with the convention $T_{el}^* = 0 \rightarrow s_3 = 0$, the basic nonlinear state controller was proposed in [21] as:

$$\begin{bmatrix} i_\gamma \\ i_\delta \end{bmatrix} = \frac{\tau_r}{M} \begin{bmatrix} \frac{1}{\tau_r} + s_3 k_p \tilde{T}_{el} & -\left(\frac{s_3}{\tau_r} + k_p \tilde{T}_{el}\right) \\ \frac{s_3}{\tau_r} + k_p \tilde{T}_{el} & \frac{1}{\tau_r} + s_3 k_p \tilde{T}_{el} \end{bmatrix} \begin{bmatrix} \psi_\gamma \\ \psi_\delta \end{bmatrix} \tag{4}$$

where $k_p > 0$ denotes a controller gain. The terms that are output error independent compensate the drift terms and provide purely oscillatory dynamics of the rotor flux linkage vector $\boldsymbol{\psi}_{\gamma,\delta}$ as the output error $\tilde{T}_{el}$ vanishes. On the other hand, the terms that are output error dependent provide the adjusted amplitude and frequency modulation of the input current vector. The feedback system is obtained from (2) to (4) in the following form:

$$\dot{\psi}_\gamma = s_3 k_p \tilde{T}_{el} \psi_\gamma - \left(s_3 \tau_r^{-1} + k_p \tilde{T}_{el} \right) \psi_\delta$$
$$\dot{\psi}_\delta = \left(s_3 \tau_r^{-1} + k_p \tilde{T}_{el} \right) \psi_\gamma + s_3 k_p \tilde{T}_{el} \psi_\delta \qquad (5)$$
$$\dot{\hat{T}}_{el} = -\tau_f^{-1} \hat{T}_{el} + \tau_f^{-1} \frac{\tau_r}{L_r} \left(s_3 \tau_r^{-1} + k_p \tilde{T}_{el} \right) \left(\psi_\gamma^2 + \psi_\delta^2 \right)$$

In [21] it was proven that (5) has the following properties: For any constant $T_{el}^* \neq 0$, any $\hat{T}_{el}(0)$ and $\boldsymbol{\psi}_{\gamma,\delta}(0) \neq \mathbf{0}$ the system converges to a state where $\hat{T}_{el} = T_{el}^*$ and the components of $\boldsymbol{\psi}_{\gamma,\delta}(t)$ are harmonic functions with the amplitude

$$\left\| \boldsymbol{\psi}_{\gamma,\delta} \right\|_2 = \sqrt{L_r \left| T_{el}^* \right|}.$$

and the angular frequency $s_3 \tau_r^{-1}$. Furthermore, a distinctive characteristic of (5) is that, for a given piecewise constant output reference T_{el}^* the minimal norm of the control input vector $\mathbf{i}_{\gamma,\delta} = [\, i_\gamma \quad i_\delta \,]^T$ is assured in steady states

$$\left\| \mathbf{i}_{\gamma,\delta} \right\|_2 = \sqrt{2 \frac{L_r}{M^2} \left| T_{el}^* \right|}.$$

The angle between the rotor flux vector and the stator current is $s_3 \pi/4$ *permanently* as was already presented in [21]. The rotation speed of the stator current vector is obtained as $\omega_I = \omega_r + \omega_m$, where the introduced "relative" speed ω_r corresponds to the inverse time rotor constant τ_r^{-1}. The relative speed exactly matches the slip speed in steady states. Further remarks and characteristics of the controlled system (5) regarding stability, robustness, singularity avoidance, and overall performance can be found in [21,22].

3. Improved Torque Tracking Controller

Since from now on all vectors are expressed as vectors in the rotor reference frame the index combination γ, δ is omitted. The main characteristic of the presented basic controller, namely the implicit rotor flux tracking without separate control loop works perfectly for a piecewise constant torque references assuring also maximal torque-per-amp ratio. The main obstacle for a good tracking performance in general is observed especially at zero crossings of T_{el}^* with finite slope since the magnitude of the rotor flux linkage becomes too small or even zero. This results in the output $\hat{T}_{el}$ getting out of control for a short period as can easily be seen from the third equation in (5).

3.1. Controller Modification

To cope with this problem the basic controller scheme (4) has to be conceptually modified. Based on the ideas outlined in [23] the control input is composed of two orthogonal parts

$$\mathbf{i} = \mathbf{i}_\psi + \mathbf{i}_\tau \qquad (6)$$

where the vector $\mathbf{i}_\psi$ influences only the rotor flux linkage magnitude ψ, whereas the vector $\mathbf{i}_\tau$ facilitates machine torque. Considering the basic control structure it is intended that the effect of the diagonal elements in (4) is replaced by an appropriately selected vector $\mathbf{i}_\psi$. The presentation of the control design can be simplified if polar coordinates are introduced for the rotor flux vector

$$\psi_\gamma = \psi \cos \varphi \qquad \psi_\delta = \psi \sin \varphi \qquad (7)$$

with

$$\psi = \|\boldsymbol{\psi}_{\gamma,\delta}\|_2 = \sqrt{\psi_\gamma^2 + \psi_\delta^2} \quad \varphi = \arctan\left(\frac{\psi_\delta}{\psi_\gamma}\right). \tag{8}$$

The plant Equations (2) and (3) are then obtained as:

$$\dot{\psi} = -\tau_r^{-1}\psi + \tau_r^{-1}M\left(i_\gamma \cos\varphi + i_\delta \sin\varphi\right) \tag{9}$$

$$\psi\dot{\varphi} = \tau_r^{-1}M\left(-i_\gamma \sin\varphi + i_\delta \cos\varphi\right) \tag{10}$$

$$\dot{\hat{T}}_{el} = -\tau_f^{-1}\hat{T}_{el} + \tau_f^{-1}\frac{M}{L_r}\left(-i_\gamma \sin\varphi + i_\delta \cos\varphi\right)\psi \tag{11}$$

$$T_{el} = \frac{M}{L_r}\left(-i_\gamma \sin\varphi + i_\delta \cos\varphi\right)\psi \tag{12}$$

Now it is obvious that the control vector $\mathbf{i}_\psi$ must be of the form

$$\mathbf{i}_\psi = \begin{bmatrix} i_{\psi\gamma} \\ i_{\psi\delta} \end{bmatrix} = \begin{bmatrix} I \cos\varphi \\ I \sin\varphi \end{bmatrix}$$

with a suitable selected magnitude I in order to influence the flux magnitude ψ only. Introducing the reference signal ψ^* for the rotor flux linkage magnitude ψ a simple proportional control is proposed for the "magnetizing" vector $\mathbf{i}_\psi$

$$\mathbf{i}_\psi = \frac{1}{M}\begin{bmatrix} \left(\psi^* + k_\psi(\psi^* - \psi)\right)\cos\varphi \\ \left(\psi^* + k_\psi(\psi^* - \psi)\right)\sin\varphi \end{bmatrix}, \tag{13}$$

where $k_\psi > 0$ is a tuning parameter and $\psi^* > 0$ is assumed. By inserting (13) in (9) it follows that the rotor flux magnitude is ψ driven by the linear first order dynamics with unit gain:

$$\dot{\psi} = -\tau_r^{-1}(1 + k_\psi)\psi + \tau_r^{-1}(1 + k_\psi)\psi^* \tag{14}$$

The time constant of (14) can be simply influenced by an appropriate selection of the parameter k_ψ. It is important to note that (14) is completely decoupled from the other two states φ, $\hat{T}_{el}$.

Since $\mathbf{i}_\psi$ should replace the effect of the diagonal elements in (4) the second control vector $\mathbf{i}_\tau$ influencing only the output error results from the remaining part of (4):

$$\mathbf{i}_\tau = \frac{\tau_r}{M}\begin{bmatrix} -\left(s_3\tau_r^{-1} + k_p\tilde{T}_{el}\right)\psi_\delta \\ \left(s_3\tau_r^{-1} + k_p\tilde{T}_{el}\right)\psi_\gamma \end{bmatrix}$$

$$= \frac{\tau_r}{M}\begin{bmatrix} -\left(s_3\tau_r^{-1} + k_p\tilde{T}_{el}\right)\psi\sin\varphi \\ \left(s_3\tau_r^{-1} + k_p\tilde{T}_{el}\right)\psi\cos\varphi \end{bmatrix} \tag{15}$$

Obviously $\mathbf{i}_\tau$ has no effect on the flux magnitude ψ. By inserting (15) into (10), it follows

$$\dot{\varphi} = s_3\tau_r^{-1} + k_p\tilde{T}_{el}.$$

When the torque error $\tilde{T}_{el}$ vanishes the rotor flux linkage dynamics results in a harmonic oscillator with an angular speed $\dot{\varphi} = s_3\tau_r^{-1}$. This property which is implicitly contained in the basic controller (4) is reasonable only for piecewise *constant* references T_{el}^* but it is too restrictive in general. Therefore, it is proposed to replace the term $s_3\tau_r^{-1}$ in (15) with the expression

$$R_r\frac{T_{el}^*}{(\psi^*)^2} \quad \text{(with the restriction } \psi^* > 0\text{)}.$$

One of the reasons for this replacement is the fact that in the case of a constant reference T_{el}^* and the choice $\psi^* = \sqrt{L_r\,|T_{el}^*|}$ the same steady state is obtained as by using the basic controller (4). This means that the important property of (4), i.e., producing a constant torque with minimum norm of the control input is recovered by the modified controller. Finally, by combining (13) and (15) (with the proposed modification), a new control law is obtained:

$$
\begin{aligned}
i_\gamma &= \frac{\tau_r}{M}\Big[-\big(R_r\frac{T_{el}^*}{(\psi^*)^2} + k_p\tilde{T}_{el}\big)\psi\sin\varphi. \\
&\quad + \tau_r^{-1}\big(\psi^* + k_\psi(\psi^* - \psi)\big)\cos\varphi\Big] \\
i_\delta &= \frac{\tau_r}{M}\Big[\big(R_r\frac{T_{el}^*}{(\psi^*)^2} + k_p\tilde{T}_{el}\big)\psi\cos\varphi. \\
&\quad + \tau_r^{-1}\big(\psi^* + k_\psi(\psi^* - \psi)\big)\sin\varphi\Big]
\end{aligned}
\tag{16}
$$

The resulting closed loop system built up by (9)–(11), (16) is given as:

$$
\begin{aligned}
\dot{\psi} &= -\tau_r^{-1}(1 + k_\psi)\psi + \tau_r^{-1}(1 + k_\psi)\psi^* \\
\dot{\varphi} &= R_r\frac{T_{el}^*}{(\psi^*)^2} + k_p\tilde{T}_{el} \\
\dot{\hat{T}}_{el} &= -\tau_f^{-1}(1 + \frac{k_p}{R_r}\psi^2)\hat{T}_{el} + \tau_f^{-1}\big(\frac{\psi^2}{(\psi^*)^2} + \frac{k_p}{R_r}\psi^2\big)T_{el}^*
\end{aligned}
\tag{17}
$$

Of course the control law (16) can be transformed back into the original state variable form by the inverse relations (7) and (8). Which, in combination with the plant model (2) and (3) leads to the following description of the closed loop system:

$$
\begin{aligned}
\dot{\psi}_\gamma &= -\tau_r^{-1}\psi_\gamma - \left(R_r\frac{T_{el}^*}{(\psi^*)^2} + k_p\tilde{T}_{el} \right)\psi_\delta \\
&\quad + \tau_r^{-1}\left(\psi^* + k_\psi\Big(\psi^* - \sqrt{\psi_\gamma^2 + \psi_\delta^2} \Big) \right)\frac{\psi_\gamma}{\sqrt{\psi_\gamma^2 + \psi_\delta^2}} \\
\dot{\psi}_\delta &= -\tau_r^{-1}\psi_\delta - \left(R_r\frac{T_{el}^*}{(\psi^*)^2} + k_p\tilde{T}_{el} \right)\psi_\gamma \\
&\quad + \tau_r^{-1}\left(\psi^* + k_\psi\Big(\psi^* - \sqrt{\psi_\gamma^2 + \psi_\delta^2} \Big) \right)\frac{\psi_\delta}{\sqrt{\psi_\gamma^2 + \psi_\delta^2}} \\
\dot{\hat{T}}_{el} &= -\tau_f^{-1}\hat{T}_{el} + \frac{\tau_f^{-1}}{R_r}\left(R_r\frac{T_{el}^*}{(\psi^*)^2} + k_p\tilde{T}_{el} \right)\big(\psi_\gamma^2 + \psi_\delta^2\big)
\end{aligned}
\tag{18}
$$

3.2. Stability Analysis

Since the closed loop system has two external inputs it must be proven that the norm of the state vector $\mathbf{x} = \begin{bmatrix} \psi_\gamma & \psi_\delta & \hat{T}_{el} \end{bmatrix}^T$ remains bounded whenever the norm of the input vector defined as $\mathbf{r} := \begin{bmatrix} \psi^* & T_{el}^* \end{bmatrix}^T$ and $\|\mathbf{x}(0)\|_2$ are bounded. This task becomes much easier, if the model (17) is used. Obviously we have

$$
\|\mathbf{x}\|_2^2 = \psi^2 + \hat{T}_{el}^2,
\tag{19}
$$

therefore only the first and the third equation of (17) have to be considered. Next, it is assumed that the following bounds hold for the external inputs

$$0 < \psi^*_{\min} \leq \psi^*(t) \leq \psi^*_{\max} \quad \text{for all } t \geq 0 \tag{20}$$

$$|T^*_{el}(t)| \leq T^*_{el,\max} \quad \text{for all } t \geq 0, \tag{21}$$

where $T^*_{el,\max}$, $\psi^*_{\min}$, $\psi^*_{\max}$ are some finite positive constants. With the restriction $k_\psi > 0$ it follows immediately from the first equation in (17) that

$$\psi(t) \leq e^{-\tau_r^{-1}(1+k_\psi)t}\psi(0) + \psi^*_{\max} \leq \psi(0) + \psi^*_{\max}. \tag{22}$$

Thus $\psi(t)$ remains bounded for any finite initial value $\psi(0) \geq 0$ and any external input $\psi^*(t)$ satisfying (20).

Now the third equation in (17) can be regarded as a linear *timevariant* first order system with the external input T^*_{el} and with the output $\hat{T}_{el}$. For the stability analysis of this system the homogeneous case (i.e., $T^*_{el} = 0$) is considered first which leads to:

$$\hat{T}_{el}(t) = e^{\int_{t_0}^{t} -\tau_f^{-1}(1+\frac{k_p}{R_r}\psi^2)d\tau}\hat{T}_{el}(t_0)$$

$$= e^{-\tau_f^{-1}(t-t_0)}e^{\int_{t_0}^{t} -\tau_f^{-1}\frac{k_p}{R_r}\psi^2 d\tau}\hat{T}_{el}(t_0)$$

Taking into account $k_p > 0$, it follows

$$|\hat{T}_{el}(t)| \leq e^{-\tau_f^{-1}(t-t_0)}|\hat{T}_{el}(t_0)|,$$

which shows that the homogeneous system is uniformly exponentially stable. Furthermore from (20) to (22) we can derive the result

$$\tau_f^{-1}\left|\frac{\psi^2}{(\psi^*)^2} + \frac{k_p}{R_r}\psi^2\right| \leq c \quad \text{for all } t \geq 0$$

where c denotes some finite positive constant. This inequality combined with the fact that the homogeneous part is uniformly asymptotically stable assures that $|\hat{T}_{el}(t)|$ remains bounded, whenever $|T^*_{el}(t)|$ and $\hat{T}_{el}(0)$ are bounded [24]. Finally from (19) the desired result follows that $\|x\|_2$ remains bounded. Furthermore, it is important to note that also the angular rotation speed $\dot{\varphi}$ of the rotor flux vector remains bounded (see second equation in (17)).

4. Magnetic Saturation and Flux Reference Optimization

The new control law (16) provides the flux reference signal ψ^* as an additional external input. In this section the problem how to choose ψ^* in order to achieve energy efficient tracking of a desired torque T^*_{el} will be tackled. Since we would like to take full advantage of the dynamic capabilities of the induction machine, magnetic saturation must be considered. A correct mathematical description of this phenomenon would lead to a very complicated model, useless for control design. Therefore, a simple approach proposed in [25] for field oriented control is adopted which is also applied in [26]. It is based on the magnetization curve of the induction machine given in the form

$$\psi = f(i_\psi), \tag{23}$$

where the magnetizing current i_ψ is defined as

$$i_\psi := i_\gamma \cos\varphi + i_\delta \sin\varphi. \tag{24}$$

It should be mentioned that $|i_\psi| = \|\mathbf{i}_\psi\|_2$ is valid. Simple geometric considerations based on the phasors $\mathbf{i}_{\gamma,\delta}$ and $\boldsymbol{\psi}_{\gamma,\delta}$ reveal that this relation holds, where $\mathbf{i}_\psi$ is defined in (6) and (13) respectively. The magnetization curve can be obtained by measurements (e.g., [26]). Throughout the rest of the paper M and L_r are the nominal values of the mutual inductance and rotor inductance below saturation.

The following nonlinear flux model was introduced in [25]

$$\dot{\psi} = \tau_r^{-1} M \left(-f^{-1}(\psi) + i_\psi \right), \tag{25}$$

where $f^{-1}(\psi)$ is the inverse magnetizing function. For the next considerations it is useful also to introduce the torque producing current component

$$i_\tau := -i_\gamma \sin \varphi + i_\delta \cos \varphi. \tag{26}$$

(with the property $|i_\tau| = \|\mathbf{i}_\tau\|_2$, given by the same geometric considerations as mentioned above), which allows us to write (12) in the form:

$$T_{el} = \frac{M}{L_r} i_\tau \psi \tag{27}$$

Note that the real electrical torque T_{el} is considered in the present case. The Equations (25) and (26) are the basis for the flux optimization problem. In the following subsection a piecewise constant torque reference T_{el}^* is considered and the corresponding optimization problem is solved. It is assumed that the desired torque can be produced by the IM within the given current and voltage limitations. The resulting optimal flux magnitude ψ_{opt} can be used as flux reference signal ψ^* in the proposed control structure.

4.1. Constant Reference T_{El}^*

In the first case, the following question is considered: How should the *constant* flux ψ be selected, so that a desired *constant* torque $T_{el} = T_{el}^* \neq 0$ is achieved with minimum norm of the current vector in steady state? In other words, we want to get the *maximum torque per amp* feature. If magnetic saturation is neglected (i.e., for the linear flux model) the solution is simple

$$\psi_{opt} = \sqrt{L_r \left| T_{el}^* \right|}, \tag{28}$$

was already mentioned in Section 3. In the saturation case we get from (23)

$$i_\psi = f^{-1}(\psi),$$

which is the necessary value of i_ψ in order to achieve the constant ψ in steady state. From (27) it follows

$$i_\tau = \frac{L_r}{M} \frac{T_{el}^*}{\psi}$$

and using the fact $\|\mathbf{i}\|_2^2 = i_\psi^2 + i_\tau^2$ the problem can be formulated as:

$$\psi_{opt} = \underset{\psi}{\arg\min} \left(\left(f^{-1}(\psi) \right)^2 + \left(\frac{L_r}{M} \frac{T_{el}^*}{\psi} \right)^2 \right) \tag{29}$$

By differentiating with respect to ψ the necessary condition is obtained

$$2 f^{-1}(\psi) \frac{df^{-1}(\psi)}{d\psi} - \frac{2 L_r^2}{M^2} \frac{T_{el}^{*2}}{\psi^3} \overset{!}{=} 0.$$

Unfortunately, this equation cannot be solved for ψ but this causes no severe problems since it can be written as:

$$L_r\,|T_{el}^*| = M\sqrt{\psi^3 f^{-1}(\psi)\frac{df^{-1}(\psi)}{d\psi}} =: g(\psi) \tag{30}$$

The function $g(\psi)$ can be evaluated and stored in a lookup table so that the desired result can be obtained numerically in the form

$$\psi_{opt} = g^{-1}(L_r\,|T_{el}^*|). \tag{31}$$

A comparison between the optimal flux magnitude based on (28) and (31) is shown in Figure 1b. A similar result was presented in [14] where power losses for torque steps were considered and a function, which determines the optimal constant flux magnitude for constant torque levels, is given.

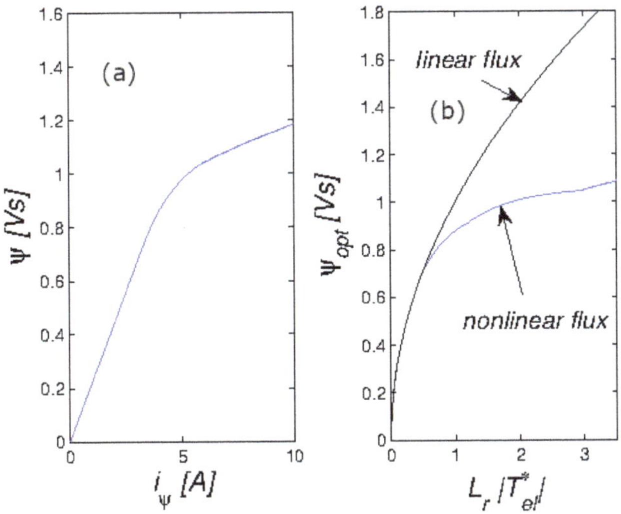

Figure 1. Magnetization curve (**a**); Optimal flux magnitude for constant torque (**b**).

4.2. General Case

In general, an arbitrary but bounded reference signal $T_{el}^*(t)$ defined over a finite time interval $0 \leq t \leq T_0$ is considered. In contrast to the forementioned case, a time function $\psi(t)$ should be determined now, so that the criterion

$$E = \int_0^{T_0} \left(i_\psi^2 + i_\tau^2\right) dt \tag{32}$$

is minimized for $T_{el}(t) = T_{el}^*(t)$. This means that the optimal flux trajectory $\psi(t)$ for an arbitrary torque reference signal should be determined. Obviously the value of E will also depend on the initial condition $\psi(0)$. In order to eliminate this influence, a periodic continuation of $T_{el}^*(t)$ is assumed and the steady state solution of the problem is considered. In this case $\psi(t)$ is a periodic time function too. From (27), it follows

$$i_\tau(t) = \frac{L_r}{M}\,\frac{T_{el}^*(t)}{\psi(t)} \tag{33}$$

and therefore the optimization problem can be formulated as:

$$\text{mimimize} \int_0^{T_0} \left(i_\psi(t)^2 + \left(\frac{L_r}{M} \frac{T_{el}^*(t)}{\psi(t)} \right)^2 \right) dt$$

subject to

$$\dot{\psi} = \tau_r^{-1} M \left(-f^{-1}(\psi) + i_\psi \right)$$

$$\psi(0) = \psi(T_0) \qquad i_\psi(0) = i_\psi(T_0).$$

$$(34)$$

The boundary conditions are given by the fact that $\psi(t)$ and $i_\psi(t)$ must also be continuous functions. This flux optimization problem can be treated by the classical methods of optimal control (see e.g., [27]). First the Hamilton function is introduced (the argument time is omitted for brevity)

$$H = i_\psi^2 + \left(\frac{L_r}{M} \frac{T_{el}^*}{\psi} \right)^2 + \lambda \tau_r^{-1} M \left(-f^{-1}(\psi) + i_\psi \right), \tag{35}$$

where λ denotes a Lagrange multiplier. Now the solution of the problem must satisfy the following necessary conditions:

$$0 = \frac{\partial H}{\partial i_\psi} = 2i_\psi + \lambda \tau_r^{-1} M \quad \rightarrow \quad i_\psi = -\frac{\lambda \tau_r^{-1} M}{2} \tag{36}$$

$$\dot{\lambda} = -\frac{\partial H}{\partial \psi} = 2 \left(\frac{L_r}{M} T_{el}^* \right)^2 \frac{1}{\psi^3} + \lambda \tau_r^{-1} M \frac{df^{-1}(\psi)}{d\psi} \tag{37}$$

$$\dot{\psi} = \frac{\partial H}{\partial \lambda} = \tau_r^{-1} M \left(-f^{-1}(\psi) + i_\psi \right) \tag{38}$$

Substituting i_ψ in Equation (38) with the expression from Equation (36), the resulting boundary value problem is obtained:

$$\dot{\lambda} = 2 \left(\frac{L_r}{M} T_{el}^* \right)^2 \frac{1}{\psi^3} + \lambda \tau_r^{-1} M \frac{df^{-1}(\psi)}{d\psi}$$

$$\dot{\psi} = \tau_r^{-1} M \left(-f^{-1}(\psi) - \frac{\lambda \tau_r^{-1} M}{2} \right)$$

$$\psi(0) = \psi(T_0) \qquad \lambda(0) = \lambda(T_0)$$

$$(39)$$

This problem can be solved numerically using appropriate software e.g., [28]. Finally, the optimal solution $\psi_{opt}(t)$ can be applied as flux reference signal $\psi^*(t)$. This solution could be implemented only if the reference trajectory is known in advance. A less restrictve assumption is discussed next.

In order to give more information about the solution of the boundary value problem (39), an example is presented. The parameters of the IM are taken from Table 1 and a reference torque $T_{el}^*(t)$ is chosen as shown in Figure 2a. The MATLAB function bvp4c was used to solve the boundary value problem. The resulting optimal flux magnitude $\psi_{opt}(t)$ is depicted in Figure 2b,c; it contains the corresponding current $i_\psi(t)$ and also the component $i_\tau(t)$ which is determined from the optimal solution by (33) with $\psi(t)$ replaced by $\psi_{opt}(t)$. The resulting value of the performance criterion (32) is given as $E = 336.04$.

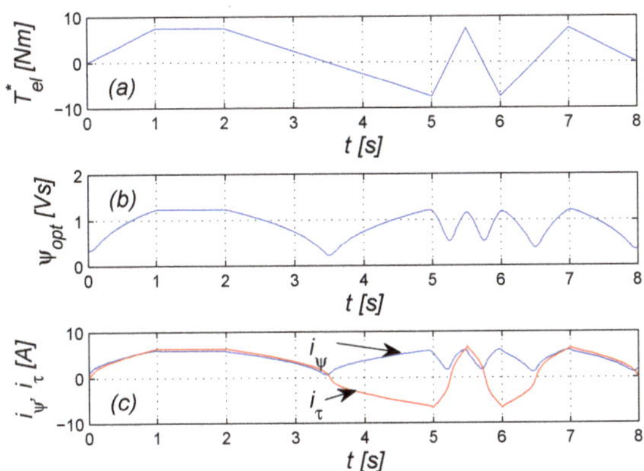

Figure 2. (a) Reference torque; (b) optimal flux magnitude; (c) optimal currents.

Table 1. Nominal motor data and controller parameters.

Symbol	Parameter	Value
P_n	nominal power	3 kW
U_n	supply voltage	220/380 V
I_n	nominal current	13/7.5 A
n_n	synchronous speed	1328 rpm
R_s	stator resistance	1.97 Ω
R_r	rotor resistance	2.91 Ω
L_s	stator inductance	0.2335 H
L_r	rotor inductance	0.2335 H
M	mutual inductance	0.223 H
L_σ	leakage inductance	0.0105 H
T_{el}	nominal torque	10 Nm
J	drive inertia	0.031 kgm^2
c	friction constant	0.025 Nms/rad
α	saturation parameter	0.13
β	saturation parameter	1.7154
τ_f	filter time-constant	0.005 s
k_ψ	flux controller gain	1.5
k_p	torque controller gain	2.5
k_p	speed controller gain	25
k_{pi}	current controller gain	20
k_{ii}	integrator gain	225.5

5. Final Controller Modifications

In the previous section two different scenarios concerning the torque reference signal T_{el}^* were investigated. The general case—being the most interesting one—needs the solution of a nonlinear boundary value problem. In this case the torque reference signal must be known in advance which drastically limits the applicability. Therefore, a suboptimal but easily implementable method for the selection of the flux reference signal ψ^* is preferred. A very simple but nevertheless reasonable possibility can be derived by a comparison of the optimization problems (29) and (34). In addition to the boundary conditions, the two problems differ mainly in the fact that the *dynamic* flux model is taken into account in (34). If replaced by the *static* flux model (23) a similar functional as in (29) is obtained. Based on the result (31) for problem (29) the following choice

$$\psi^*(t) = g^{-1}\left(L_r \left|T_{el}^*(t)\right|\right)$$

is introduced for the general case. In [14] the special case of step changes in the reference torque (or step changes filtered by a stable linear second order system) are considered. The optimization problem for the flux magnitude in order to minimize the power losses is formulated. Finally, the optimization problem is simplified in the way that only the constant power losses in steady state are taken into account. For proposed solution it should be emphasized that $\psi^*(t)$ is applied for general torque trajectories. Obviously this selection of the flux reference is optimal in the case of piecewise constant reference torques but it requires certain limitations in order to meet the requirements in the general case. Therefore, it is proposed to use

$$\psi^*(t) = \mathrm{sat}\left(g^{-1}\left(L_r \left|T_{el}^*(t)\right|\right), \psi_{min}^*, \psi_{max}^*\right), \tag{40}$$

where the saturation function is defined as

$$y = \mathrm{sat}\left(x, x_{min}, x_{max}\right) := \begin{cases} x_{min} & \text{if } x < x_{min} \\ x & \text{if } x_{min} \leq x \leq x_{max} \\ x_{max} & \text{if } x > x_{max} \end{cases}.$$

The lower bound $\psi_{min}^* > 0$ has to be introduced in order to avoid the problems at zero crossings of $T_{el}^*(t)$ as mentioned in Section 3. Of course it is not possible to give an exact bound on the "suboptimality" of the proposed flux reference strategy for an arbitrary torque reference signal which is not given in advance. However, the choice of the lower bound ψ_{min}^* offers a simple way to adjust the concept to the expected reference torque trajectories. Especially in the case of slowly varying reference torques the efficiency improvement can be significantly compared to the constant flux reference case as can be seen in Figure 2. Upper bound ψ_{max}^* should prohibit unduly high magnetization or can be used in the field weakening range being defined as a function of the rotor speed ω_m. The introduction of the lower bound ψ_{min}^* for the flux reference signal is helpful to meet different drive requirements. In cases where the dynamics of the torque response is of main interest an operation with constant flux magnitude may be preferable which can be obtained by setting $\psi_{min}^* = \psi_{max}^*$. On the other hand, if efficiency of the drive is emphasized a small value of ψ_{min}^* will be a suitable choice. Of course, in cases where the required torque $T_{el}^*(t)$ is known in advance (e.g., in robotic applications) the optimal flux reference $\psi^*(t)$ resulting from the boundary value problem (39) can be applied directly, i.e., the first equation of the controller (41) can be omitted.

The results obtained by this simple approach are shown in Figure 3. The same torque reference signal $T_{el}^*(t)$ as used in the previous example leads to a flux reference signal $\psi^*(t)$ nearly coinciding with the optimal flux $\psi_{opt}(t)$ most of the time. Here the bounds were chosen somewhat arbitrarily as $\psi_{min}^* = 0.35$ and $\psi_{max}^* = 1.4$. Interestingly, the performance criterion only marginally deteriorates compared to the optimal solution. In this case the value is given as $E = 337.6$. Considering the torque reference signal in Figure 2a and the deviation from the optimal flux in Figure 3c it follows that high dynamic changes in the reference torque require an increasing value of the lower bound ψ_{min}^* in order to get a good approximation of the optimal flux magnitude.

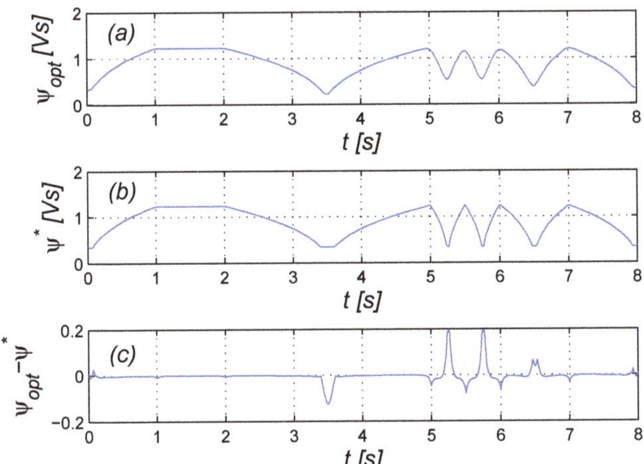

Figure 3. (**a**) Optimal flux magnitude; (**b**) flux reference resulting from (40); (**c**) deviation from optimal flux.

The inclusion of the nonlinear flux model (25) into the control concept requires a minor modification of (13). The feedforward part ψ^*/M in this scheme must be replaced by the inverse magnetization function in order to achieve the correct flux magnitude.

Thus, the torque controller is proposed in its final form as follows: given the reference torque $T_{el}^*(t)$ as external input, the estimated torque $\hat{T}_{el}(t)$ and the flux coordinates $\psi(t)$, $\varphi(t)$ (or their estimated values $\hat{\psi}(t)$, $\hat{\varphi}(t)$ from an observer/estimator) the controller outputs $i_\gamma(t)$, $i_\delta(t)$ (i.e., the reference signals for the current controller) are determined by the relations

$$
\psi^*(t) = \mathrm{sat}\left(g^{-1}\left(L_r \,|T_{el}^*(t)| \right), \psi_{\min}^*, \psi_{\max}^* \right)
$$

$$
i_\psi(t) = f^{-1}(\psi^*(t)) + \frac{k_\psi}{M}\left(\psi^*(t) - \hat{\psi}(t) \right)
$$

$$
i_\tau(t) = \left(\frac{L_r}{M}\frac{T_{el}^*(t)}{\psi^*(t)^2} + \frac{k_p L_r}{R_r M}\left(T_{el}^*(t) - \hat{T}_{el}(t) \right) \right) \hat{\psi}(t) \tag{41}
$$

$$
i_\gamma(t) = i_\psi(t)\cos\hat{\varphi}(t) - i_\tau(t)\sin\hat{\varphi}(t)
$$

$$
i_\delta(t) = i_\psi(t)\sin\hat{\varphi}(t) + i_\tau(t)\cos\hat{\varphi}(t).
$$

Remark 1. *If the nonlinear flux model (25) is used in combination with the modified equation for i_ψ from (41) the first equation of the closed loop model (17) changes to*

$$
\dot{\psi} = -\tau_r^{-1}(Mf^{-1}(\psi) + k_\psi\psi) + \tau_r^{-1}(Mf^{-1}(\psi^*) + k_\psi\psi^*). \tag{42}
$$

Since the inequality $Mf^{-1}(\psi) \geq \psi$ holds for all $\psi \geq 0$ it follows immediately that $\psi = 0$ is an asymptotically stable equilibrium of the unforced (i.e., for $\psi^*(t) = 0$) system (42). Additionally, using the Lyapunov functon $V = \frac{1}{2}\psi^2$ it can be shown that (42) has the so-called input to state stability property which means that $\psi(t)$ remains bounded whenever $\psi(0)$ and $\psi^*(t)$ are bounded [29]. Therefore, all results from the stability analysis of Section 3.2 remain valid.

5.1. Guidelines for Selection of Controller Parameters

The final control law (41) contains two parameters k_ψ and k_p which must have positive values in order to achieve stability of the closed loop system. Considering the dynamic capabilities of the controlled induction machine it would be desirable to select large values for k_ψ and k_p but of course this choice is limited by the constraint posed on the stator current vector. Therefore, some guidelines should be given next. Starting from the given maximum possible flux magnitude $\psi_{\max}$ we can determine the associated magnetization current amplitude $i_{\psi,m}$ via the measured magnetization curve, i.e., $i_{\psi,m} = f^{-1}(\psi_{\max})$. In order to achieve a desired dynamic flux control loop we may put the following constraint on the magnetization current

$$i_\psi \leq i_{\psi,\max} := 2i_{\psi,m}.$$

From (41) we have

$$i_\psi = f^{-1}(\psi^*) + \frac{k_\psi}{M}(\psi^* - \hat{\psi})$$

which immediately leads to the conservative constraint

$$f^{-1}(\psi_{\max}) + \frac{k_\psi}{M}\psi_{\max} \leq i_{\psi,\max}.$$

From this inequality we can derive a bound for the parameter k_ψ as

$$k_\psi \leq \frac{M}{\psi_{\max}}(i_{\psi,\max} - f^{-1}(\psi_{\max})) = \frac{M}{\psi_{\max}}i_{\psi,m}. \tag{43}$$

Given the maximum possible stator current amplitude $i_{\max}$ we can determine the maximum amplitude $i_{\tau,\max}$ of the torque producing current by

$$i_{\tau,\max} = \sqrt{i_{\max}^2 - i_{\psi,\max}^2}.$$

Now considering (15) it follows

$$|i_\tau| = \|\mathbf{i}_\tau\|_2 = \frac{\tau_r}{M}\left(s_3\tau_r^{-1} + k_p\tilde{T}_{el}\right)\psi$$

from which the inequality

$$(1 + k_p\tau_r\tilde{T}_{el}) \leq i_{\tau,\max}\frac{M}{\psi_{\max}}$$

can be derived. Under the assumption

$$|\tilde{T}_{el}| \leq 2T_{el,\max},$$

where $T_{el,\max}$ denotes the maximum possible torque, we get the desired bound

$$k_p \leq \frac{1}{2\tau_r T_{el,\max}}\left(i_{\tau,\max}\frac{M}{\psi_{\max}} - 1\right). \tag{44}$$

It should be emphasized that (43) and (44) are conservative bounds; therefore, the applied parameters may be increased to some extent.

5.2. Speed Control

The proposed torque control concept can be easily extended to speed control of an IM. In this case the mathematical model of the plant has to be augmented with the differential equation for the mechanical part of the system

$$J\dot{\omega}_m = -c\omega_m + T_{el} - T_l. \tag{45}$$

The inertia of the drive is denoted by J, c is a constant for viscous friction, and T_l is an unknown load torque. It is assumed that the mechanical speed ω_m can be measured. Now the proposed torque controller (41) can be simply extended by a speed controller whose output serves as reference torque input T_{el}^* for the torque controller. A standard PI-controller may be used for this purpose. The resulting structure of the controller including the flux and torque observer/estimator is shown in Figure 4. The reference signal for the mechanical speed is denoted by ω_m^*.

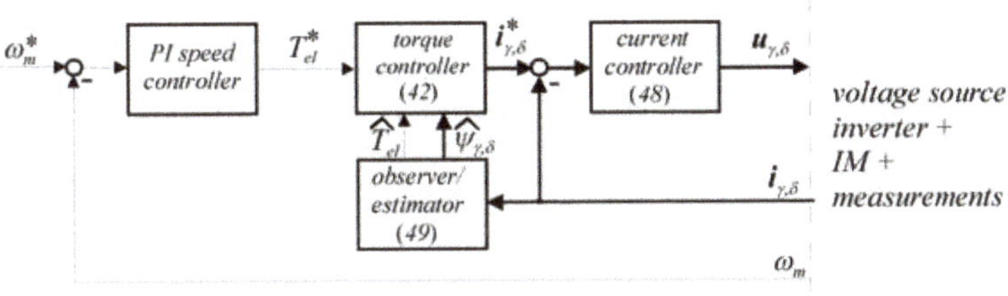

Figure 4. Controller structure for speed control.

For high performance servo applications it is preferable to choose ψ^* below the saturation point to ensure the required dynamic response. On the other hand, for slowly changing or constant speed references ω^* the flux command ψ^* could be optimized as proposed.

If we dispense with the maximum torque per ampere feature a simplified version of the speed controller can be derived from (41) in the following way: First (3) of the plant model is replaced by (45), next a suitable *constant* flux reference value ψ^* is chosen and finally the current component i_τ in (41) has to be modified such that it depends on the speed error. In this way the speed controller is given as:

$$i_\psi(t) = f^{-1}(\psi^*) + \frac{k_\psi}{M}\left(\psi^* - \hat{\psi}(t)\right)$$
$$i_\tau(t) = \left(\frac{cL_r}{M}\frac{\omega_m^*(t)}{\psi^{*2}} + \frac{k_pL_r}{R_rM}\left(\omega_m^*(t) - \omega_m(t)\right)\right)\hat{\psi}(t) \tag{46}$$
$$i_\gamma(t) = i_\psi(t)\cos\hat{\varphi}(t) - i_\tau(t)\sin\hat{\varphi}(t)$$
$$i_\delta(t) = i_\psi(t)\sin\hat{\varphi}(t) + i_\tau(t)\cos\hat{\varphi}(t).$$

The closed loop system composed of (2), (45), and (46) is described by the relation (42) and

$$\dot{\varphi} = cR_r\frac{\omega_m^*}{\psi^{*2}} + k_p\left(\omega_m^* - \omega_m\right)$$

$$\dot{\omega}_m = -\frac{1}{J}\left(c + \frac{k_p}{R_r}\psi^2\right)\omega_m$$
$$+ \frac{1}{J}\left(c\frac{\psi^2}{\psi^{*2}} + \frac{k_p}{R_r}\psi^2\right)\omega_m^* - \frac{1}{J}T_l.$$

6. Control Implementation and Experimental Results

In the following, the practical implementation of the controller combined with the appropriate inner current control along with a rotor flux linkage estimator is considered. The current control that provides reference tracking, disturbance suppression, and voltage drop compensation of the Voltage Source Inverter (VSI) was realized in the rotor reference frame. The current error

$$\tilde{\mathbf{i}}_{\gamma,\delta} = \mathbf{i}^*_{\gamma,\delta} - \mathbf{i}_{\gamma,\delta}$$

was calculated from the reference currents obtained from the torque controller and the machine stator currents, that were directly measured. Based on an internal model principle a PI controller, that provides reference tracking and compensation of the back-EMF, was used in the following form:

$$\mathbf{u}_{\gamma,\delta} = k_{pi}\tilde{\mathbf{i}}_{\gamma,\delta} + k_{pi}k_{ii}\int_0^t \left(\frac{\mathbf{J}\omega_m}{k_{ii}} + \mathbf{I}\right)\tilde{\mathbf{i}}_{\gamma,\delta}d\tau \tag{47}$$

The constants $k_{pi}, k_{ii} > 0$ are free design parameters. In this application $k_{ii} = \tau_\sigma^{-1}$ was set. The implementation of the proposed control law clearly requires a rotor flux linkage estimator or observer. In experiments, a simplified magnetically nonlinear estimator was used based on the rotor model (25) in the following form

$$i_\psi = i_\gamma \cos\hat{\varphi} + i_\delta \sin\hat{\varphi}$$
$$i_\tau = -i_\gamma \sin\hat{\varphi} + i_\delta \cos\hat{\varphi}$$
$$\dot{\hat{\psi}} = \tau_r^{-1}M\left(-f^{-1}(\hat{\psi}) + i_\psi\right) \tag{48}$$
$$\dot{\hat{\varphi}} = \tau_r^{-1}M\frac{i_\tau}{\hat{\psi}}$$
$$\dot{\hat{T}}_{el} = -\tau_f^{-1}\hat{T}_{el} + \tau_f^{-1}\frac{M}{L_r}\hat{\psi}i_\tau.$$

The magnetization curve $\hat{\psi} = f(i_\psi)$ was implemented as lookup table.

The experimental setup consisted of the tested IM coupled with a separately controlled DC motor serving as variable load, a dSPACE PPC Controller board, a host PC with installed development environment, an incremental encoder with 5000 pulses per revolution, current sensors, voltage inverter, and a torque transducer. During tests, data acquisition, transformations, the flux, and torque estimator and the control algorithm were executed on the PowerPC with a sampling time of 250 µs, while a slave DSP was used for the vector modulation executed at 4 kHz. The program codes for the PowerPC and for the slave DSP were developed with the Real Time Interface and Simulink. Characteristic motor data and controller parameters for torque control are given in Table 1.

The first experiment presented in Figure 5 shows the machine response for a torque control with a piecewise constant torque reference and with adjusted rotor flux linkage magnitude as proposed in (31). The diagram (a) shows the estimated torque, the reference torque, and the filtered measured torque. For comparison on diagram (b), the same experiment is repeated using just the basic FOC (constant flux command, no compensation of magnetic nonlinearity, constant parameter flux estimator, current controllers in d-q reference frame, PI speed controller). Since no compensation of nonlinearity was used, some mismatch between observed, measured, and reference torque can be observed at higher torque values. Diagrams (c) and (d) show measured and reference currents while on the fifth diagram (e) the rotor flux linkage vector trajectory is shown. The last diagram (f) shows the reference and adjusted magnitude of the rotor flux linkage vector considering the nonlinear magnetizing characteristic.

Considering our previous proposition of the basic controller in (4), the advantages of the proposed extended control become evident if the reference changes sign with finite slope. This is illustrated for the reference profile that includes torque reversal in Figure 6. The proposed controller enables precise torque tracking also at zero crossings (a). On diagram (b) the comparison with FOC shows some deviation in the measured torque at higher reference values. Slightly better performance can be observed for the proposed control mostly due to the fact that just basic FOC was implemented. As in the previous experiment reference and measured currents (c,d), rotor flux linkage vector trajectory (e) and adjusted magnitude of the rotor flux linkage vector (f) are shown for the proposed control. Diagrams (g,h) show the stator current vector norms and corresponding integrals for a constant and optimized rotor flux. The current norm and the quantity E were obtained numerically from the measurements. For a constant rotor flux obtained with the proposed control and FOC the difference in E is a consequence of slightly bigger magnetizing current that was applied for a proposed control (curves 2 and 3). As expected, E is minimized for optimized rotor flux (curve 1). Similar performance could be achieved also with FOC for optimized rotor flux.

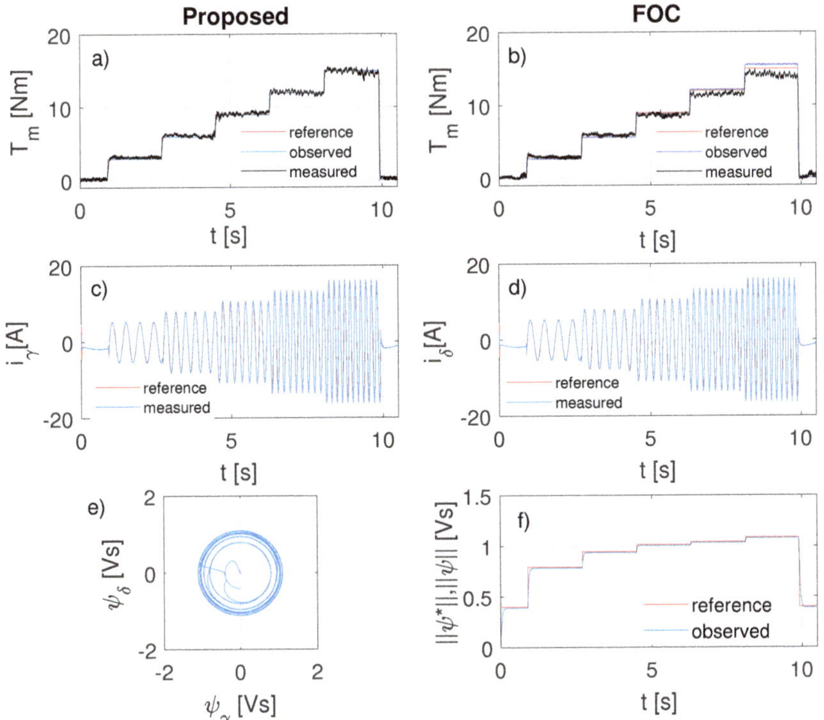

Figure 5. Torque control with adjusted rotor flux magnitude command and piecewise constant torque reference: torque response for proposed method (**a**), torque response for FOC (**b**), stator current components gamma and delta (**c,d**), rotor flux vector trajectory (**e**), rotor flux vector magnitude (**f**).

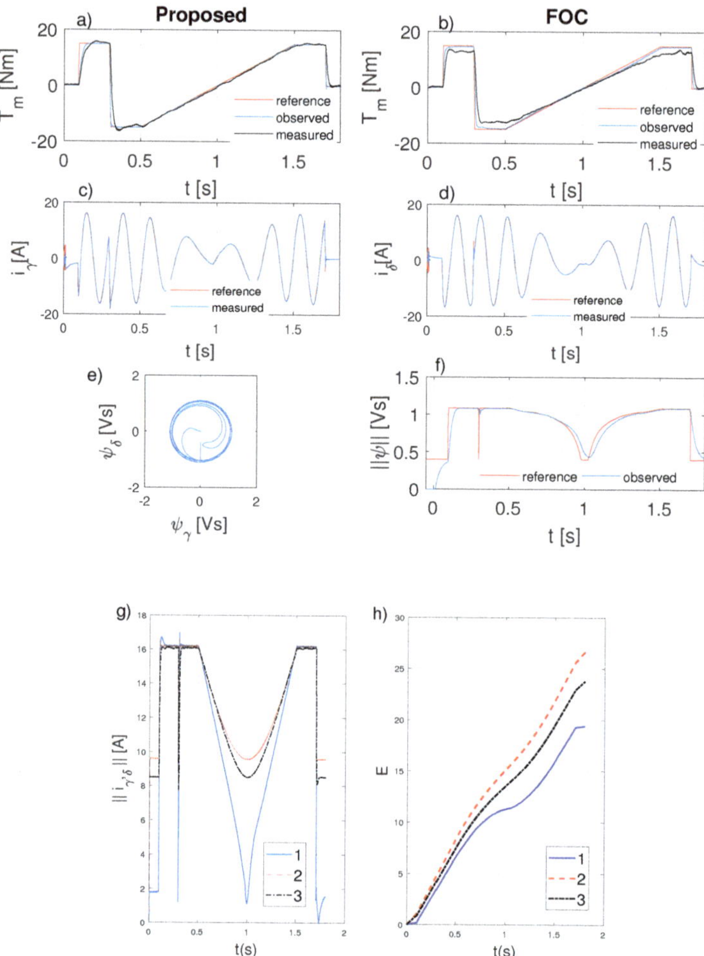

Figure 6. Torque control with optimized rotor flux magnitude command for the torque reversal including zero crossing with a finite slope (**a**–**f**) and comparison of stator current norm and corresponding integrals (**g**,**h**) for the proposed controller with optimized rotor flux (1) and with constant rotor flux (2) as well as for field oriented control (FOC) with constant rotor flux (3).

In Figure 7 the performance comparison for low speed reference is shown. No significant differences can be observed between proposed control and FOC. In the experiment shown in Figure 8 tracking performance for a harmonic speed reference with simultaneous constant load torque $T_l = -8$ Nm starting at $t = 1.9$ s is given. Actual and reference speed are shown in diagram (a), measured torque is given in diagram (b), while in diagrams (c) and (d) reference and measured currents i_γ and i_δ are shown. In Figure 9, the same variables as in the previous experiment are shown for a speed reversal command.

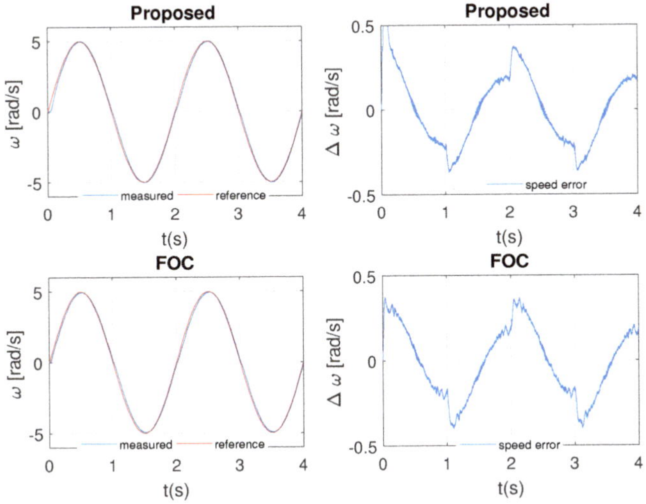

Figure 7. Performance comparison at low speed.

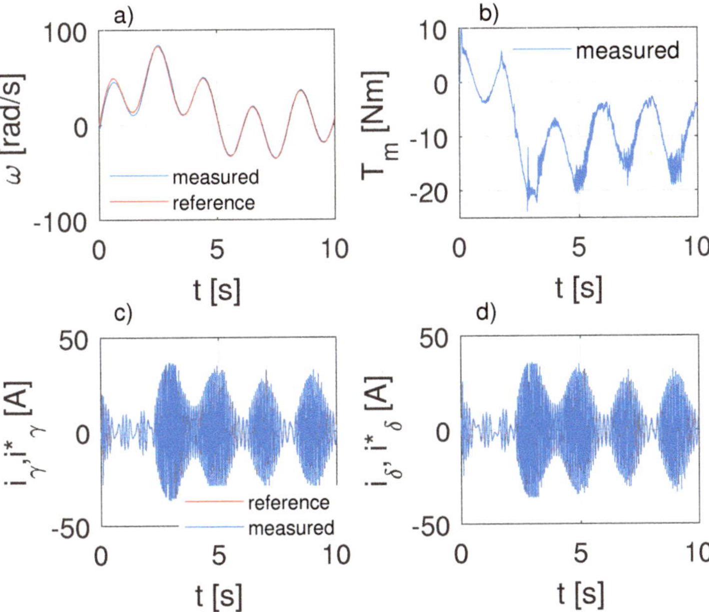

Figure 8. Speed response for harmonic speed reference and a constant load torque starting at $t = 1.9\ s$: reference and measured speed (**a**), measured torque (**b**), reference and measured stator current vector components (**c**,**d**).

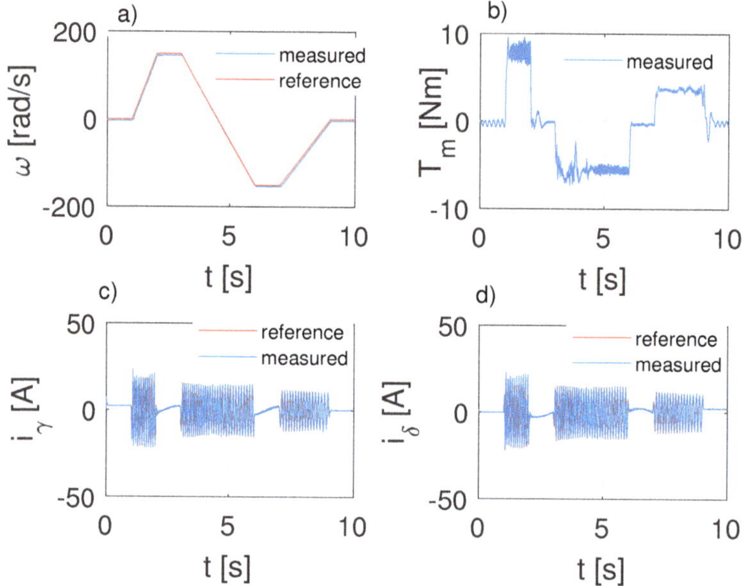

Figure 9. Control performance for the speed reversal command: speed response (**a**), measured torques (**b**), and reference and measured stator current vector components (**c,d**).

7. Conclusions

In this paper we proposed an extended controller for an induction machine based on the 3-D non-holonomic integrator that assures tracking of a general torque reference. This was achieved by extending a basic non-linear state controller with a separate control of the rotor flux linkage magnitude. The modified control results in a structure that is similar to FOC, although a completely different approach was used. It should be pointed out that brief additional results obtained with classical FOC are given exclusively to show that similar overall performance could be obtained with the proposed approach and are not meant for detail performance evaluation or comparison. The additional control loop in proposed control scheme provides extra flexibility to meet performance requirements posed on the drive. As occasion demands the rotor flux linkage magnitude can be optimized to assure energy efficiency or can be kept constant to provide high dynamic tracking performance in servo applications. It follows from our experience that both requirements are hard to satisfy simultaneously without prior information about the class of reference signals. Simultaneous online optimization of the flux magnitude for completely arbitrary torque reference is computationally still too demanding for existing hardware. With the proposed controller, singularity problems of the original controller (4) are completely removed. Since the nonlinear state controller requires a rotor flux linkage estimate, further improvement could be achieved by implementing well known estimation techniques in the proposed control.

Author Contributions: Conceptualization, B.G. and A.H.; writing—original draft preparation, B.G. and A.H.; experimental validation, G.Š.; writing—review and editing, G.Š. All authors have read and agreed to the published version of the manuscript.

Funding: This work was supported in part by the Slovenian Research Agency, project no. P2-0115 and J2-1742.

Conflicts of Interest: The authors declare no conflict of interest.

References

1. Marino, R.; Peresada, S.; Tomei, P. Global Adaptive Output Feedback Control of Induction Motors with Uncertain Rotor Resistance. *IEEE Trans. Autom. Control* **1999**, *44*, 964–980. [CrossRef]
2. Bodson, M.; Chiasson, J. Differential-Geometric Methods for Control of Electric Motors. *Int. J. Robust Nonlinear Control* **1998**, *8*, 927–952. [CrossRef]
3. Oteafy, A.; Chiasson, J. Lyapunov stability of an open-loop induction machine. In Proceedings of the American Control Conference, St. Louis, MO, USA, 10–12 June 2009; pp. 3452–3457.
4. Ortega, R.; Loria, A.; Nicklasson, P.J.; Siera-Ramirez, H. Current fed induction motors. In *Passivity-Based Control of Euler-Lagrange Systems*; Springer: Berlin, Germany, 1998; pp. 381–439.
5. Peresada, S.; Tilli, A.; Tonielli, A.Current regulation strategies for vector-controlled induction motor drives. *IEEE Trans. Power Electron.* **2003**, *18*, 151–163. [CrossRef]
6. Holmes, D.G.; McGrath, B.P.; Parker, S. Current regulation strategies for vector-controlled induction motor drives. *IEEE Trans. Ind. Electron.* **2012**, *59*, 3680–3689. [CrossRef]
7. Hajian, M.; Soltani, J.; Markadeh, G.A.; Hosseinnia, S. Adaptive nonlinear direct torque control of sensorless IM drives with efficiency optimization. *IEEE Trans. Ind. Electron.* **2010**, *57*, 975–985. [CrossRef]
8. Casadei, D.; Member, S.; Serra, G.; Stefani, A.; Tani, A.L. Zarri, L. DTC drives for wide speed range applications using a robust flux-weakening algorithm. *IEEE Trans. Ind. Electron.* **2007**, *54*, 2451–2461. [CrossRef]
9. Alexandridis, A.T.; Konstantopoulos, G.C.; Zhong, Q.-C. Advanced integrated modeling and analysis for adjustable speed drives of induction motors operating with minimum losses. *IEEE Trans. Energy Convers.* to be published. [CrossRef]
10. Stojić, D.; Milinković, M.; Veinović, S.; Klasnić, I. Improved stator flux estimator for speed sensorless induction motor drives. *Ieee Trans. Ind. Electron.* **2015**, *30*, 2363–2371. [CrossRef]
11. Canudas de Wit, C.; Ramirez, J. Optimal Torque Control for Current-Fed Induction Motors. *IEEE Trans. Autom. Control* **1999**, *44*, 1084–1089. [CrossRef]
12. Fernández-Bernal, F.; García-Cerrada, A.; Faure, R. Model-Based Loss Minimization for DC and AC Vector-Controlled Motors Including Core Saturation. *IEEE Trans. Ind. Appl.* **2000**, *36*, 755–763. [CrossRef]
13. Stumper, J.F.; Dötlinger, A.; Kennel, R. Loss Minimization of Induction Machines in Dynamic Operation. *IEEE Trans. Energy Convers.* **2013**, *28*, 726–735. [CrossRef]
14. Borisevich, A.; Schullerus, G. Energy Efficient Control of an Induction Machine Under Torque Step Changes. *IEEE Trans. Energy Convers.* **2016**, *31*, 1295–1303. [CrossRef]
15. Schnurr, C.; Hohmann, S.; Kolb, J. Non-linear MPC for winding loss optimised torque control of anisotropic PMSM. *J. Eng.* **2018**, *17*, 4252–4256. [CrossRef]
16. Brockett, R. Characteristic phenomena and model problems in nonlinear control. *IEEE Trans. Autom. Control* **1998**, *48*, 391–396. [CrossRef]
17. Hespanha, J.P.; Morse, A.S. Stabilization of nonholonomic integrators via logic-based switching. *Automatica* **1999**, *35*, 385–393. [CrossRef]
18. Morgansen, K.A. Controllability and trajectory tracking for classes of cascade-form second order nonholonomic systems. In Proceedings of the 40th IEEE Conference on Decision and Control, Orlando, FL, USA, 4–7 December 2001; pp. 3031–3036.
19. Bloch, A.; Drakunov, S. Stabilization and tracking control in the nonholonomic integrator via sliding modes. *Syst. Control Lett.* **1996**, *29*, 91–99. [CrossRef]
20. Xua, W.L.; Huob, W. Variable structure exponential stabilization of chained systems based on the extended nonholonomic integrator. *Syst. Control Lett.* **2000**, *41*, 225–235. [CrossRef]
21. Grčar, B.; Cafuta, P.; Štumberger, G.; Stankovič, A.; Hofer, A. Non-holonomy in Induction Torque Control. *IEEE Trans. Control Syst. Technol.* **2011**, *19*, 367–375. [CrossRef]
22. Grčar, B.; Hofer, A.; Cafuta, P.; Štumberger, G. A contribution to the control of the non-holonomic integrator including drift. *Automatica* **2012**, *48*, 2888–2893. [CrossRef]
23. Hofer, A.; Grčar, B.; Štumberger, G.; Horn, M.; Reichhartinger, M. Führungsregelung für den nichtholonomen Integrator mit Drift. *Automatisierungstechnik* **2015**, *63*, 700–712.
24. Chen, C.T. *Linear System Theory and Design*; Oxford University Press: Oxford, UK, 1984.

25. Heinemann, G.; Leonhard, W. Self-tuning field-oriented control of an induction motor drive. In Proceedings of the International Power Electronics Conference, Tokyo, Japan, 2–6 April 1990.

26. Novotnak, R.T.; Chiasson, J.; Bodson, M. High-Performance Motion Control of an Induction Motor with Magnetic Saturation. *IEEE Trans. Control Syst. Technol.* **1999**, *7*, 315–327. [CrossRef]

27. Atans, M.; Falb, P.L. *Optimal Control*; McGraw-Hill Book Company: New York, NY, USA, 1966.

28. Shampine, L.F.; Kierzenka, J.; Reichelr, M.W. *Solving Boundary Value Problems for Ordinary Differential Equations in MATLAB with bvp4c*; The MathWorks Inc.: Natick, MA, USA, 2000.

29. Sontag, E.D. Smooth stabilization implies coprime factorization. *IEEE Trans. Autom. Control* **1989**, *34*, 435–443. [CrossRef]

Article

Water and Energy Efficiency Improvement of Steel Wire Manufacturing by Circuit Modelling and Optimisation

Muriel Iten [1,*], Miguel Oliveira [2], Diogo Costa [3] and Jochen Michels [4]

[1]   Instituto de Soldadura e Qualidade, 4415-491 Grijó, Portugal
[2]   Departamento de Engenharia Química, Instituto Superior Técnico, 1049-001 Lisboa, Portugal; miguelmcoliveira@tecnico.ulisboa.pt
[3]   Departamento de Engenharia Mecânica, Faculdade de Engenharia da Universidade do Porto, 4200-465 Porto, Portugal; up201406806@fe.up.pt
[4]   Dechema Research Institute, 60486 Frankfurt am Main, Germany; michels@dechema.de
*   Correspondence: mciten@isq.pt; Tel.: +351-22-747-1950

Received: 29 November 2018; Accepted: 3 January 2019; Published: 11 January 2019

Abstract: Industrial water circuits (IWC) are frequently neglected as they are auxiliary circuits of industrial processes, leading to a missing awareness of their energy- and water-saving potential. Industrial sectors such as steel, chemicals, paper and food processing are notable in their water-related energy requirements. Improvement of energy efficiency in industrial processes saves resources and reduces manufacturing costs. The paper presents a cooling IWC of a steel wire processing plant in which steel billets are transformed into wire. The circuit was built in object-oriented language in OpenModelica and validated with real plant data. Several improvement measures have been identified and an optimisation methodology has been proposed. A techno-economic analysis has been carried out to estimate the energy savings and payback time for the proposed improvement measures. The suggested measures allow energy savings up to 29% in less than 3 years' payback time and water consumption savings of approximately 7.5%.

Keywords: energy efficiency; industry; water circuits; OpenModelica; optimisation

---

1. Introduction

Industrial Water Circuits (IWCs) encompass all the water streams of a factory unit, including the most common uses of water industrially such as water cooling and water treatment. Such processes are responsible for significant consumption of water as well as energy. The industrial sector is responsible for 20% of the total water consumption in the world [1], while in terms of energy it represents 30% of the global final end-use of energy [2]. The European industry in particular is responsible for 40% of the total water abstractions [3] and 25% of the final end-use of energy in the European Union is attributed to this sector [4]. Electric motors, in turn, represent 65% of the total electric energy consumption in European industry [5] and pumping systems in particular account for 21% of the aforementioned share of electricity consumption [6]. These figures show that it is crucial to improve the energy efficiency of IWCs.

The 2012 Energy Efficiency Directive [7] and implementing directive for the Ecodesign requirements for water pumps [8] focus on reducing energy consumption and consequently, water consumption. These directives emerged in the scope of the climate change and energy targets of the Europe 2020 Strategy [9]. Hence, increasing attention have been given to alternative strategies for the reduction of energy and water consumptions.

The European project WaterWatt [10] aims to develop new tools and guidelines to reduce energy consumption in IWCs. IWCs are generally treated as secondary streams in the production process of a plant and therefore neglected. Furthermore, the associated energy consumption is frequently despised, as the circuits only represent a minor part of the total electricity consumption. In this alignment, the present paper presents an object-oriented model, developed in the open source OpenModelica Software in order to analyse and improve the energy consumption of an IWC case study part of the WaterWatt project [11].

The optimisation problems approached in this work are based on the application of improvement measures to reduce electric energy consumption and water consumption in IWC. The measures for energy efficiency improvement in IWC, and similar systems such water networks and water distribution systems, have been proposed by several authors. Cabrera et al. [11] proposed two sets of strategies for energy efficiency improvement, an operational-type set and structural-type set. In operational group, it was proposed: the operation of the pumping systems at the best efficient point (BEP), the improvement of the system regulation to avoid surplus energy, the minimization of leaks and the minimization of friction losses. For the structural group, it were proposed: the use of more efficient pumps, the decoupling of energy sectors and the improvement of designs and layouts.

As part of the WaterWatt project and this work specifically, the optimization based on energy efficiency improvement of the IWC will be performed taking into account the knowledge about the proposed measures for this type of systems and the knowledge about the achievement of energy savings by the adjustment of the operation of electric motors. The application of variable speed drives (VSD) in electric motors is proposed to attain considerable reductions in both water and energy consumptions, which in the case of the pumping systems it has been revealing as an excellent alternative to currently applied improvement measures.

In this paper, different improvement measures are proposed, and energy savings and payback associated to these are estimated. The case study corresponds to water cooling circuit of a steel wire processing plant in Germany in which steel billets are transformed into wire. In general, iron and steel industry present an electrical energy consumption of 2327 ktoe/year, which represents about 12% of the total energy consumption of the industrial sector namely in Germany [12]. Moreover, the production of basic metals in the manufacturing industry in Germany has a total water consumption of about 581 dam^3, in which 85% corresponds to water cooling [13].

2. Model Assembling

2.1. Circuit Description

The cooling circuit under analysis includes a hot roiling mil, four cooling towers, three pump groups, four sand filters, one oil separator and a cyclone separator (Figure 1). The pump group 1 is constituted by two pumps that transport cooled water from a tank (Cold Water Tank) to the rolling mill. Then, the water stream flows down the channel to an oil separator and cyclone (Cyclone Tank) for the pre-treatment of water. The pre-treated water is distributed to two sand filters by the action of pump group 2, constituted by two pumps in operation. From the sand filters, the water stream flows back to the cooling tower tank. The designated pump group 3 is part of a secondary circuit, constituted by two pumps. Pump group 3 corresponds to the same type as pump groups 1 and 2 to reduce maintenance and logistic effort. This separate water stream is transported from the cooling tower tank to the four cooling towers and the outlet water streams from the cooling tower are then, transported back to the tank. This secondary circuit is part of a recent modernisation drive in the factory. Such modernization aims to improve energy efficiency. It was installed to allow the operation of the cooling towers only when required (i.e., when the temperature at the cold water tank increases above the operational target-value by 0.5 °C) and as soon as temperature drops the circuit stops. For instance, this may constitute a constraint for the process optimization based on energy efficiency improvement, in which the temperature of the cold water tank must not be exceeded by 0.5 °C relatively to its target-value. Otherwise, this operational requirement is considered to be surpassed. The separation of the cooling towers from the main circuit also allows a closer control on the cooling process as the water demand in the circuit is variable. This is because the billets are produced in batches and due to the existence of variabilities in the material and the rolling mill operational conditions, such as rolling speed and cooling speed.

The inlet water stream in the rolling mill is used for the main purpose of direct and indirect cooling. Nonetheless, it used in the remaining processes of hot forming, such as descaling and scale transport. For that reason, a water treatment section is installed after the hot forming process. This section is constituted by the filtration units, constituted each one, by two sand filters. Moreover, it is preceded by an oil separator and a cyclone, constituting a pre-treatment section.

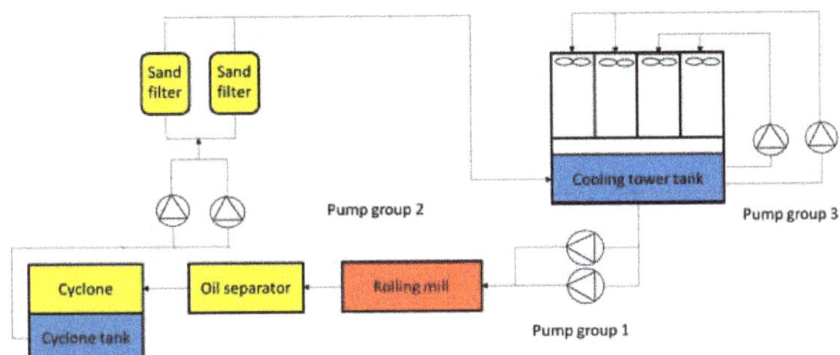

Figure 1. Schematics of the case study cooling circuit.

Table 1 summarizes the specifications for the design and assembling of the model circuit (described Section 2.2).

Table 1. Specifications for the assembling of the cooling circuit.

	Annual Operational Time (h/year)		6600
	Main Circuit Nominal Flow rate (m^3/h)		1000
	Secondary Circuit Nominal Flow rate (m^3/h)		1200
	Main Circuit Operational Flow Rate (m^3/h)		900
	Secondary Circuit Operational Flow Rate (m^3/h)		900
	Latitude of the Plant		51°36′ N
	Pump Group 1	Nominal Flow rate (m^3/h)	500
		Nominal Head (m)	48
		Motor Velocity (rpm)	1450
		Motor Efficiency (%)	90
		Design Power (kW)	110
		Running Power (kW)	90
	Rolling Mill Pipe	Diameter (m)	0.5
		Length (m)	200
		Pressure Drop (bar)	4.6
		Outlet Temperature (°C)—$T_{in,1}$	24.5
	Pump Group 2	Nominal Flow rate (m^3/h)	500
		Nominal Head (m)	48
Main circuit		Motor Velocity (rpm)	1450
		Motor Efficiency (%)	90
		Design Power (kW)	110
		Running Power (kW)	90
	Cyclone Tank	Area (m^2)	45
		Height of water (m)	2
	Pipe (Pump Group 2 to Sand Filters)	Diameter (m)	0.5
		Length (m)	200
		Pressure Drop (bar)	3.1837
	Sand Filters	Nominal Pressure Loss (bar)	1.3
		Area (m^2)	78.5
		Diameter (m)	5
	Pipe (Sand Filters to Cold Water Tank)	Diameter (m)	0.5
		Length (m)	20
		Pressure Drop (bar)	0.3184

Table 1. *Cont.*

		Nominal Flow rate (m^3/h)	600
		Nominal Head (m)	47.5
	Pump Group 3	Motor Velocity (rpm)	1450
		Motor Efficiency (%)	90
		Design Power (kW)	110
		Running Power (kW)	104
Secondary circuit		Motor Velocity (rpm)	1500
		Installed Power (kW)	22
	Cooling Towers	Inlet Temperature (°C)—$T_{out,2}$	22
		Outlet Temperature (°C)—$T_{in,2}$	20
		Ambient Temperature (°C)	11
		Wet-bulb Temperature (°C)	8
		Diameter (m)	0.5
	Pipe (Cold Water Tank to Cooling Towers)	Length (m)	7
		Pressure Drop (bar)	3.96
Main and Secondary circuit	Cold Water Tank	Area (m^2)	75
		Height of water (m)	1
		Material	Stainless Steel
	Additional pipe parameters	Roughness (mm)	0.015
		R_e	24300
		Friction Factor	0.025

The heat input at the hot rolling mill ($q_{input,RM}$) is determined by:

$$q_{RMinput} = Q_1 \times \rho_w \times c_{pw} \times (T_{RMinlet} - T_{RMoutlet}) \tag{1}$$

where Q_1 and Q_2 corresponds to the water flowrate of the main and secondary circuit respectively (Table 2). The density of the water (ρ_w) corresponds to 999 kg/m^3 and the specific heat (c_{pw}) to 4.186 kJ/kg°C. The T_{inlet_RM} is determined by:

$$Q_1 \times \rho_W \times C_{P,w} \times (T_{in,1} - T_{out,1}) = Q_2 \times \rho_W \times C_{P,w} \times (T_{in,2} - T_{out,2}) \tag{2}$$

whereas:

$$T_{out,1} = T_{RMinlet} \tag{3}$$

These two additional inputs of the circuit are listed in Table 2.

Table 2. Additional inputs in the IWC.

Parameter	Results
$q_{input,RM}$ (kW)	1380
$T_{RMinlet}$ (°C)	22.1

The measurements at the plant included the water temperature, water flowrate and pressure drop across pumps and sand filters (Table 3). In order to measure the flowrate, ultrasonic signals have been used, employing the transit time difference principle. The ultrasonic signals are emitted by a transducer installed on the pipe and received by a second transducer. The transit time difference is measured and allows the flowmeter to determine the average flow velocity along the propagation path of the ultrasonic signals. A flow profile correction is then performed to obtain the area averaged flow velocity, which is proportional to the volumetric flow rate. Two integrated microprocessors control the entire measuring process, removing disturbance signals, and checking each received ultrasonic wave for its validity which reduces noise. Moreover, the temperature of the water circulating within the pipes was measured by coupling two resistance thermometer (RTD) temperature probe (applicable to the supply and return pipes simultaneously) whose outputs are integrated with the flow measurements at the meter. The pressure drops across pumps and filters have been measured by the coupling of manometers before and after each component. Also, the power of pumps have been recorded during representative period of time. This has been performed by a power analyser connected to a data logger, taking three phase power measurements by coupling clamp meters into the phases. Table 1 summarizes the specifications for the design and assembling of the model circuit (described Section 2.2).

Table 3. Specifications for each type of measurement.

Measurement	Flowrate	Temperature	Pressure	Power
Range	0.03 to 82 ft/s	−238 to +1040 °F	0 to 100 psi	10 W to 10 GW
Resolution	0.15% of reading ±0.03 ft/s	0.01 K	2 psi	±1% ± 0.3% of nominal power
Accuracy	±1.6% of reading ±0.03 ft/s	±0.01% of reading ±0.03 K	3/2/3% of reading	0.01 kW

2.2. Design and Assembling of IWC Model

The IWC was built in the object-oriented language in OpenModelica, used for the modelling of such physical systems and supports most of the equipment (Figure 2).

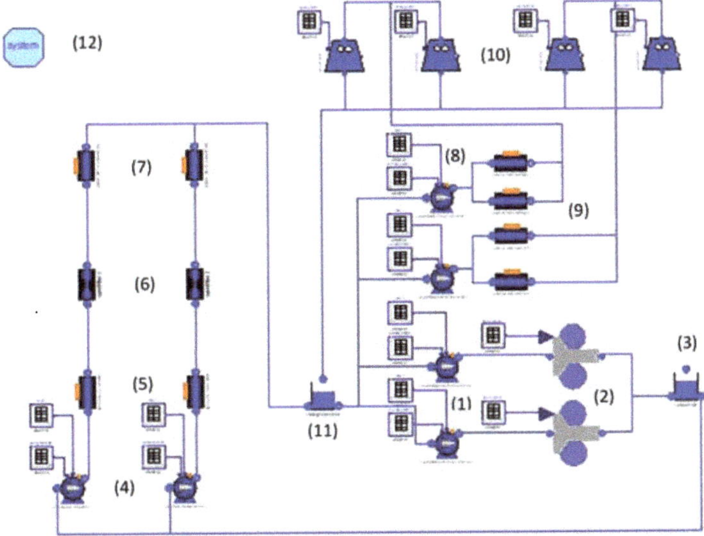

Figure 2. Diagram of the circuit of a steel wire processing plant in OpenModelica (Legend: (1) Pump Group 1, (2) Hot Rolling Mills, (3) Cyclone Tank, (4) Pump Groups 2, (5) Pipes from pumps to filters, (6) Sand Filters, (7) Pipes from filters to tank, (8) Pump Group 3, (9) Pipes from pumps to cooling towers, (10) Cooling Towers, (11) Cold Water Tank, (12) Reference System).

Currently, a number of free and commercial component libraries in different domains are available, including electrical, mechanical, thermo-fluid and physical-chemical. The current case study was built using the ThermoPower library [14] that has been developed with the goal of providing a solid foundation for dynamic thermal and power modelling. It includes a list of components, such as pumps, tanks, cooling towers, pipes and filters, that can be "drag and drop" for the modelling assembling. Each component corresponds to singular model to be setup accordingly with the required parameters and therefore, replicability is assured.

The selected components and respective models correspond to the main components of the circuit namely: tank model, pump model, pipe model, filter model, cooling tower model and rolling mill as further described. Each model was accordingly setup using the measured data collected in the case study circuit and presented in the beginning of this section.

2.2.1. Tank Model

The model represents a free-surface water tank corresponding to the vertical axis in the region where the fluid level is supposed to lie within. The tank geometry is described the input parameters: V_0 (volume of the fluid when the level is at reference zero height), A (cross section of the free surface) and y_0 (height of "zero-level"). The model also requires the input parameters *ymin* and *ymax*, which should be appropriate to avoid underflow or overflow of the tank. The pressure at the inlet and outlet ports accounts for the ambient pressure and the static head due to the level, while the pressure at the inlet port corresponds to the ambient pressure.

Concerning the mass and energy balances the model assumes there is no heat transfer except through the inlet and outlet flows. It is to note that, due to the inexistence of the performance of heat transfer directly in the modelling of the cold water tank in the present IWC, the sand filters outlet water streams are cooled through the contact with the water stream from the secondary circuit at a lower temperature.

2.2.2. Pump Model

The model represents a centrifugal pump, or rather a group of centrifugal pumps in parallel. The input parameters are related with the pump curves therefore defined in the model for each single pump or pump group. The component characteristics are given for nominal conditions of flowrate and rotational speed. These characteristics may be then adapted according to similarity equations, namely the affinity laws for the pumps to different flow conditions (Equation (7)).

There are two input functions: *flowCharacteristic* that corresponds to the relationship of the volumetric flowrate and the head, while the *efficiencyCharacteristic*, corresponds to the relationship between the volumetric flowrate and the hydraulic efficiency. One of the output parameter is the electric power of the pump (Pump). This also accounts for the mechanical efficiency of the pump motor (*motorEfficiency*) also set as input.

In addition to the inlet and outlet ports of the water stream, the pump models also include inlet ports for the motor speed (*in_n*), the electricity tariff (*tax*) and the number of pumps in parallel (*in_NP*).

2.2.3. Pipe Model

The model represents the piping system of the circuit, either as a single tube or by N parallel tubes, representing the water flow streams of the factory unit. It is built with certain assumptions, such as: one-phase fluid state, uniform velocity across the section, constant turbulent friction, longitudinal heat diffusion is neglected and uniform pressure distribution in the energy balance equation.

The finite volume method is used to discretize the mass, momentum and energy balance equation, taking as the state variables one pressure, one flowrate and N-1 specific enthalpies. The dynamic momentum is not accounted by default.

The input parameters for the pipe model are: the pressure losses on the pipes due to the friction phenomena (*dpnom*) and the static head (*h*) as well as the mass flowrate (*wnom*). In addition to the inlet and outlet ports of the water stream, the pipe model also contains a wall-type inlet (*wall*), which can be specified for a certain heat input. Taking into account that the circuit pipes are all installed at the same level, this is, it does not exist pressure losses due to elevation, the total pressure drop of the sections of the piping system are determined considering the head losses due the friction of the pipes, according to:

$$\Delta P_f = \rho_w g f \frac{L}{D} \frac{u^2}{2g} \qquad (4)$$

The friction factor (*f*) was determined using the Moody pipe friction chart, according to a transitional flow regime ($R_e = 24{,}300$). Moreover, for the total head losses the minor losses are also considered. The minor losses consist of the local pressure drop (Equation (5)) due to sudden or gradual expansion or contraction, the presence of bends, elbows, tees and valves (open or partially closed).

$$\Delta P_L = \rho_w g f K_L \frac{u^2}{2g} \tag{5}$$

The K_L is determined according to tabulated values for each type of local pressure drop [15].

2.2.4. Filter Model

The model is set to induce and determine the pressure loss in the circuit. The pressure drop across the inlet and outlet connectors of the components is proportional to the square velocity of the fluid, defined according to a turbulent friction model. The input parameter corresponds to the pressure drop across de filter (*dpnom*).

2.2.5. Cooling Tower Model

The model represents an open cooling tower with fans and it is filled with metallic packaging. The humid ambient air is used to cool the incoming water stream. The fluid is modelled using the *IF-97 water-steam model* as an ideal mixture of dry air and steam.

The hold-up of water in the packaging is modelled assuming a linear relationship between the hold-up in each volume and the corresponding outgoing flow. The finite volume method is used to discretize the corresponding 1-D counter-current equations. The Merkel's equation [16] is used for the mass and heat transfer phenomena from the hot water to the humid air (Equation (6)):

$$\frac{k\,S}{Q_{Mw}} = C_{pw} \int_{T_{w,out}}^{T_{w,in}} \frac{1}{(H_s - H_{air})} dT \tag{6}$$

Note that k designates the overall heat and mass transfer coefficient and S the surface of packing. The required input parameters correspond to: the overall mass and heat transfer (*k_wa_tot*), nominal water mass flowrate (*wlnom*), nominal air volume flow rate (*qanom*), normal air density (*rhoanom*), surface of packing (*S*), nominal fan rotational speed (*rpm_nom*) and the nominal power consumption (*Wnom*). The fan behaviour is modelled by kinematic similarity: the air flow is proportional to the fan speed (*rpm_nom*) and the electric power consumption (*Wnom*) is proportional to the cube of the speed, according to the affinity laws for the fans (Equations (7) and (8)).

The cooling tower model also contain an inlet port for the fan speed (*fanRpm*). In which, similarly to the pump motor rotational speed, the fan speed can be changed relatively to the nominal value (*in_n*). The outcome parameters include the inlet ($T_{CTinlet}$) and outlet ($T_{CToutlet}$) ports of the water stream, there is also an outlet port for the electric power consumption of the fan (*powerConsumption*).

2.2.6. Rolling Mill

The rolling mill connects two models from the ThermoPower library, namely the heat source model and the pipe model already described above. Regarding the heat source model it corresponds to an ideal tubular heat flow source, with uniform heat flux. The actual heating power (P) is provided as input by the power signal connector (*in_p*). The finite volume method is used for discretisation. The output parameter corresponds to the water outlet temperature ($T_{RMoutlet}$).

While the comprehension of the thermal phenomena intrinsic in the analysis of this type of circuits have been performed for the design and calibration assuming the more accurate values possible, it is to note that the cooling system of a rolling mill component has been simplified. This represents a thermal power input in the circuit, allowing the increase of temperature proportional to that heat input and water flow rate. Therefore, it still lacks a more complex understanding of the occurrence of phenomena at a more dynamic basis.

2.2.7. System and Time Tables

The System component must be added in any circuit in order to define the system condition, namely water temperature and ambient pressure. In this study, the temperature was considered approximately 15 °C according to the average water temperature provided by the utility company in Germany. The Time Table component allows the user to specify a certain input value over time for a specific component. In the current case study, they are connected to the pump motor and to the fan of the cooling tower to specify the rotational speed of these equipment(rpm). Furthermore, it can also be connected to the rolling mill to specify the thermal input change with time. Additionally, a time table can be connected to specify the electricity tariff price with time.

3. Model Validation

The model results have been validated with real casa data from the case study. For the energy and water consumption the following parameters have been identified for validation: pumps power (*Pump1*, *Pump4*, *Pump8*), the inlet and outlet water temperature at the cooling towers ($T_{CTinlet}$ and $T_{CToutlet}$) and the water outlet temperature of the rolling mills ($T_{RMoutlet}$). The outputs of the numerical results for these parameters are presented from Figures 3–5.

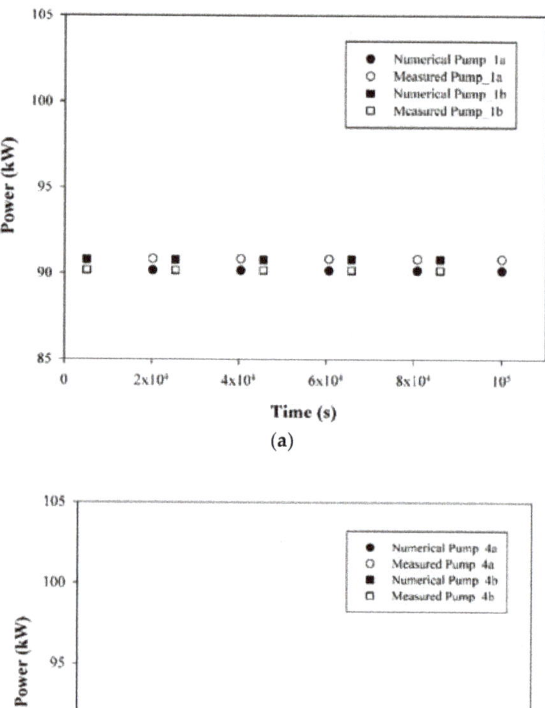

Figure 3. *Cont.*

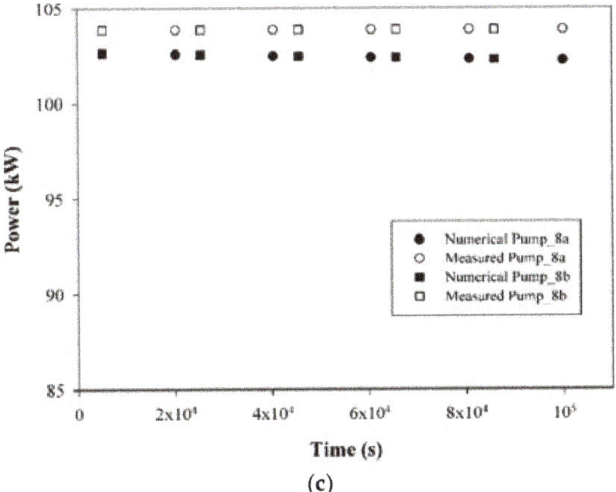

(c)

Figure 3. Numerical and measured pump power (**a**) pump group 1, (**b**) pump group 2 and (**c**) pump group 3.

Furthermore, it is also relevant to analyse the values obtained for the temperatures in different points of the circuit, namely the inlet and outlet water stream temperature at the cooling tower (Figure 4) and outlet water temperature at the rolling mill (Figure 5).

The deviation that has been observed for each pump group to comparison to the nominal value was listed in Table 3 above. The measured electric power used for the operation of the pumps was analysed and validated against plant data and summarized in Table 1. It is observed that the values obtained in the simulation are overall consistent with plant data. Moreover, the value outlet water stream temperature of the hot rolling mill and at the inlet of the cooling towers have minor deviations compared to the real values. The slightly major deviation is related with the outlet temperature at the cooling towers (Table 4). This may be related with the fact that the water circuit in the cooling towers corresponds to a separated circuit from the main circuit. Overall, the thermal phenomena are consistent with plant data and the required cooling load is given by the model.

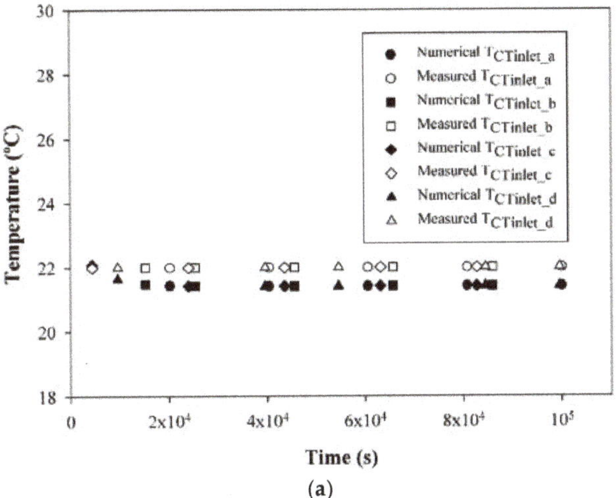

(a)

Figure 4. *Cont.*

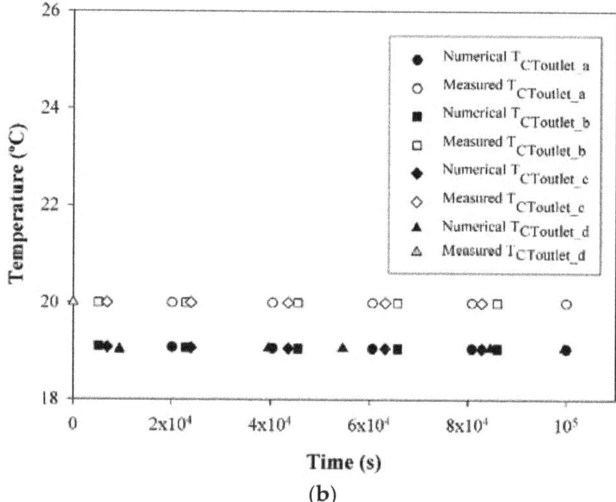

Figure 4. Water temperature at cooling tower (**a**) inlet: $T_{CTinlet}$ and (**b**) outlet: $T_{CToutlet}$.

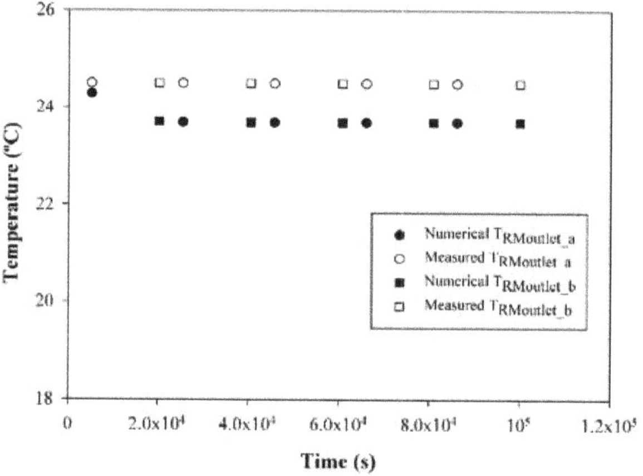

Figure 5. Water outlet temperature of rolling mill ($T_{RMoutlet}$).

Table 4. Deviations obtained for the pumps measured power.

Parameter	Deviation (%)
Power of Pump Group 1 (Pump_1)	0.74
Power of Pump Group 2 (Pump_4)	0.77
Power of Pump Group 3 (Pump_8)	1.56
Inlet Temperature of Cooling Tower ($T_{CTinlet}$)	2.62
Outlet Temperature of Cooling Tower ($T_{CToutlet}$)	4.75
Outlet Temperature ($T_{RMoutlet}$)	3.29

4. Sensitivity Analysis

A sensitivity analysis of the model was performed. It corresponds to a crucial analysis for the solving of the optimisation problems. The methodology to perform this analysis is distinct from the ones elaborated in dynamic modelling and simulation. As there is a high number of required inputs to be given to the model in order to run, a pre-selection of variables to analysis is required. The selected variables for the analysis correspond to the speeds of pump motor and the cooling tower fan which have a direct influence on power consumption and operational requirements of the IWC. Different scenarios have been defined for these variables (Tables 5 and 6) to understand their influence on the power consumption (Figure 6). The scenarios have been assembled following a methodology similar to the Monte Carlo method. For this study, is based on random sampling, in which the results for power consumption are obtained for random rotational speeds values. Thus, different values for power consumption are obtained for different scenarios. To note that the best scenarios have been selected based on the lower power consumption of all the pump groups and also without surpassing the operational requirements.

Table 5. Scenarios for the variation of Pump Rotational Speed.

Scenario	Pump Groups 1 and 2 Rotational Speed (nPG1 and nPG2)	Pump Group 3 Rotational Speed (nPG3)
1	1444	1401
2	1413	1434
3	1408	1373
4	1405	1342
5	1372	1328
6	1418	1401
7	1411	1392
8	1350	1350
9	1442	1403
10	1341	1400

Moreover, it is also necessary to evaluate the influence of those input parameters on simulations results related to temperature and hydraulic parameters. In this study, the influence of the change of the motors speed of pumps and fans (cooling tower) on the water temperature at the cold water tank will be analysed (Figures 7 and 8).

For the sensitivity analysis related with the change of the pump motor rotational speeds, the power consumptions of both pump groups 1 and 2 were analysed simultaneously, since both are part of the main circuit, and having the same technical characteristics. Figure 6 presents the results ten scenarios for the power consumption of pump groups 1, 2 and 3.

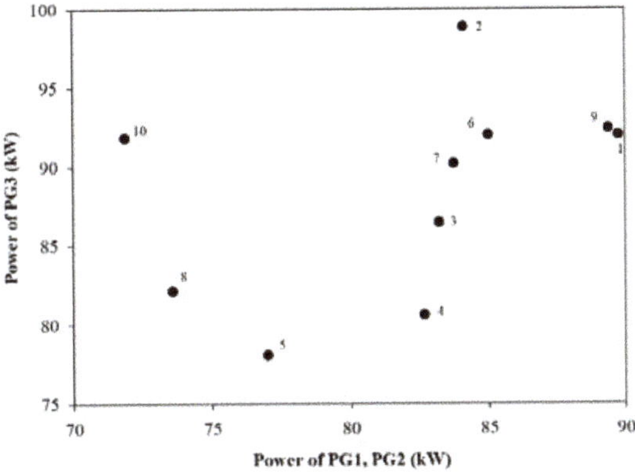

Figure 6. Influence of the rotational speed of the pump motors on the power consumption.

From the ten created scenarios, it is necessary to select the ones associated to higher reduction in power consumption of both pump group 1 and 2 and pump group 3. This analysis focuses on the favorable scenarios on the viewpoint of energy efficiency and the possibility of its application. Therefore, observing Figure 6, the three scenarios correspond to 5, 8 and 10 and to be selected to proceed with the sensitivity analysis (Figure 7). The remaining scenarios were neglected, as they do not present significant reduction in power consumption.

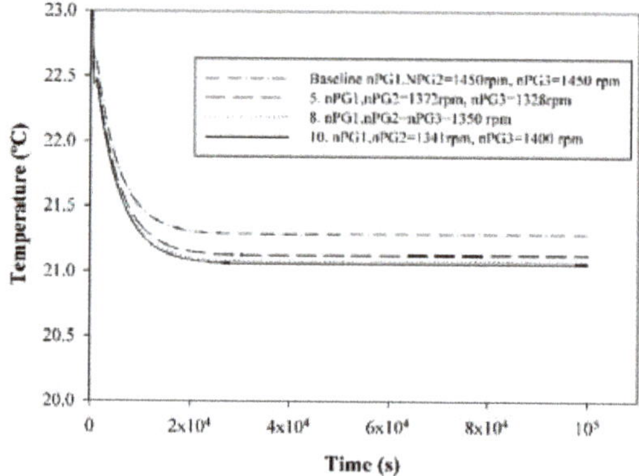

Figure 7. Influence of rotational speed of the pump motors on the water temperature of the cold water.

Table 6. Scenarios for the variation of Cooling Tower Fans Rotational Speed.

Scenario	Cooling Towers Fans Rotational Speed (nCT)
1	1392.48
2	1350.50
3	1190.55

The scenarios for the cooling towers (Table 6) were created considering the share of reduction in power consumption by the decrease of the rotational speed of the cooling tower fans, following the affinity laws of the fans (Equation (8)). Thus, three scenarios designated 1, 2 and 3 were created bearing in mind shares of reduction in power consumption of, respectively, 20%, 27% and 50%.

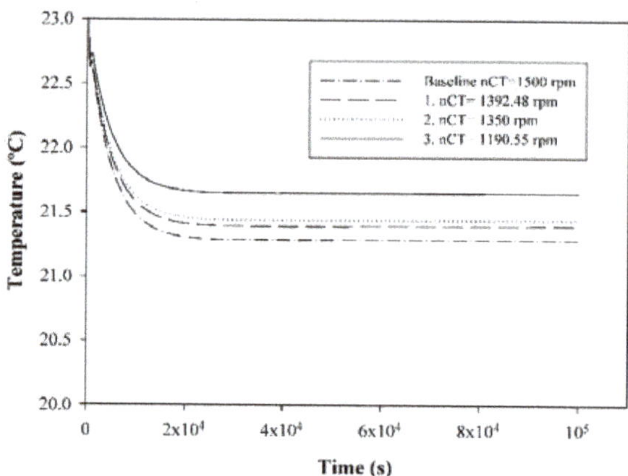

Figure 8. Influence of rotational speed of the cooling tower fans (nCT) on the water temperature of the cold water tank (T).

The most efficient scenarios for each case: promoting the major energy reductions in power consumption respecting the operational requirements in temperature, will be analysed furtherly.

5. Identification of Improvement Measures

The efficiency of water circuits can be reached by the dynamic adjustment of the circuit operation to the production process, through the reduction of pressure demand and pressure losses in the circuit and by improvement of energy efficiency of pumps and cooling tower. Such improvements include matching the scale of the motor service to the work demand (i.e., speed control devices such as adjustable speed drives); reducing demand for energy services (e.g., substituting a blower for compressed air or turning off steam supplied to inactive equipment) and the replacement to high efficiency motors. The improvement measures should account for the technical specifications of the process: avoiding interruptions, improvement of production reliability (maintaining required flow, pressure and temperature) and improved maintenance practices. The measures do also account for economic and environmental aspects namely cost reduction, energy savings and reduction of CO_2 impact which are the primary aspect that industrials consider regarding energy efficiency.

The present model allows a prior analysis of efficiency measures to be implemented in the IWC avoiding extra costs with technical work. To note that the structural related input variables such as: length and diameter of the pipes, heat inputs in the case of non-looped circuits, despite being highly influential and part of the formulation of certain optimisation problems, they are set as design variables, meaning constant parameters of the assembling of the circuits. Therefore, the analysis of such measures, as related to energy efficiency measures at the design stage and therefore not in the scope of this work.

Furthermore, aim of this work does not focuses on the analysis of process control strategies. Noteworthy such strategies are intrinsic in the operation of all the processes in industry, including the water circuits. Thus, a more accurate analysis of these models would require the comprehension of the models related to types of valves, flowmeters, electric meters and temperature sensors. Such extensions, would possibly assist the operational improvement measures, as well as the application of control related measures. Hence, this paper analyses the application of five improvement measures to the case study circuit in order to reduce the energy and water consumption as well as guaranteeing the required operation conditions.

5.1. Variable Speed Drives (VSD) in Pumps

The coupling of VSD's allow energy savings of 20% to 25% [17]. The installation of VSD in pump motors allows the automatic flow adjustment to the needs of the process. The change of the pump speed adjusts the pump rotation frequency to the optimal efficiency point. The performance of this measure relies on the affinity laws for the pumps. The automatic flow adjustment is described by Equation (7) [18]:

$$\frac{Q_{w,nom}}{Q_{w,opt}} = \frac{Q_{pump,nom}}{Q_{pump,opt}} \tag{7}$$

For the cooling circuit under analysis, the operation of the pump motors was initially set at rotational speed of 1450 rpm for the nominal flowrate. Applying the Equation (7) for the real running flowrate, the optimised rotational speed corresponds to 1350 rpm. New simulations have been performed for the optimal point speed and savings have been estimated and presented in Table 7.

5.2. Refurbishment of Pumps

The refurbishment of pumps corresponds to the mechanical cleaning and overall of a pump to restore its initial functioning, namely its energy efficiency. Pumps without maintenance over the years generate a lower flow. The improvement measure for the pumping system of the case study circuit considered the refurbishment of all pumps. For the calculations, the authors have assumed an improvement of 10% in the hydraulic efficiency of the pump, set in IMechE [19].

5.3. Change of Filters

In water treatment process, sand filters present considerable impact on energy efficiency due to lower pressure drop induced into the water. The improvement measures for the treatment section was to. In replace the two sand filters with a pressure drop of 1.3 bar to new ones with a pressure drop of 0.5 bar. The results are presented in Table 7.

5.4. Change of Electric Motors of the Pumps

The change of electric motors in pumping systems, specifically to electric motors categorized as IE3 Premium Efficiency allow high energy savings. A higher efficiency motor is more expensive than a conventional motor however its lifespan is much longer as it operates at a lower temperature and hence heat losses are lower [20]. In the current circuit, the pump motors with an initial mechanical efficiency of 90% were exchanged to IE3 electric motors with a mechanical efficiency of 95.4%. The savings are presented in Table 7.

5.5. VSD in the Cooling Tower Fans

The installation of VSDs in the cooling tower fans allow a dynamic adjustment of the airflow respecting the required temperature values in the circuit. The power consumption associated to the operation of the fan is directly linked to the fan speed, therefore high energy savings can be achieved with adjustments decreasing the fan speed [21]. The performance of this measure relies on the affinity laws for the fans. The automatic air flow adjustment is described by Equation (7) [22]. Moreover, the change on the power of the fans is described by Equation (8) [18]:

$$\frac{Q_{air,nom}}{Q_{air,opt}} = \frac{\eta_{fan,nom}}{\eta_{fan,opt}} \tag{8}$$

$$\frac{P_{fan,nom}}{P_{fans,opt}} = \left(\frac{\eta_{fan,nom}}{\eta_{fan,nom}}\right)^3 \tag{9}$$

In the circuit under analysis the operation of the four cooling tower fans operates at an initial rotational speed of 1450 rpm. This has been changed to an optimal speed of 1350 rpm. The results are presented in the Table 7.

6. Definition of Optimization Methodology

An optimisation methodology must be developed with the aim to improve energy efficiency in this IWC. This methodology is primarily based on the assembling of several settings for each case, in which a certain input parameter is modified with the aim to achieve lower energy consumption. These are primarily related with the operational conditions of the pumps and cooling towers. The definition of this methodology is necessary for the techno-economic evaluation of the analysed improvement measures. Based on the scenarios assembled in the sensitivity analysis, the objective functions are attained by the changes on the decision variables, although that change must respect the operational requirement of the circuit which are translated by the constraints.

The main objective of the presented optimisation problems is the reduction on electric energy consumption. Although thermal energy consumption is also a relevant concern, and heat is a direct input in certain components of these models, the thermal phenomena are mostly treated as imposed variables of the model, namely as input parameters. Hence, the optimisation methodology includes the definition of objective functions, decision variables and constraints.

The secondary circuit is only switched on when the temperature of the water inside the cold water tank (T_{CWT}) reaches more than 0.5 °C relatively to the target value (22.1 °C). From this, the constraint is defined as a requirement of not exceeding 0.5 °C of the temperature at the cold water tank. This allows to maintain the cooled water temperature close to the required in the cooling process.

The objective functions correspond to the reduction of energy consumption. The decision variables are the same as the ones selected for the sensitivity analysis: rotational speeds of pump motors (n_{PG1}, n_{PG2} and n_{PG3}), cooling tower fans (n_{CT}) and the hydraulic and mechanical efficiencies of the pumps (designated by η_{hydr} and η_{mech}, respectively) as well as the pressure drop in the filters. Table 7 summarizes the objective functions, decision variables and constraints of the optimisation problems.

Table 7. Deviations obtained for the pumps measured power.

Objective Functions	Decision Variables	Constraints
$min\ (ELC_{PG1} + ELC_{PG2} + ELC_{PG3} + ELC_{CT})$	n_{PG1} n_{PG2} n_{PG3} n_{CT} η_{hydr} η_{mech} ΔP_{filter}	$T_{CWT}\ (n_{nom}) - T_{CWT}\ (n) < 0.5\ {}^{\circ}C$

The scenarios presented in the sensitivity analysis enables high energy savings by respecting the circuit operational conditions. For the solving of optimisation problems, the scenarios associated to the lower energy consumption respecting the process limit conditions, this is, its constraints, will be selected and analysed for energy efficiency improvements. The application of VDS's in pump motors is related to scenario 8 for the variation of pump motor speed. The application of VSD's in cooling tower fans are related with the scenario 2 for the variation of fan speed.

7. Techno-Economic Evaluation

A techno-economic analysis was undertaken for the implementation of specific improvement measures. The economic savings of an improvement measure, determined in a per year basis, are, in general, calculated according to Equation (9), in which *Savings* designates power savings, t_{op} the annual operational time, C the electricity cost per energy unit and P_{nom} and P_{opt} the power at initial conditions and optimized conditions, respectively:

$$Savings = t_{op} \times C_{elec} \times (P_{nom} - P_{opt}) \tag{10}$$

In the case of the implementation of electric motors with higher efficiency, the method to calculate economic savings by Equation (10) takes the form of Equation (11):

$$Savings = t_{op} \times C \times \left(\frac{P_{IE1}}{\eta_{IE1}} - \frac{P_{IE3}}{\eta_{IE3}} \right) \tag{11}$$

The P_{IE1} and η_{IE1} correspond to the pump power and its overall efficiency at initial conditions of a standard efficiency motor E1. While P_{IE3} and η_{IE3} correspond to the pump power and its overall efficiency of a premium efficiency motor IE3.

In the case of the application of VSD's in electric motors, the savings are calculated according to Equation (12), in which P_{nom} and η_{nom} designate, the power of a pump and its overall efficiency at initial conditions respectively; P_{opt} and η_{opt} correspond to the power of a pump and its mechanical efficiency, respectively, by the installation of VSD's and i to the numeral designation of a load regime:

$$Savings = \sum_{i} t_{op} \times C \times \left(\frac{P_{nom}}{\eta_{nom}} - \frac{P_{VSD}}{\eta_{VSD}} \right) \tag{12}$$

Since the electric motors in question, namely pump and fan motors, are operated in a continuous load regime, Equation (12) is simplified to:

$$Savings = t_{op} \times C \times \left(\frac{P_{nom}}{\eta_{nom}} - \frac{P_{VSD}}{\eta_{VSD}} \right) \tag{13}$$

The investment payback period of a given measure is calculated according to Equation (14), in which *PB* designates the payback period and *Inv* the investment cost:

$$PB = \frac{Inv}{Savings} \tag{14}$$

The implementation of an improvement measure is considered economically viable under the patterns of European industry if the payback period is below 2–3 years. The economic evaluation was proceeded by considering an average electricity price in Portugal of 0.0923 €/kWh [23] and a simple payback, not considering an inflation rate of the investment costs. The cost of the electric motors has been obtained from the manufacturer catalogue [24] as well as the cost of the variable speed drives [25]. The cost of the refurbishment of each pump was estimated to be 2 full days of working hours of a maintenance technician with a monthly salary of 800 Euros. It is assumed that the material required for repairs and exchanges of equipment are intrinsic to the company hence no cost was considered. The energy savings, a share of energy savings, investment cost and the payback values are presented in Table 8.

Table 8. Energy savings, investment cost and payback period of proposed measures.

Improvement Measure	Annual Energy Savings (MWh/year)	Share of Energy Savings	Investment Cost (€)	Payback (Years)
(i.1) Couple VSD in pump group 1	231.74	19%	21,622	1.0
(i.2) Couple VSD in pump group 2	231.05	19%	21,622	1.0
(i.3) Couple VSD in pump group 3	266.24	20%	21,622	0.9
(ii) Refurbishment of the pumps	524.18	12%	436	0.01
(i.1) and (ii) in pump group 1	400.60	29%	21,786	0.6
(i.2) and (ii) in pump group 2	399.83	29%	21,786	0.6
(i.3) and (ii) in pump group 3	452.77	29%	21,786	0.5
(iii) Change filters (1.3 bar to 0.5 bar)	146.67	4%	42,196	3.1
(iv) Change motors to high-efficiency (IE3)	212.46	6%	134,916	6.9
(i.1) and (iv) in pump group 1	286.51	24%	66,594	2.5
(i.2) and (iv) in pump group 2	285.84	24%	66,594	2.5
(i.3) and (iv) in pump group 3	327.86	24%	66,594	2.2
(v) Couple VSD in fans	157.40	27%	14,872	1.0

Furthermore, the application of VSD's in pump motors also allows the achievement of savings in water consumption by the decrease of water flow rate (Table 9).

Table 9. Savings in Water Consumption.

Improvement Measure	Savings in Water Consumption (m^3/h)	Share of Water Savings (%)
(i.1) Couple VSD in pump group 1 (i.2) Couple VSD in pump group 2	35	7.5
(i.3) Couple VSD in pump group 3	45	
Total	80	

8. Discussion

Overall all the applied improvement measures allowed energy savings. The payback period was generally lower than 3 years, with exception to measure iv) and largely overcoming that period and therefore not considered as an efficient economic measure.

In general, the application of VSD in the pumps (measure i.1, i.2, i.3) allows energy consumption reductions from 19% to 20%. This converges with the typical values of energy savings with such application [19]. Furthermore, it reveals itself as favorable measure due to its low payback time of approximately from 11 mouths to 12 months for all pump groups. The coupling of VSD in the cooling tower fans (measure v) is equally an attractive measure, with energy savings of 27% and payback period of nearly a year.

The refurbishment of pumps (measure ii) is peculiarly advantageous because despite saving less than the applying of VSD it presents a low payback time. This is due to the low investment cost of the implementation of such measure. Adding the application of VSD in pumps and its refurbishment is shown as the most propitious measures since it produces the larger reduction in energy consumption associated to an acceptable payback period.

The replacement of conventional pump motors to the IE3 premium efficiency pumps has not the same propitiousness as the aforementioned measures. This is also applied to the change of filters. However, to note that motors and filters are changed at the end of their lifecycle and savings can be expected at that time. If the replacement is joined with the coupling of VSD however it becomes a very attractive measure presenting a payback of 2.5 years for pump group 1 and 2 and 2.2 years for pump group 3. Such approach is also beneficial on a technical point of view as the application of VSD reduces of the motors lifetime and replacing to a high efficiency motor enables to overcome such limitation [17].

From an environmental perspective, the application of VSD in pumps allows to reduce fresh water consumption. The current nominal water flow rate corresponds to 600 m^3/h for the pump group 3. From the simulation, it was observed that required the cooling load was achieved for a flowrate of 555 m^3/h. Taking into account that average annual water consumption per inhabitant in Germany in 2016 was 45 m^3 [26] the water savings could meet the needs of 6601 inhabitants.

9. Conclusions

The present work presents the modelling and optimisation of an IWC, with the following steps: model assembling, modelling, simulation results, model validation, sensitivity analysis, formulation of optimisation problems, implementation of energy efficiency improvement measures and a techno-economic evaluation.

The simulation results for the presented IWC have been successfully calibrated against real measured data. The deviations corresponded to: pump group 1 (0.74%), pump group 2 (0.77%), pump group 3 (1.56%), the cooling tower inlet (2.62%) and outlet (4.75%) and the rolling mill water outlet temperature (3.29%).

Overall, the hydraulic phenomena, namely the pump power demonstrate lower deviations than the thermal phenomena. This is due the hydraulic phenomena being directly related to energy use of the pumping systems that correspond to the main power consumers of the IWC. Thus, they are also associated to the main improvement measures. While the thermal phenomena are related with temperatures requirements that are associated to the limit conditions. Furthermore, several improvement measures have been identified and their implementation allowed up to 29% share of energy savings. For this case study, indeed there is a large potential for energy savings and improvement of the efficiency in related industrial water circuits (IWC). This has been envisaged in general by the WaterWatt project [7] and demonstrated through the case studies part of the project.

The presented methodology has been successfully replicated for the other steel sector case-studies of the WaterWatt project [7] including: closed cooling circuit of an inductive furnace and closed cooling circuit of a blast furnace. Similarly, for each case-study, simulation results have been validated against real measured data, optimisation problems have been formulated and sets of scenarios have been defined for the improvement measures as well as for the techno-economic analysis.

Author Contributions: M.I., M.O. and D.C. conceived and designed the experiments; M.O. and D.C. performed the experiments; M.I. and J.M. analyzed the data; J.M. contributed with materials and analysis tools; M.I. and M.O. wrote the paper.

Funding: This project has received funding from the European Union's Horizon 2020 research and innovation programme under grant agreement "No 695820".

Conflicts of Interest: The authors declare no conflict of interest.

Nomenclature

Index

1	Main Circuit
2	Secondary Circuit
air	Air
CT	Cooling Tower
CWT	Cold Water Tank
elec	Electricity
G	Group
i	Load regime
in	Inlet
IE1	IE1 Standard Efficiency Motors Installed
IE3	IE3 Premium Efficiency Motors Installed
L	Minor Losses
nom	Nominal conditions
out	Outlet
opt	Optimised conditions
op	Operational
RM	Rolling mill
S	Saturated Air
VSD	Variable Speed Drives Installed
w	Water
filter	filter

Abbreviations

Baseline	Operational conditions for nominal pump motor and fans speed
EEM	Energy Efficiency Improvement Measure
ELC	Electric Energy Consumption
IWC	Industrial Water Circuits
Measured	Measured plant data
RTD	Resistance thermometer
VSD	Variable Speed Drive

Parameters

$C_{P,w}$	Water heat capacity (kJ $^\circ$C^{-1} kg^{-1})
C_{elec}	Cost of electricity (€/kWh)
D	Pipe diameter (m)
ELC	Energy Consumption (MWh/year)
f	Darcy friction factor
g	Gravitic acceleration (m s^{-2})
H	Enthalpy (J/kg)
k	Overall heat mass transfer coefficient (kg m^{-2} s^{-1})
K_L	Coefficient of Minor Losses
L	Pipe length (m)
n	Rotational Speed (rpm)
P	Power (kW)
Q	Water flowrate (m^3/s)
Q_M	Water mass flowrate (kg/s)
R_e	Reynolds Number

Savings	Economic Savings (€/year)
S	Cooling tower surface of packing (m^2)
T	Water temperature (°C)
t_{op}	Operational Time (h/year)
Z	Depth of a vertical pipe (m)
ΔP	Pipe pressure drop (Pa)
ρ_w	Water density (kg/m^3)
η_{hydr}	Hydraulic efficiency
η_{mech}	Mechanical efficiency

References

1. Hoekstra, A.Y.; Mekonnen, M.M. The Water Footprint in Humanity. *Proc. Natl. Acad. Sci. USA* **2012**, *109*, 3232–3237. Available online: https://doi.org/10.1073/pnas.1109936109 (accessed on 18 December 2018). [CrossRef]

2. Banerjee, R.; Cong, Y.; Gielen, D.; Jannuzzi, G.; Maréchal, F.; McKane, A.T.; Rosen, M.A.; van Es, D.; Worrell, E. Chapter 8—Energy End-Use: Industry. In *Global Energy Assessment—Toward a Sustainable Future*; Cambridge University Press: Cambridge, UK; New York, NY, USA; The International Institute for Applied Systems Analysis: Laxenburg, Austria, 2012; pp. 513–574.

3. Eurostat. Eurostat Statistics Explained, Water Use in Industry. 2014. Available online: http://ec.europa.eu/eurostat/statisticsexplained/index.php/Archive:Water_use_in_industry (accessed on 26 July 2018).

4. Eurostat. Eurostat Statistics Explained, Consumption of Energy. 2017. Available online: http://ec.europa.eu/eurostat/statistics-explained/index.php/Consumption_of_energy (accessed on 26 July 2018).

5. Tolvanen, J. LCC approach for big motor-driven systems savings. *World Pumps* **2008**, *2008*, 24–27. Available online: https://doi.org/10.1016/S0262-1762(08)70314-6 (accessed on 26 July 2018). [CrossRef]

6. Almeida, A.T.; Fonseca, P.; Bertoldi, P. Energy-efficient motor systems in the industrial and in the services sectors in the European Union: Characterisation, potentials, barriers and policies. *Energy* **2013**, *28*, 673–690. Available online: https://doi.org/10.1016/S0360-5442(02)00160-3 (accessed on 26 July 2018). [CrossRef]

7. EUR-Lex. Directive 2012/27/EU of the European Parliament and of the Council of 25 October 2012 on Energy Efficiency, Amending Directives 2009/125/EC and 2010/30/EU and Repealing Directives 2004/8/EC and 2006/32/EC Text with EEA Relevance. Official Journal of the European Union. Available online: http://data.europa.eu/eli/dir/2012/27/oj (accessed on 18 December 2018).

8. EUR-Lex. Commission Regulation (EU) No 547/2012 of 25 June 2012 Implementing Directive 2009/125/EC of the European Parliament and of the Council with Regard to Ecodesign Requirements for Water Pumps. Official Journal of the European Union. Available online: http://data.europa.eu/eli/reg/2012/547/oj (accessed on 26 December 2018).

9. European Commission. Europe 2020 Strategy. 2016. Available online: https://ec.europa.eu/info/business-economy-euro/economic-and-fiscal-policy-coordination/eu-economic-governance-monitoring-prevention-correction/european-semester/framework/europe-2020-strategy_en (accessed on 26 July 2018).

10. WaterWatt. Improvement of Energy Efficiency in Industrial Water Circuits Using Gamification for Online Self-Assessment, Benchmarking and Economic Decision Support. Funded under: H2020-EU. Available online: https://www.waterwatt.eu/index.php?page=about-project (accessed on 26 July 2018).

11. Cabrera, E.; Gómez, E.; Espert, V.; Cabrera, E., Jr. Strategies to improve the energy efficiency of pressurized water systems. *Procedia Eng.* **2017**, *186*, 294–302. Available online: https://doi.org/10.1016/j.proeng.2017.03.248 (accessed on 18 December 2018). [CrossRef]

12. Eurostat. Energy Balances, Energy Data. Available online: http://ec.europa.eu/eurostat/web/energy/data/energy-balances (accessed on 26 July 2018).

13. Statistisches Bundesamt. *Nichtöffentliche Wasserversorgung und nichtöffentliche Abwasserentsorgung*; Fachserie 19, Reihe 2.2; Statistisches Bundesamt: Wiesbaden, Germany, 2013.

14. Politecnico di Milano. ThermoPower Home Page. 2009. Available online: http://home.deib.polimi.it/casella/thermopower/help/ThermoPower.html/ (accessed on 28 September 2018).

15. Çengel, Y.A.; Cimbala, J.M. *Fluid Mechanics: Fundamentals and Applications*, 2nd ed.; McGraw Hill: New York, NY, USA, 2010.

16. Shah, P.; Tailor, N. Merkel's Method for designing induced draft cooling tower. *J. Impact Factor* **2015**, *6*, 63–70. Available online: https://www.researchgate.net/publication/299411018_MERKEL\T1\textquoterightS_METHOD_FOR_DESIGNING_INDUCED_DRAFT_COOLING_TOWER (accessed on 18 December 2018).
17. Fernandes, M.C.; Matos, H.A.; Nunes, C.P.; Cabrita, J.C.; Cabrita, I.; Martins, P.; Cardoso, C.; Partidário, P.; Gomes, P. *Medidas Transversais de Eficiência Energética Para a Indústria*, 1st ed.; Tecnico Lisboa: Lisbon, Portugal, 2016; pp. 31–52, ISBN 978-972-8268-41-1.
18. GAMBICA—Automation, Instrumentation and Control Laboratory Technology. *Variable Speed Driven Pumps—Best Practice Guide*; BPMA: West Bromwich, UK, 2003.
19. IMechE—Institution of Mechanical Engineers. *Pumping Cost Savings in the Water Supply Industry*; Fluid Machinery Committee of the Power Industries Division: London, UK, 1989.
20. ADENE. *Cursos de Utilização Racional de Energia—Eficiência Energética na Indústria*; ADENE: Gaia, Portugal, 2014.
21. Almeida, A.T.; Ferreira, F.J.T.E.; Bock, D. Technical and economical considerations in the apllication of Variable-Speed Drives with Electric Motor Systems. *IEEE Trans. Ind. Appl.* **2005**, *41*, 188–199. Available online: https://ieeexplore.ieee.org/document/1388677/ (accessed on 26 July 2018). [CrossRef]
22. Viljoen, J.H.; Muller, C.J.; Craig, I.K. Dynamic modelling of induced draft cooling towers with parallel heat exchangers, pumps and cooling water network. *J. Process Control* **2018**, *68*, 34–51. Available online: https://doi.org/10.1016/j.jprocont.2018.04.005 (accessed on 26 July 2018). [CrossRef]
23. ERSE—Entidade Reguladora dos Serviços Energéticos. Preços das Tarifas Transitórias de Venda a Clientes Finais em Portugal Continental em 2018; Eletricidade, Tarifas e Preços. Tarifas Reguladas em 2017. Available online: http://www.erse.pt/pt/electricidade/tarifaseprecos/2017/Paginas/TtransVCFPortugalcont2017.aspx (accessed on 26 July 2018).
24. Siemens Industry. *SIMOTICS Low-Voltage Motors*; Type Series 1LA, 1LE, 1LG, 1LL, 1LP, 1MA, 1MB, 1PC, 1PP, 1PQ, Frame Sizes 63 to 450, Power Range 0.09 to 1250 kW, Price List D 81.1 P; Siemens Industry: Munich, Germany, 2015.
25. ABB Group. *ACS55, ACS150, ACS310, ACS355, ACS550, 2014. Conversores de Frequência Para Controlo Preciso do Motor e Poupança Energética*; Tabela de Preços; ABB: Zürich, Switzerland, 2014.
26. Statista. Entwicklung des Wasserverbrauchs pro Kopf und Tag in Deutschland in den Jahren 1990 bis 2016. Available online: https://de.statista.com/statistik/daten/studie/12353/umfrage/wasserverbrauch-pro-einwohner-und-tag-seit-1990/ (accessed on 26 July 2018).

MDPI
St. Alban-Anlage 66
4052 Basel
Switzerland
Tel. +41 61 683 77 34
Fax +41 61 302 89 18
www.mdpi.com

Energies Editorial Office
E-mail: energies@mdpi.com
www.mdpi.com/journal/energies